ENGINEERS' HANDBOOK

of

INDUSTRIAL MICROWAVE HEATING

Roger Meredith

The Institution of Electrical Engineers

Published by: The Institution of Electrical Engineers, London,
United Kingdom

The Institution of Electrical Engineers,
Michael Faraday House,
Six Hills Way, Stevenage,
Herts. SG1 2AY, United Kingdom

British Library Cataloguing in Publication Data

A CIP catalogue record for this book
is available from the British Library

ISBN 0 85296 916 3

Typeset by On-Screen Typography, Cheshunt

Printed in England by Short Run Press Ltd., Exeter

To Judy,

without whose support and encouragement

this would not have happened

Contents

Preface

Most people who become involved in microwave heating are from disciplines other than electrical engineering. Usually their introduction is through the domestic oven they have used for a simple experiment which has demonstrated to them the feasibility of a process. Others may become involved because their company plans to install a major plant, or it forms part of a research project. Electrical engineers themselves often have little in-depth knowledge of the subject. Although there are many technical papers written on specialist topics, a few general books, and a plethora of symposia having their customary 'short courses' on microwaves, there is a clear absence of a book with an engineering base concerned with the fundamentals of the subject.

There are, however, many books on the subject of microwaves, starting with the famous MIT series of the 1940s with many published since, but they are mostly radar oriented and at a high mathematical level difficult or impossible to apply to the problems arising in microwave heating. They do not form an easy route to this technology

After nearly four decades of involvement in industrial microwaves, I have become ever more acutely aware that there is a generation of engineers and physicists working in this field, few of whom have had the benefit of an initial training in microwave engineering. They are unfamiliar with the fundamentals, and in some cases are unaware of the existence of a technology beyond the domestic microwave oven.

My purpose in writing this book is to present the theory and practice of microwave engineering relevant to its use as a processing tool, starting from very basic concepts of electric and magnetic fields, progressing to waveguide theory and practice and consideration of applicator design and practice. I have attempted to be descriptive as far as possible, but have not eschewed mathematical analysis where a quantified presentation is possible and helpful. The mathematical content should be well within the range of capability of the majority of readers, and there is much helpful descriptive discussion for those to whom mathematics is anathema. I have included a chapter on heat flow within a workload, a topic frequently overlooked, which is fundamental to any volumetric heating process.

Two chapters are concerned with microwave generators. The first is mainly concerned with magnetrons and their performance characteristics, the second with power conversion from input AC power to high-voltage DC for the bulk power supply to the magnetron and the auxiliary supplies. Both the conventional and the new switch-mode power supplies are considered, the latter of great importance, heralding a new era in efficient, reliable power conversion.

I recognise a possible criticism in omitting a substantial section on the numerical analysis of Maxwell's equations. This is intentional because the subject is still new and is developing rapidly. It is a topic requiring great professional skill and generally remains within the ambit of university research. It is a potential trap for the unwary, readily producing incorrect but plausible results. Those wishing to embark on this route should consult specialists prominent in the field. I would particularly like to commend Drs. Ricky Metaxas and David Dibben of the University of Cambridge and Dr. Raymond Lau for their immense contribution to the subject.

A further possible criticism is the absence of detailed discussion and photographs of specific microwave installations. This is deliberate because such information rapidly becomes dated, and it is more helpful to advise on practical techniques.

I owe a debt of acknowledgement to many, but first to my wife, Judy, whose enduring patience I have stretched almost to the limit in the long hours this work has taken. In writing a book on fundamentals, the pioneers of the subject are necessarily prominent and I would particularly mention the staff of the former B.T-H Research Laboratory, Rugby, UK, in the 1950s, especially H.B. Taylor, the late Dr. J.E. Curran and J.R.G. Twistleton who designed the first industrial L-band magnetron which forms, with only detail changes, the basis of the present industry-standard magnetrons, also to Tony Wynn of California Tube Laboratory Inc. who recognised early on the commercial importance of rebuilding magnetrons, and continues ably to pioneer their further development to even higher output powers. To them, and many others including those listed in the References, I say thank you for your contributions to the development of this technology.

May I also thank my publishers for their professional approach, and in particular Jonathan Simpson and Fiona MacDonald.

In a work of this scale there are inevitably some, hopefully only a few, errors and omissions: the author and publisher would greatly appreciate being notified of them.

Roger Meredith
Cold Overton, Leicestershire

October 1997

Chapter 1

Introduction and fundamental concepts

1.1 Introduction

Of all the processes used in manufacturing industry there can be little doubt that heating is the most commonplace, widely used in the food, chemical, textile and engineering industries for drying, promoting chemical or physical change, and many other purposes. Yet it remains one of the most difficult techniques to control, being slow and imprecise when practised in the usual way of heating the surface of the workpiece by radiation, convection or conduction or, commonly, a poorly controlled mixture of all three; and even if perfection were achieved in surface heating, the process time is limited by the rate of heat flow into the body of the workpiece from the surface, which is determined entirely by the physical properties of the workpiece: its specific heat, thermal conductivity and density. These last three are combined into one parameter, the thermal diffusivity (Carslaw and Jaeger, 1959), which uniquely determines the temperature rise within a material as a function of time and depth from the surface, subject to a given set of conditions at the surface. Nothing can be done in the application of surface heating to accelerate heating once the surface has reached a specified maximum temperature. Internal temperature distribution is then limited by the thermal diffusivity of the material.

Because, in conventional heating, all the heat energy required in the workpiece must pass through its surface, and the rate of heat flow to within is limited by temperature and the thermal diffusivity, the larger the workpiece, the longer the heating takes.

Everyone's experience is that surface heating is especially slow in a thick material. It is not only slow but nonuniform, with the surfaces, and in particular edges and corners, being much hotter than the inside. Consequently, the quality of the treated workload is variable and frequently inferior to what is desired.

Imperfect heating resulting from these difficulties is a frequent cause of product reject and energy waste; above all the extended process time results in large production areas devoted to ovens. Large ovens are slow to respond to impressed temperature changes, take a long time to warm up and have high heat capacities. Their sluggish performance results in a failure to respond to sudden changes in production requirements: management control becomes difficult, subjective and expensive.

1.2 Electrical volumetric heating

By electrical means, volumetric heating is possible wherein all the infinitesimal elements constituting the volume of a workload are each heated individually, ideally at substantially the same rate. The heat energy injected into the material is transferred through the surface electromagnetically, and does not flow as a heat flux, as in conventional heating. The rate of heating is no longer limited by thermal diffusivity and surface temperature, and the uniformity of heat distribution is greatly improved. Heating times can often be reduced to less than 1% of that required using conventional techniques, with effective energy variation within the workload less than 10%.

Any material can be heated directly by electrical volumetric heating provided that it is neither a perfect electrical conductor nor a perfect insulator, implying that the range extends from metals to dielectric materials which could be considered quite good insulators. No single electrical technique is effective in all cases and there are four methods used in practice, classified by the effective electrical resistivity and physical properties of the workpiece (Metaxas, 1996).

1.2.1 Conduction and induction heating

These processes are used for heating metals with low resistivity and involve passing a heavy current through the workload to cause I^2R heating. The current may pass between physical electrical connections to the workload (conduction, or resistance, heating). Alternatively the workload may form the secondary of a step-down transformer in which the induced EMF causes the heating current to circulate (induction heating). The electrical frequency used in these techniques ranges from DC to 60 Hz (conduction heating), and 50 Hz to about 30 kHz for induction heating. Davies (1990), provides an excellent presentation of the theory and practice of these methods.

Conduction and induction heating are used widely in the metal-forming industries for welding, annealing, hardening and brazing, often under vacuum to prevent oxidation.

1.2.2 Ohmic heating

Ohmic heating is a conduction heating technique for liquids and pumpable slurries; it consists of equipment for passing an alternating current through the liquid between electrodes. Aqueous solutions, in particular, are almost always sufficiently conductive to permit a high power density to be dissipated, because dissolved salts provide ions as charge carriers.

Ohmic heating invariably uses a power-frequency supply (50 or 60 Hz) and is extremely efficient as a converter of energy to heat in the workload, efficiency of conversion being over 95%. It is capable of handling liquids containing solids, e.g. diced meat or vegetables or fruit segments, as the solids usually have similar conductivity to the gravy or juice. Large amounts of power are readily dissipated (>300 kW) in continuous-flow systems of high mass-flow throughput.

Ohmic heating is gaining popularity rapidly in installations in the food industry and is particularly valued for sterilisation in aseptic processing.

1.2.3 Radio frequency heating

When the workload has high resistivity, the voltage required to pass sufficient current for a practical power-dissipation density becomes prohibitive at the low frequency used for conduction heating. This problem can be overcome by increasing the frequency to the range 1–100 MHz, most often 27.12 MHz, one of several internationally agreed frequencies for the purpose.

Typical applications are plastics (welding and forming), wood (seasoning and gluing), textiles, paper and board (drying), food (post-baking/drying) and ceramics (drying).

The workload is placed between electrodes in the form of plates or rods, to which is applied a high voltage (usually several kilovolts) at the chosen high frequency. This structure forms a capacitor in which the workload is part of the dielectric, and a charge/discharge current flows between the electrodes. Because the workload is an imperfect dielectric, the structure can be considered as an ideal capacitor in series with a small resistor which represents the source of heating within the workload.

Radio-frequency (RF) heating has been used in industry since the 1930s and has grown to a substantial and important industry.

1.2.4 Microwave heating

Intensive research during the Second World War into high-definition radar led to the development of microwave frequencies (500 MHz to 100 GHz), and in particular the magnetron valve as a microwave generator of very high power output with exceptional efficiency (University of Birmingham, 1940). In the post-war years further development resulted in microwaves being used for heating, especially for domestic purposes, but also significantly in industry, where there are some important advantages compared with processing at lower frequencies. These are enlarged upon in later sections.

Modern industrial-microwave-heating systems are used for a diversity of processes in the food industry, tempering and thawing, continuous baking, vacuum drying, pasteurisation and sterilisation, and in the ceramics, rubber and plastics industries, as well as many specialised processes in the chemical industry where there is great interest in vacuum processing.

Contemporary equipment has very high reliability and running costs are competitive with other heating methods, especially when the advantages of volumetric heating are included; moreover high-power magnetrons, though initially expensive, are now rebuilt after normal end of life at a cost representing less than 10% of the energy they use.

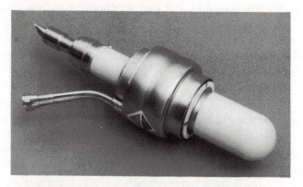

a

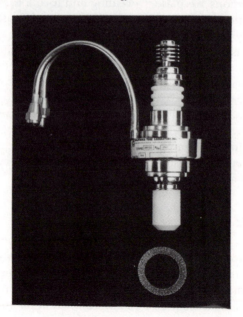

b

Figure 1.1   Photographs of industrial high-power magnetrons

    a   75 kW, 896/915 MHz
    b   15 kW, 2450 MHz
    (Courtesy California Tube Laboratory Inc.)

In industry, microwave heating is performed at either a frequency close to 900 MHz or at 2450 MHz, frequencies which are chosen by international agreement with the principal aim of minimising interference with communication services.

Most of the materials which can be heated at RF can be also be treated at microwave frequencies together with some others which are difficult with RF because of their low loss factor. Because microwave heating operates at a much higher frequency than RF, the applied electric field strength is less, so the risk of arcing is less, as discussed in detail in later sections. The higher power density of heating can also be used, resulting in physically smaller plant. However, the penetration depth (Section 2.5) is less at microwave frequencies than for RF, and, with the shorter wavelength giving greater prominence to standing waves (Section 4.3), the uniformity of heating may be inferior.

The overall efficiency of microwave heating systems is usually very high because of the exceptional efficiency of high-power magnetrons (85% at 900 MHz, 80% at 2450 MHz).

Because microwave frequencies have very short wavelengths (33.3 cm at 900 Mhz, and 12.2 cm at 2450 MHz), the electrical techniques used differ greatly from RF heating: RF equipment uses conventional electrical components such as inductors and capacitors, with open conductors for the electrical connections. Microwave equipment cannot use these components because their size is comparable with the operating wavelength; under these conditions the components behave anomalously and the circuits would radiate most of the energy into space. Instead, microwave heating uses waveguides (hollow metal tubes) to convey power from the magnetron(s) to the heating oven, frequently called the applicator. The applicator may have many forms but is almost invariably based on a closed metal structure, with an access door or small open ports to allow the workload to pass through in a continuous flow.

The engineering theory and practice of industrial microwave systems is the primary topic of this book.

1.3 The electromagnetic spectrum

All the electroheating techniques described above have particular operating frequencies for the electrical energy applied to the workload. This energy, which comprises combined electric and magnetic fields, is the same as that used for broadcasting, television, radar and satellite communications. Clearly, great care must be taken to avoid interference between the electroheat systems and these services, which essentially radiate and receive signals from locations spaced far apart. Whereas in communications the purpose is to radiate power into space, in electroheat the power is often generated at much higher power levels, but must be contained within the treatment equipment.

Indeed the amount leaking away has to be controlled to within specified limits for safety of personnel, and to avoid radio-frequency interference (RFI) to other services.

Figure 1.2 shows, on a logarithmic frequency scale, the frequencies of the electroheat techniques relative to those of other users. Frequencies allocated for electroheat are often designated ISM 'industrial, scientific and medical' frequencies. The frequencies chosen for RF and microwave heating are the result of historical evolution and a complexity of international committees which constantly review the use of the electromagnetic spectrum. Unfortunately, there is little logic in the frequency choice and most of the frequencies were determined in the period 1930 to 1970, before electroheat technology had developed significantly. Moreover, some frequencies were not included in particular geographical areas, largely because the determining body had no representation on electroheat matters, and frequency allocations were made wholly on the requirements of communications. In particular, there is at present (1997) no allocation of a 900 MHz frequency in continental Europe, which is a serious shortcoming: mainland European installations at 900 MHz have to be specially screened to meet exacting specifications for residual leakage, far lower than required for safety.

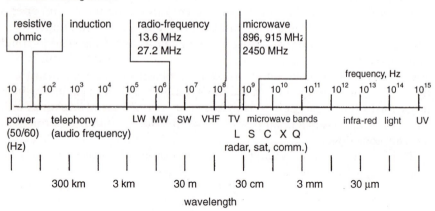

Figure 1.2   The electromagnetic spectrum

Note that the absence of an allocated frequency band for ISM does not mean total exclusion, and provided that the equipment is designed to emit not more than a legally prescribed limit of RF or MW power, it is acceptable. In practice, this means that great care has to be taken in equipment design, which inevitably involves a cost penalty; the matter is discussed at length in Chapter 8.

Harmonics of the fundamental frequency of ISM equipment must also be at very low leakage amplitude because they fall at frequencies generally allocated for other purposes. This is a particular problem at RF, where the principal harmonics are at frequencies within bands devoted to civil and military communication and navigation aids.

1.4 Electromagnetic fields

In 1873 the British physicist J.C. Maxwell published his famous 'Treatise on electricity and magnetism', including the Maxwell equations which embody and describe mathematically all phenomena of electromagnetism. There is much literature on this subject and references 6,7,8 are a good introduction for those wishing to inquire deeply. Here, we shall confine our attention to those basic aspects necessary for a proper understanding of electroheat technology, and in particular microwave heating.

Qualitatively we are all familiar with electromagnetic (EM) waves propagating in space not only as in broadcasting and radar but also as radiant heat and visible, infra-red and ultra-violet light. Two fundamental components are necessary for a propagating EM wave to exist: an electric field (*E*, volts per metre) and a magnetic field (*H*, ampères per metre). The electric field, in particular, is a recurring parameter in microwave and RF heating, being the prime source of energy transfer to the workload, and familiarity with it is essential its nature is discussed, together with the magnetic field and its relationship to the electric field, in subsequent sections.

In subsequent sections there are frequent references to the permittivity ($\varepsilon_0 = 8.854 \times 10^{-12}$ F/m) and the permeability ($\mu_0 = 1.257 \times 10^{-6}$ H/m) of free space. These universal and fundamental constants relate the mechanical forces existing between two electrostatic charges spaced apart in vacuo (permittivity) and two current loops similarly spaced apart (permeability). Their detailed definitions are fully developed in most texts on electricity and magnetism.

1.4.1 Electric fields

The electric field is familiar as that in electrostatics, and is a field intensity, or stress, at a point in space due to the presence of a charged body. Figure 1.3 shows a simple arrangement of two perfectly conducting plates spaced apart by a distance *d* (metres), small compared with the size of the plates, connected to a voltage source *V* (volts).

The electric field intensity between the plates and away from the edges is simply *V/d* volts per metre. In this example the electric field everywhere between the plates is substantially constant, but the geometric discontinuity at the edges causes a local field distortion which increases the field intensity. If the edges have sharp corners, the intensity is increased considerably, possibly to cause corona or voltage breakdown, this being the reason why very high voltage circuits have radius edges to conductors.

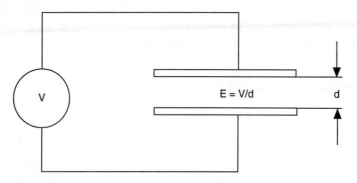

Figure 1.3   Illustrating an electric field

Electric field intensity is a vector. It is immediately apparent that the value of electric field tangential to the plates in Figure 1.3 (i.e. grazing the surface) must be zero because they are perfect conductors and by definition there can therefore be no voltage drop across their surfaces. In this example the electric field is thus a vector normal to the facing surfaces of the plates. The electric field is frequently drawn diagramatically as 'lines of force' representing the path along which a charged particle would travel due to the force on it from the field, and the direction of the vector.

1.4.2 Magnetic fields

Hans Christian Œrsted announced in 1820 that he had discovered that a conductor carrying a current produces a magnetic field in its vicinity; he was followed by André Marie Ampère, who quantified the relationship.

The magnetic field intensity is defined as the magnetic-field intensity H (amperes per metre) at the centre of a circular loop of conductor of radius r (metres) carrying a current I (amperes) where

$$H = \frac{I}{2\pi r} \tag{1.1}$$

A circular loop is chosen because this is the simplest practical geometry which takes account of the magnetic field of the return path of the current, and conforms with an easily realised experiment. Figure 1.4 shows the configuration of the definition.

The magnetic field can also be produced by a permanent magnet; this is due to the equivalent current loops caused by spinning and orbiting electrons within the molecules which, in a magnet, are aligned so as to add in their magnetic effect.

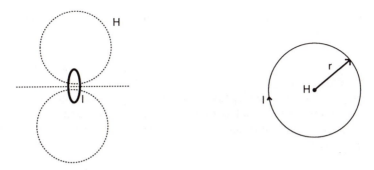

Figure 1.4   *Magnetic field of a current loop*

Maxwell realised that eqn. 1.1 is incomplete because it postulates a closed conducting electric circuit, and does not explain the flow of current through a capacitor when connected to a time-varying voltage source. Moreover, there is a magnetic field associated with the current passing through the capacitor, in the absence of a conducting path. To include this phenomenon, Maxwell introduced the concept of displacement current, which is a current, or more especially a magnetic field, produced by a time-varying electric field in the absence of a conductor. The magnitude of the displacement current is readily quantified. Consider the capacitor in Figure 1.3; its capacitance is given, as for a parallel plate capacitor, as

$$C = \frac{\varepsilon_0 \varepsilon' A}{d} \tag{1.2}$$

where A is the area of one plate (square metres), ε_0 is the permittivity of free space (8.854×10^{-12} F/m) and ε' is the relative permittivity of the material between the plates of the capacitor. For air, $\varepsilon' = 1.00$.

Also, the charge on the capacitor Q (coulombs or amp seconds) is

$$C = \frac{Q}{V} = \int \frac{i dt}{V} \qquad \text{or, differentiating,} \qquad i = C\frac{dV}{dt} \tag{1.3}$$

Substituting for C,

$$i = \frac{A\varepsilon_0 \varepsilon'}{d}\frac{dV}{dt} = A\varepsilon_0 \varepsilon'\frac{dE}{dt} = A\frac{dD}{dt} \tag{1.4}$$

where $D = \varepsilon' \varepsilon_0 E$ is the electric displacement density.

In this last expression the total voltage difference between the capacitor plates has been divided by their separation distance to give an electric field stress E in the space between them. Clearly, the time-varying electric field stress dE/dt results in a flow of current in the space between the plates, which, in turn, creates a magnetic field H in accordance with the argument of Section 1.4.2.

By writing $D = \varepsilon_o\varepsilon'E$ the concept of Maxwell's displacement current density D is introduced, which is analogous to the current density in a conventional circuit, resulting from an applied EMF. Like the current flowing in a conductor, the displacement current has an associated magnetic field surrounding it. When electric circuits have dimensions comparable with the wavelength corresponding to the frequency of excitation, the displacement current becomes substantial compared with the conduction current, and leads to the possibility of radiation of energy into space away from the circuit as a radio wave.

1.4.3 Electromagnetic induction

In 1831 another British physicist, Michael Faraday, discovered that when a magnet is moved in the vicinity of an electric circuit a voltage is induced in the circuit, which causes a current to flow, so bringing together the hitherto separate subjects of electrostatics and magnetism. Movement, in fact, means time change in the magnetic field as 'seen' by the circuit, and comes about either by movement in a steady field, or by time change in amplitude without movement. It is the latter which is of prime interest in electroheat.

Quantitatively, the intensity of electric field is related to the magnetic field by

$$V = \int EdS = \mu_0\mu' A\frac{dH}{dt} \tag{1.5}$$

This equation simply states that the voltage developed around the circuit (the line integral of E around the circuit) is the rate of change of magnetic-field intensity multiplied by the area of the circuit and the coefficient $\mu_0\mu'$. This coefficient has two parts: μ_0 which is the permeability of free space ($4\pi \times 10^7$ H/m), and a dimensionless number μ' which is the relative permeability of the surrounding medium. For air, $\mu' = 1.00$, but it has higher values for ferromagnetic materials. In practice, the relative permeability is written ($\mu' + j\mu''$), where μ'' represents losses in the magnetic material.

1.4.4 Maxwell's equations

In Sections 1.4.1–1.4.3 we have considered some particular cases of the interaction between electric and magnetic fields. Maxwell expressed these relationships with complete generality and conciseness in his famous Maxwell's equations within which all electrical phenomena are determined. They can be presented in many forms, and are presented here for the sake of completeness in vector form for sinusoidal, time-varying fields with an angular frequency $\omega = 2\pi f$ (radians per second), and for the case of no sources of electric charge or magnetic dipoles:

$$\text{div } \boldsymbol{D} = 0 \tag{1.6}$$

$$\text{div } \boldsymbol{B} = 0 \tag{1.7}$$

$$\text{curl } \boldsymbol{E} = -j\omega\boldsymbol{B} \tag{1.8}$$

$$\text{curl } \boldsymbol{H} = \boldsymbol{J} + j\omega\boldsymbol{D} \tag{1.9}$$

together with the auxiliary equations

$$\boldsymbol{J} = \sigma\,\boldsymbol{E} \tag{1.10}$$

$$\boldsymbol{D} = \varepsilon_o\varepsilon'\boldsymbol{E} \tag{1.11}$$

$$\boldsymbol{B} = \mu_o\mu'\boldsymbol{H} \tag{1.12}$$

This set of differential equations can be solved only for particular configurations of simple geometry for the boundaries of the fields, e.g. rectangular or spherical structures. For more complicated shapes numerical methods must be used, summarised by Metaxas (1996). Prominent amongst these is the finite-difference time-domain (FDTD) technique (Lau and Sheppard, 1986; Lau, 1995; Dibben and Metaxas, 1995).

Essentially, Maxwell's equations are solved by finding solutions to the fields to match the requirements of the field intensities which must exist at the boundaries of the structure. The principal boundary conditions are discussed in Section 1.4.4.1.

1.4.4.1 Boundary conditions

(*a*) The electric field intensity at the surface of the conductor, in a direction parallel to the surface, i.e. grazing the surface, is zero. This is to be expected because for a perfect conductor there can be no potential difference between two points however great the current flowing. This condition is frequently written

$$E_{tan} = 0 \tag{1.13}$$

However the component of electric field normal to a conducting surface has, in general a nonzero value.

(*b*) The component of magnetic field normal to a conducting surface is zero because it is not otherwise possible to create a magnetic loop, i.e.

$$H_{norm} = 0 \tag{1.14}$$

Analogously but orthogonally to the electric field, the magnetic field grazing the surface has a nonzero value, and has with it an associated current flowing in the surface.

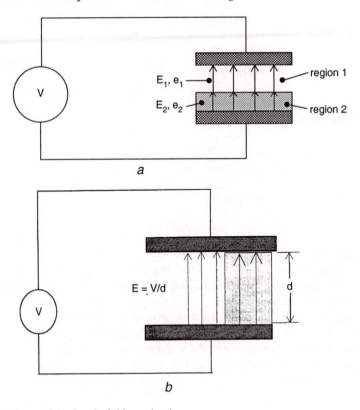

E_1, e_1 — region 1

E_2, e_2 — region 2

a

$E = V/d$ *d*

b

Figure 1.5   Illustrating electric-field continuity

    a Dielectric block partly filling the height, normal to the dielectric interface

    b Dielectric block partly filling the width, parallel to the dielectric interface

(*c*) Clearly, there must be continuity of displacement current across the boundary between two dielectric regions. This is a particularly important condition in electroheat because in part it determines the relative values of electric field inside and outside the workload. The electric field inside the workload is the source of heating. Consider a parallel-plate capacitor as shown in Figure 1.5*a* in which a dielectric slab (relative permittivity ε_2') partly fills the space between the plates to form region 2, whilst the air space ($\varepsilon_1' = 1$) is designated region 1. Continuity of displacement current means that *D* has the same amplitude in both regions. Thus:

$$D = \varepsilon_0 \varepsilon_1' E_1 = \varepsilon_0 \varepsilon_2' E_2 \qquad (1.15)$$

whence

$$E_2 = \frac{\varepsilon_1}{\varepsilon_2} E_1 \qquad (1.16)$$

Thus the electric field is less in the workload, in this simple geometry, than in the air space by the factor $1/\varepsilon'_2$, which may be a large factor, particularly with wet loads, because of the high permittivity of water.

Continuity of displacement current across a dielectric boundary strictly means the component of displacement current normal to the boundary interface, and in the above example is illustrated to its maximum effect: such a situation arises in parallel-plate RF heaters.

More usually, the displacement-current vector is inclined to the boundary so that only the component resolved normal to the interface follows the above analysis. For the component tangential to the interface, the electric field E must be continuous, as can be seen in Figure 1.5b where a dielectric post is erected between the plates extending to the full height. Clearly, the electric field intensity is the same inside the post as just outside, since the voltage and the separation between the plates is the same for both regions.

Thus in the general case the ratio of the electric field inside the workload to that in the surroundings lies in the range 1 to $1/\varepsilon'$; generally and in practice it is a variable function of position within the workload. For the special case of a sphere embedded in an otherwise uniform electric field, the internal and external fields are related (Stratton, 1941; Ramo *et al.*, 1965) by

$$E_2 = \frac{3}{2+\varepsilon'} E_1 \qquad (1.17)$$

In this case the field is uniform throughout the sphere.

1.5 Plane waves

The propagation of radio waves as plane waves in unbounded loss-free space is the simplest solution of the Maxwell equations and is well documented (e.g. Stratton, 1941; Ramo *et al.*, 1965; Carter, 1954).

Plane waves are very useful in helping to visualise practical field configurations. For example, they can be used, by addition of waves propagating along inclined axes, to synthesise the fields existing in a waveguide or cavity, illustrating and quantifying some of their important features (Metaxas and Meredith, 1983). True plane waves are, by definition, infinite in extent and therefore imply infinite power flow and cannot exist in practice. Nevertheless, radio waves at a long distance (compared with their wavelength) from the source are a close approximation. In studying plane waves it is not important or relevant to ask where they came from.

The solution of Maxwell's equations for plane waves shows they consist of an electric field vector and a complementary magnetic field vector, orthogonal to each other, and to the direction of propagation as shown in Figure 1.6 for a wave travelling in the direction of the Oz axis. The wave has its electric-field vector in the direction of the Oy axis so that Ey has a finite

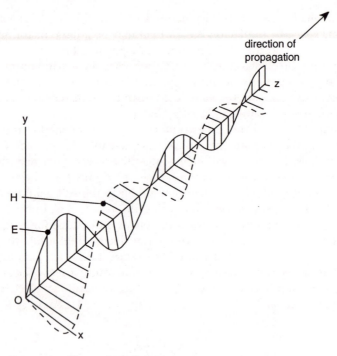

Figure 1.6   Perspective illustration of plane-wave components

value, but E_x and E_z are zero. Similarly, H_x has finite value but H_y and H_z are zero. Note that there is no field component in the direction of propagation; the wave is called transverse electromagnetic (TEM) because all the field components are transverse to its direction of travel.

The field components E_y and H_x vary sinusoidally in time and also in space: a 'snapshot' would reveal a sinusoidal variation in the direction of travel, as shown in Figure 1.6. Formally, the wave is represented by:

$$E_y = E_0 \sin(\omega t - \beta z) \tag{1.18}$$

$$H_x = \frac{E_0}{Z_0} \sin(\omega t - \beta z) \tag{1.19}$$

where E_0 is an amplitude factor, and

$$Z_0 = \sqrt{\frac{\mu_0}{\varepsilon_0}} = 377\,\Omega \tag{1.20}$$

and is called the 'characteristic impedance'

$$\beta = \frac{2\pi}{\lambda_0} \tag{1.21}$$

and is called the propagation constant (radians metre^{-1})

The characteristic impedance is a very important parameter, giving the ratio between the electric and magnetic fields; although for free space it has the value 377 Ω as shown above, in dielectric or ferromagnetic material the value is modified as

$$\sqrt{\frac{1}{\varepsilon'}} \quad \text{or} \quad \sqrt{\mu'}$$

for the relative permittivity or permeability of a lossless medium. If the medium is lossy, these parameters become complex numbers. As is shown in Chapter 4, when a wave passes from one medium into another with a different value of characteristic impedance from the first, a wave is reflected from the interface between the two. The magnitude of the reflected wave increases with the mismatch between the two characteristic impedances, and 'impedance matching' is an important procedure, discussed in later chapters, designed to minimise the reflected wave. The reflected wave is undesirable, first because it represents power wasted, and secondly because it can affect adversely the performance of the generator, in the extreme causing damage.

Some further important features of plane waves are:

(*a*) They travel at a velocity *c* metres per second where

$$c = \sqrt{\frac{1}{\mu_0 \varepsilon_0}} = \sqrt{\frac{1}{4\pi \times 10^{-7} x 8.854 \times 10^{-12}}} = 3.00 \times 10^8 \text{ m/s} \tag{1.22}$$

Note that this velocity is independent of frequency, and is the same for the whole of the electromagnetic spectrum, including visible light. However, the velocity is slowed in a dielectric medium by the factor

(*b*) In the direction of propagation the wave has a wavelength, defined as the

$$\sqrt{\frac{1}{\mu' \varepsilon'}}$$

distance between two planes in which the vectors are in the same time phase. For example, if the wave were 'frozen' in time, the wavelength would be the distance between two adjacent maxima of electric field intensity. The wavelength is related to frequency and is given by

$$c = f\lambda_0 \tag{1.23}$$

Note that if the medium is a dielectric material, because the velocity slows as described in (*a*) above due to its permittivity, the wavelength will shorten as the frequency remains the same.

(*c*) In the example described so far, the electric- and magnetic-field vectors are aligned parallel to the *Oy* and *Ox* axes, respectively, and there are no components along any other axes. The wave is said to be 'plane polarised' in the *Oy–Oz* plane, i.e. the plane of the *E*-field vector, or vertically polarised in the diagram (Figure 1.6).

It is equally possible for a wave to be polarised with the *E*-field vector in the *Ox–Oz* plane and the *H*-field vector in the *Oy–Oz* plane, or horizontally polarised. Moreover, it is possible to have the two waves present simultaneously at the same frequency; if they are in time-phase, they combine by vector addition to form a resultant wave polarised in a plane inclined to the *Ox–Oy* axes by an angle depending on their relative amplitudes.

(*d*) If the two waves in (*c*) are in time quadrature and of equal amplitude, the resultant wave becomes circularly polarised; a view of the wave in the plane transverse to the propagation axis would show the electric- and magnetic-field vectors rotating at constant amplitude, with a rotation speed equal to the frequency. The direction of rotation depends on whether the vertically polarised wave leads or lags the horizontal in time-phase.

1.6 Power flow and the Poynting vector

Analogously with ordinary electric circuits, in which the power (watts) is simply the product voltage × current, the power flow associated with a plane wave is *E* (volts per metre) × *H* (amperes per metre). Dimensionally, this gives the power flow as a power flux density *p* (watts per square metre).

Already (Section 1.5) we have noted that the electric- and magnetic-field vectors in a plane wave lie in a plane normal to the direction of propagation travel of the wave, and obviously the power flow must also be in the direction of travel. This result was formally established by Poynting circa 1884, and the vector relationship

$$\mathbf{p} = \overline{E} \times \overline{H} \text{ watts per square metre} \qquad (1.24)$$

is called the Poynting vector, where the electric and magnetic fields are shown as vectors, multiplied mathematically as a vector product (Ramo *et al.*, 1965), giving the resultant power flow mutually orthogonal to the field vectors.

Pursuing further the analogy with ordinary electric circuits, there are, from Ohm's law, the following relations equating the field intensity to the power flux density:

$$E = Z_0 H \tag{1.25}$$

and

$$p = \frac{E^2}{Z_0} = H^2 Z_0 \text{ watts per square metre} \tag{1.26}$$

Table 1.1 shows the field intensities in a plane wave in air ($Z_0 = 377\ \Omega$) for a range of power-flux densities in watts per square metre, and watts per square centimetre, the latter being more convenient to visualise.

Table 1.1   Power-flux density and E- and H-Field intensities

Power-flux density	Power-flux density	E	H
(W/m^2)	(W/cm^2)	(V/m)	(A/m)
1×10^6	1×10^2	19.4×10^3	51.5
3×10^5	30	10.6×10^3	28.2
1×10^5	10	6.1×10^3	16.3
3×10^4	3.0	3.4×10^3	8.9
1×10^4	1.0	1.9×10^3	5.1
3×10^3	0.3	1.06×10^3	2.82
1×10^3	0.1	6.14×10^2	1.63
1×10^2	1×10^{-2}	1.94×10^2	0.51
10	1×10^{-3}	61.4	0.16
1×10^{-2}	1×10^{-6}	1.94	5.15×10^{-3}
$1 \vee 10^{-8}$	1×10^{-12}	1.9×10^{-3}	5.15×10^{-6}
2.6×10^{-11}	2.6×10^{-15}	1.0×10^{-4}	0.26×10^{-6}

In electroheat applications the power flux density is usually in the range 0.1–100 W/cm^2 with a corresponding electric-field-stress intensity of 6–200 V/cm; however, this can be exceeded particularly at lower frequencies (e.g. RF) where resonant circuits are often used which involve stored energy which increases the voltage stress. These voltage stresses are modest compared with the breakdown strength of air at atmospheric pressure in a uniform field of 30 kV/cm; however, in practice there are many factors which erode this apparently large margin, as discussed in detail in Section 2.6.2.

Note that the power-flux density in a plane wave is not the power actually dissipated in an absorbing medium, i.e. a workload; it is the power incident on the workload. The power-dissipation density is the power absorbed per unit volume of the workload and has the dimension watts per cubic metre; it is dependent on the electrical properties of the workload (Section 2.4)

Table 1.1 also shows field intensities corresponding to low power densities, where considerations of personal safety and radio interference are important. Personal safety limits are in the range 1.0–10.0 mW/cm^2 for which the electric field stress is 0.6–2.0 volt/cm.

For radio-interference limitation, the field intensity allowed out-of-band is of the order of 100 μV/m, corresponding to a power-flux density of 2.6×10^{-11} W/m, a level close to the noise limit of most antenna–receiver systems.

These quantified field intensities and power-flux densities are given to provide a perspective on the magnitude of signals encountered in electroheat systems, discussed in detail in later chapters.

Chapter 2

Microwave interaction with dielectric materials

2.1 Introduction

Heating of a dielectric material by a microwave field is a process which, for purity, should be discussed in detail at molecular level, and the engineer interested in the mechanism in depth is referred to the many texts (e.g. Von Hippel, 1954; Frohlich, 1958; Metaxas and Meredith, 1983) on the subject. Usually, however, it is sufficient to know that there are two primary heating mechanisms, one in which the dielectric material behaves like a poor electrical conductor, having a finite resistivity measurable at DC, and which is usually substantially constant as the frequency extends upwards into the microwave region.

In the other, dipolar components of molecules in many dielectric materials couple electrostatically to the microwave electric field (a vector in space) and tend to align themselves with it mechanically. Since the microwave field is alternating in time, the dipoles will attempt to realign as the field reverses, and so are in a constant state of mechanical oscillation at the microwave frequency. Frictional forces within the molecule cause heat to be developed due to the motion of the dipoles. This effect is fundamentally frequency sensitive because the amount of energy dissipated is constant per cycle of applied alternating field; the rate of dissipation of energy (i.e. power dissipation) therefore increases as the frequency is raised. However, this assumes that the mechanical displacement of the dipoles remains constant with changing frequency; in practice there are mechanical resonances within the molecule in the microwave-frequency region which result in peaks of power absorption in the frequency spectrum. Individual chemical compounds have well defined spectral responses in the microwave-frequency region and microwave spectroscopy is a well known analytic technique making use of this phenomenon.

Many dielectric materials display polar characteristics, the most common being water which absorbs microwave energy very strongly even when highly pure, although it then has a very high DC electrical resistivity (approx. $10^5\ \Omega$ m). However, when even small quantities of solids are dissolved the DC resistivity falls and conduction via movement of ionic charge carriers may become a significant component in heat dissipation. It is interesting to note that conduction heating tends to fall away with rising frequency beyond approximately 100 MHz, compared with dipolar heating, because the mass of the ions is such that their movement is curtailed. Heating water by microwave energy usually involves both dipolar and conduction effects. Many industrial microwave processes are dependent on the presence of water in the workload.

In Section 2.4 there is a list of a wide range of materials commonly present in workloads, showing their dielectric properties which determine their rate of heating.

There is also a range of dielectric materials which is not polar and has high values of DC resistivity, and which does not heat significantly in a microwave field. These form a very important group because without them microwave insulators and transparent windows would not be realisable. The best and most common insulators are quartz, alumina, PTFE (Teflon®), polycarbonate, polyethylene and polypropylene.

2.2 Relative permittivity

The permittivity of free-space in vacuo has already been defined in Chapter 1 in terms of the mechanical force exerted between two charged particles spaced apart. If the space is filled with a dielectric material the force is increased by a factor called the relative permittivity for that material (ε'). Relative permittivity is a dimensionless number which for a lossless homogeneous material is the same in all three rectangular coordinate directions. It is simply related to refractive index for optical materials, and analogously with optics the lines of electric field entering the surface of the dielectric material are refracted (i.e. direction changed as well as magnitude), and also partially reflected. Other important related effects are the change in propagation velocity of a plane wave entering the material, and the change in characteristic impedance; see Section 1.5.

2.2.1 Complex relative permittivity and dielectric loss factor

Relative permittivity, as presented in Section 2.2, is a multiplying factor by which the capacitance of a vacuum capacitor would be increased if instead it were filled with a dielectric material of that permittivity; but it says nothing about the power absorption of the material.

Simple AC theory shows that a capacitor of value C connected to a sinusoidal generator of RMS voltage V at angular frequency ω will draw a current I where

$$I = j\, V \omega\, C \tag{2.1}$$

The operator symbol j signifies that the sinusoidal current leads the voltage by precisely 90°, and because there is no component of the current flowing in phase with the voltage there is no net power dissipation.

Consider now the capacitor C as a parallel-plate capacitor for which the capacitance value is well known and given by

$$C = \frac{A \varepsilon_0 \varepsilon^*}{d} \tag{2.2}$$

where ε^* is the relative permittivity of the dielectric material filling the space between the plates, A is the area of one plate and d is the spacing between the plates.

Suppose the permittivity itself has an imaginary component and is written as

$$\varepsilon^* = \varepsilon' - j\, \varepsilon'' \tag{2.3}$$

Substituting eqns. 2.2 and 2.3 in 2.1, the current I:

$$I = V\omega\, \frac{\varepsilon_0 A}{d} \left(j\varepsilon' + \varepsilon'' \right) \tag{2.4}$$

The first term in parentheses shows the component of current in phase quadrature with the voltage as it would be with a lossless capacitor, but the second term is a component in phase with the applied voltage, and therefore representing power dissipation. This power dissipation P is given by

$$P = \mathrm{Re}(VI) = V^2 \omega \frac{\varepsilon_0 A}{d} \varepsilon'' \tag{2.5}$$

The term ε'' clearly quantifies the power dissipation in the capacitor having a dielectric filling with relative permittivity defined by eqn. 2.1. It is called the 'loss factor' of the dielectric.

2.2.2 Dielectric loss angle

Another term sometimes used to quantify the 'lossyness' of a dielectric is the 'loss angle' (δ). This is the angle by which the resultant current differs from the ideal 90° phase angle relative to the voltage. Inspection of eqn. 2.4 reveals that

$$\delta = \tan^{-1} \frac{\varepsilon''}{\varepsilon'} \tag{2.6}$$

2.3 The dielectric heating equation

From eqn. 2.5 it is now a simple step to form a relation between the power-dissipation density, the loss factor and the electric field intensity and its frequency. Clearly:

(i)   The volume of dielectric material between the plates is Ad (cubic metres)
(ii)   The voltage stress in the dielectric is $E_i = V/d$ (volts per metre)
(iii) The total power dissipated (P watts) gives the power dissipation density p (watts per cubic metre) as $p = P/Ad$ (watts per cubic metre).
(iv)   Angular frequency $\omega = 2\pi f$ radians per seconds where f is in hertz.

Substituting the above into Eq. 2.5 and rearranging yields the dielectric heating equation:

$$p = 2\pi f \, \varepsilon_0 \, \varepsilon'' \, E_i^2 \qquad (2.7)$$

This derivation of the dielectric heating equation is not rigorous, but it gives the engineer a clear and simple argument based on the familiar hardware of a parallel-plate capacitor. Those seeking a more rigorous proof starting from Maxwell's equations are referred to other sources (e.g. Metaxas and Meredith, 1983).

Some very important features of dielectric heating are at once evident from eqn. 2.7:

(*a*) The power density dissipated in the workload is proportional to frequency where the other parameters are constant. This means that the volume of workload in the oven may be reduced as the frequency rises, resulting in a more compact oven.
(*b*) The power density is proportional to the loss factor.
(*c*) For a constant power dissipation density the electric field stress $\mathbf{E}_i$ reduces with $\sqrt{f}$. This means that, if ε'' remains constant with frequency, the risk of voltage breakdown reduces as the chosen operating frequency rises.
(*d*) Note that ε'' usually varies with frequency (Section 2.4), especially in materials where dipolar loss dominates. Generally, but not always, ε'' rises with frequency, adding to the effects (i) and (iv) above.
(*e*) The electric-field stress $\mathbf{E}_i$ is the field *within the dielectric*. The electric field just outside the dielectric $\mathbf{E}_{ext}$ is at least equal to $\mathbf{E}_i$ and may greatly exceed it, depending on local geometry and ε'. Voltage breakdown is a possible limitation due to high value of $\mathbf{E}_{ext}$. The value of $\mathbf{E}_{ext}$ is considered in Sections. 1.4.4 *et seq.* and 2.3.1
(*f*)   In practice the value of ε'' varies not only with frequency, but also with temperature, moisture content, physical state (solid or liquid) and composition. All these may change during processing, so it is important to consider ε'', and also ε', as variables during the process.

The dielectric heating equation is central to the decision of which is the best frequency to use for a specific application of RF or microwave heating.

The total power requirement from the heat balance (Section 3.5), and the available space for the heating oven and therefore the volume of the workload, having been determined, a preliminary figure for the power density is established. The power density calculated in this way must then be reviewed in terms of the physical properties of the workload. For example, will steam be generated internally to cause foaming or fracture? Sometimes controlled foaming is a desired result. This review may result in a need to reduce power density, or perhaps to allow an increase, resulting in a space saving.

With a knowledge of ε'', the internal voltage stress can be calculated for the proposed frequencies.

Next, the external voltage stress is estimated, and its magnitude may determine the optimum frequency if voltage breakdown is a possibility. It is usually found that a wide spectrum of frequencies is possible and the decision is then based on economic factors, and sometimes national regulations where there are restrictions on allocated frequencies.

2.3.1 Internal and external electric-field stress

In Section 2.3 the power dissipation density is given in terms of the electric-field stress within the workload. It is very important to know the magnitude of the electric-field stress *external* to the workload because it may reach an intensity sufficient to cause voltage breakdown of the gases in the vicinity; often this is a cocktail of hot processing gases.

Unfortunately the external voltage stress can be calculated analytically only for simple geometries, e.g. spheres, ellipses or flat plates remote from their edges. For others, it is necessary to use numerical methods, or, more usually, to rely on experience and judgement. As shown in Section 1.4, the value of E_{ext}/E_{int} is, depending on orientation of the E-field vector to the surface of the dielectric $1 < E_{ext}/E_{int} < \varepsilon'$. For the special case of a sphere, it is given by eqn. 1.17.

In practice the fields present in microwave or RF applicators are not uniform for many reasons:

(i)   The workload has boundary edges and corners, and its surfaces may be irregular, which cause field concentration by 'fringing'. Fringing is a well known effect in static field distributions of all kinds where local concentrations of stress, e.g. in mechanical, fluid-flow or electrical fields, are created by discontinuities.

(ii)   The workload may not be homogeneous, with considerable variations in permittivity, e.g. large volumes of lean meat and fat in a frozen block.

(iii) Metal structures may be present within the heating chamber, especially with sharp edges (they should be 'rounded').

(iv)   Dielectric structures may be present within the heating chamber to support the workload etc.

(v)   There may be standing waves due to reflections within the heating oven.

(vi)   Attenuation of energy propagating into the workload may occur (Section 2.5, penetration depth).

(vii) Field concentration may exist adjacent to power injection points where the power flux density is high.

These effects are often present simultaneously, and so the problem of estimating E_{ext} is compounded. Numerical methods may be used but they need care in interpretation, particularly to ensure that the mesh pitch is appropriate to the configuration.

In practice, the highest level of confidence is secured through experience and experiment.

2.3.2 Tables of E_i and E_{ext} at ISM frequencies for a range of dielectric properties

Tables 2.1–2.3 give values of E_i and E_{ext} for the three principal ISM frequencies of 27 MHz, 915 MHz, and 2450 MHz for a range of power-dissipation densities and a typical range of dielectric properties of the workload ε' and ε". Values of E_i are obtained directly from eqn. 2.7 and a uniform field is assumed; whilst values for E_{ext} are for workloads in the form of:

(*a*) a slab of dielectric material with E-field normal to the surface;
(*b*) a dielectric sphere.

Also shown in Table 2.1 is the rate of rise of temperature of a typical workload with density 0.8 gm/cm^3 and specific heat 0.5 cal/gm/deg C.

2.4 Dielectric properties of materials

The dielectric properties of materials vary widely, not only with composition, but also with density, temperature and frequency, and there is an extensive literature on methods of measurement (Von Hippel, 1954; Gallone *et al.*, 1966) and on the results obtained (Von Hippel, 1954; Bengtsson and Risman, 1971). Knowledge of dielectric data is essential in the design of heating systems because it enables estimates to be made of the power density and associated electric-field stress as presented in Sections 2.3 *et seq.*, and equally important to the microwave-penetration depth in the material (Section 2.5). It will be appreciated from the following that there can be substantial variations from the norm for a given material, and so it is desirable that measurements be made of each specific material to be processed; the following data should be used for guidance purposes only.

Table 2.1   Values of E_i and E_{ext} for 27 MHz

Power density (W/m³)	(W/cm³)	dT/dt (degC/min)	Dielectric loss factor 0.01 (V/cm)	0.03 (V/cm)	0.1 (V/cm)	0.3 (V/cm)	1 (V/cm)	3 (V/cm)	10 (V/cm)
(a) Values of E_i									
1.00E+05	0.1	4	8.16E+02	4.71E+02	2.58E+02	1.49E+02	8.16E+01	4.71E+01	2.58E+01
3.00E+05	0.3	11	1.41E+03	8.16E+02	4.47E+02	2.58E+02	1.41E+02	8.16E+01	4.47E+01
1.00E+06	1	36	2.58E+03	1.49E+03	8.16E+02	4.71E+02	2.58E+02	1.49E+02	8.16E+01
3.00E+06	3	107	4.47E+03	2.58E+03	1.41E+03	8.16E+02	4.47E+02	2.58E+02	1.41E+02
1.00E+07	10	357	8.16E+03	4.71E+03	2.58E+03	1.49E+03	8.16E+02	4.71E+02	2.58E+02
(b) Values of E_{ext} for slab workload, E_{ext} perpendicular to surface									
Relative permittivity									
1			2.58E+03	1.49E+03	8.16E+02	4.71E+02	2.58E+02	1.49E+02	8.16E+01
3			7.74E+03	4.47E+03	2.45E+03	1.41E+02	7.74E+02	4.47E+02	2.45E+02
10			2.58E+04	1.49E+03	8.16E+03	4.71E+03	2.58E+03	1.49E+03	8.16E+02
30			7.74E+04	4.47E+04	2.45E+04	1.41E+04	7.74E+03	4.47E+03	2.45E+03
(c) Values of E_{ext} for spherical workload									
1			2.58E+03	1.49E+03	8.16E+02	4.71E+02	2.58E+02	1.49E+02	8.16E+01
3			4.30E+03	2.48E+03	1.36E+03	7.85E+02	4.30E+02	2.48E+02	1.36E+02
10			1.03E+04	5.96E+03	3.26E+03	1.88E+03	1.03E+03	5.96E+02	3.26E+02
30			2.75E+04	1.59E+04	8.70E+03	5.02E+03	2.75E+03	1.59E+03	8.70E+02

dT/dt is the rate of rise of temperature for a typical workload with density 0.8 g/cm^3 and specific heat 0.5 $cal/g/degC$. For E_{ext} power density = 1.0 W/cm^2, scale pro rata to the voltage stresses of Table 2.1a for other power densities.

Table 2.2 Values of E_i and E_{ext} for 900 MHz

Power density		dT/dt	Dielectric loss factor						
			0.01	0.03	0.1	0.3	1	3	10
(W/m³)	(W/cm³)	(degC/min)	(V/cm)	(V/cm)	(V/cm)	(V/cm)	(V/cm)	(V/cm)	(V/cm)

(a) Values of E_i

(W/m³)	(W/cm³)	(degC/min)	0.01	0.03	0.1	0.3	1	3	10
1.00E+05	0.1	4	1.41E+02	8.16E+01	4.47E+01	2.58E+01	1.41E+01	8.16E+00	4.47E+00
3.00E+05	0.3	11	2.45E+02	1.41E+02	7.74E+01	4.47E+01	2.45E+01	1.41E+01	7.74E+00
1.00E+06	1	36	4.47E+02	2.58E+02	1.41E+02	8.16E+01	4.47E+01	2.58E+01	1.41E+01
3.00E+06	3	107	7.74E+02	4.47E+02	2.45E+02	1.41E+02	7.74E+01	4.47E+01	2.45E+01
1.00E+07	10	357	1.41E+03	8.16E+02	4.47E+02	2.58E+02	1.41E+02	8.16E+01	4.47E+01

(b) Values of E_{ext} for slab workload, E_{ext} perpendicular to surface

Relative permittivity

	0.01	0.03	0.1	0.3	1	3	10
1	4.47E+02	2.58E+02	1.41E+02	8.16E+01	4.47E+01	2.58E+01	1.41E+01
3	1.34E+03	7.74E+02	4.24E+02	2.45E+02	1.34E+02	7.74E+01	4.24E+01
10	4.47E+03	2.58E+03	1.41E+03	8.16E+02	4.47E+02	2.58E+02	1.41E+02
30	1.34E+04	7.74E+03	4.24E+03	2.45E+03	1.34E+03	7.74E+02	4.24E+02

(c) Values of E_{ext} for spherical workload

	0.01	0.03	0.1	0.3	1	3	10
1	4.47E+02	2.58E+02	1.41E+02	8.16E+01	4.47E+01	2.58E+01	1.41E+01
3	7.45E+02	4.30E+02	2.36E+02	1.36E+02	7.45E+01	4.30E+01	2.36E+01
10	1.79E+03	1.03E+03	5.65E+02	3.26E+02	1.79E+02	1.03E+02	5.65E+01
30	4.77E+03	2.75E+03	1.51E+03	8.70E+02	4.77E+02	2.75E+02	1.51E+02

dT/dt is the rate of rise of temperature for a typical workload with density 0.8 g/cm^3 and specific heat 0.5 $cal/g/degC$. For E_{ext}, power density = 1.0 W/cm^2, scale pro rata to the voltage stresses of Table 2.1a for other power densities.

Table 2.3   Values of E_i and E_{ext} for 2450 MHz

Power density (W/m³)	(W/cm³)	dT/dt (degC/min)	Dielectric loss factor 0.01 (V/cm)	0.03 (V/cm)	0.1 (V/cm)	0.3 (V/cm)	1 (V/cm)	3 (V/cm)	10 (V/cm)
(a) Values of E_i									
1.00E+05	0.1	4	8.57E+01	4.95E+01	2.71E+01	1.56E+01	8.57E+00	4.95E+00	2.71E+00
3.00E+05	0.3	11	1.48E+02	8.57E+01	4.69E+01	2.71E+01	1.48E+01	8.57E+00	4.69E+00
1.00E+06	1	36	2.71E+02	1.56E+02	8.57E+01	4.95E+01	2.71E+01	1.56E+01	8.57E+00
3.00E+06	3	107	4.69E+02	2.71E+02	1.48E+02	8.57E+01	4.69E+01	2.71E+01	1.48E+01
1.00E+07	10	357	8.57E+02	4.95E+02	2.71E+02	1.56E+02	8.57E+01	4.95E+01	2.71E+01
(b) Values of E_{ext} for slab workload, E_{ext} perpendicular to surface									
Relative permittivity									
1			2.71E+02	1.56E+02	8.57E+01	4.95E+01	2.71E+01	1.56E+01	8.57E+00
3			8.13E+02	4.69E+02	2.57E+02	1.48E+02	8.13E+01	4.69E+01	2.57E+01
10			2.71E+03	1.56E+03	8.57E+02	4.95E+02	2.71E+02	1.56E+02	8.57E+01
30			8.13E+03	4.69E+03	2.57E+03	1.48E+03	8.13E+02	4.69E+02	2.57E+02
(c) Values of E_{ext} for spherical workload									
1			2.71E+02	1.56E+02	8.57E+01	4.95E+01	2.71E+01	1.56E+01	8.57E+00
3			4.51E+02	2.61E+02	1.43E+02	8.24E+01	4.51E+01	2.61E+01	1.43E+01
10			1.08E+03	6.26E+02	3.43E+02	1.98E+02	1.08E+02	6.26E+01	3.43E+01
30			2.89E+03	1.67E+03	9.14E+02	5.28E+02	2.89E+02	1.67E+02	9.14E+01

- dT/dt is the rate of rise of temperature for a typical workload with density 0.8 g/cm^3 and specific heat 0.5 $cal/g/degC$.
For E_{ext}, power density = 1.0 W/cm, scale pro rata to the voltage stresses of Table 2.1a for other power densities.

Table 2.4   Dielectric properties of selected materials

Material	Temperature (°C)	$\varepsilon'/\varepsilon''$		
		30 MHz	1 GHz	2.5 GHz
a Liquids				
Water, ice	−12	3.8/0.7	3/0.004	3.2/0.003
Snow	−20	1.2/0.01	1.2/0.001	1.2<0.001
Water, distilled	+25	78/0.4	77/5.2	77/13
Water, distilled	+85	58/0.3	56/1	56/3
Acqueous NaCl				
0.1 molal solution	+25	76/480	76/30	76/20
0.3 molal solution	+25	76/1000	70/70	70/17
0.5 molal solution	+25	75/2400	68/140	68/54
Amyl alcohol	+25	—	—	3.5/1.2
Carbon tetrachloride	+25	2.2/<0.001	2.2/<0.001	2.17/0.0087
Ethyl alcohol	+25	23/3	12/3	7.0/6.5
Ethylene glycol	+25	41/3.5	20/8	12/12
Methyl alcohol	+25	31/1.0	27/6	25/15
N-Butyl alcohol	+25	—	—	3.60/1.96
N-Propyl alcohol	+25	—	—	4.74/2.94
Vaseline	+25	2.2/<0.001	2.2<0.001	2.2/0.0014
b Inorganic solids				
Alumina ceramic Al_2O_3	+25	8.9/0.0013	8.9/0.008	8.9/0.009
Beryllium oxide	+25	4.2/0.008	4.2/0.003	4.2/0.002
Borosilicate glass (Corning 7740)	+25	4.05/0.004	4.05/0.004	4.05/0.005
Borosilicate glass (Corning 7740)	+750	—	—	5.9/0.1
Brick, blue 'engineering'	+15	—	4.7/0.4	—
Clay, fine, 6% MC dwb	+15	—	1.8/0.34	—
Feldspar, 1.25 g/cm^3	+24	—	2.61/0.024	2.61/0.020
Fused quartz	+25	3.78/<0.001	3.78/<0.001	3.78/<0.001
Mica (muscovite) 0.45 g/cm^3	+24	—	1.63/0.006	1.62/0.005
Mica (phlogopite) 0.7 g/cm^3	+24	—	2.1/0.014	2.04/0.11
Silicon carbide	+20	—	—	30/11
Silicon, semiconductor grade	+25	—	4.3/<0.05	—
Silicon, semiconductor grade	+300	—	9/1.3	—

continued...

Table 2.4 continued

Material	Tempera-ture (°C)	$\varepsilon'/\varepsilon''$ 30 MHz	1 GHz	2.5 GHz
Soda–silica glass, 20% Na_2O, 80% SiO_2	+25	6.2/0.1	5.8/0.09	5.7/0.09
Soil, loamy, 0% MC dwb	+25	2.48/0.03	2.46/0.008	2.44/0.004
Soil, loamy, 14% MC dwb	+25	17/10	20/2.5	20/2.5
Soil, sandy, 0% MC dwb	+25	2.55/0.033	2.55/0.012	2.55/0.007
Soil, sandy, 17% MC dwb	+25	20/30	20/0.3	17/0.3
c Organic solids				
Bakelite	+24	4.6/0.34	3.8/0.26	3.7/0.23
Butyl rubber (<98% isobutylene)	+25	2.35/<0.002	2.35/<0.002	2.35/<0.002
Ethyl cellulose	+25	2.9/0.045	2.8/0.048	2.7/0.051
Methyl cellulose	+22	4.6/0.46	3.7/0.26	3.4/0.20
Nylon 66	+25	3.2/0.072	3.08/0.049	3.02/0.041
Polycarbonate thermoplastic	+15	—	—	2.72/0.0034 at 9 GHz
Polypropylene				
Polystyrene	+25	2.56/<0.003	2.55/0.0008	2.55/0.0008
Polystyrene	+80	2.54/<0.0008	2.54/0.0009	2.54/0.001
Polyethylene	+25	2.25/<0.0004	2.25/0.0005	2.25/0.0007
Polyvinyl chloride (PVC)	+20	2.86/0.029	2.85/0.016	2.85/0.016
Polyvinyl chloride (PVC)	+47	3.0/0.043	2.9/0.025	2.8/0.021
Polyvinyl chloride (PVC)	+76	2.83/0.065	2.8/0.049	2.7/0.036
Polyvinyl chloride (PVC)	+96	2.75/0.011	2.7/0.066	2.7/0.058
Polyvinyl chloride (PVC)	+80	5.3/0.34	4.5/0.49	4.4/0.43
PTFE (Teflon®)	+25	—	—	2.04 to 2.08/ 3.1×10^{-4} to 5×10^{-4}
Rubber nitril (natural)	+25	3.2/0.24	2.83/0.06	2.8/0.05
Rubber, EDPM, 57% silicious chalk	+100	—	—	3.1/0.05
Rubber, natural (pale crepe)	+25	2.4/0.01	2.4/0.008	2.2/0.006
Rubber, neoprene compound. 0.4% carbon black plus inert fillers	+24	5.0/0.52	4.1/0.16	4.0/0.14
Rubber, silicone SiO_2 filler	+20	3.2/0.012	3.2/0.03	3.2/0.03
Sodium benzoate, 5% MC	+15	—	2.0/0.1 —	—

continued. . .

Table 2.4 continued

Material	Temperature (°C)	$\varepsilon'/\varepsilon''$ 30 MHz	1 GHz	2.5 GHz
Sodium benzoate, 13% MC	+15	—	3.3/7.7	—
Talc	+15	—	1.7/0.04	—
d Foods and Agricultural Materials				
Apple	+19	—	61/9	57/12
Apple (Granny Smith), 88.4% MC wwb, 0.76 g/cm³	+23	—	57/9	53/12
Banana (Cavendish) 78.1% MC wwb, 0.94 g/cm³	+23	—	64/20	57/19
Beef, lean, vacuum freeze dried	+80	—	1.64/0.032	1.5/0.018
Beef, cooked, 8.6% fat	−20	—	—	4.8/0.44
Beef, cooked, 8.6% fat	+3	—	—	31.6/12.2
Beef, cooked, 8.6% fat	+60	—	—	28.7/9.2
Beef, raw 8.6% fat	−10	—	—	5.9/1.03
Beef, raw 8.6% fat	+40	—	—	41.8/12.6
Beef, raw 8.6% fat	+60	—	—	39.6/13.1
Beef, raw 1.07% g/cm³	+25	—	56/22	52/17
Beef, raw 1.07% g/cm³	+65	—	51/32	49/19
Beef, raw 1.07% g/cm³	+5	—	58/20	54/20
Beefsteak, round	−15	3.53/0.37	—	—
Beefsteak, round	−9.5	3.76/0.87	5.29/0.21	5.29/0.21
Beefsteak, round	−4	8.77/1.95	7.43/1.44	8.34/0.66
Bread, starch-reduced	+15	—	9.4/4.0	—
Bread dough	c+15	—	24/12	20/10
Bread, (10 min bake)	c+15	—	18/8	12/4.5
Bread, (20 min bake)	c+15	—	12/4	10/2
Bread, (30 min bake)	c+15	—	8/2	4.5/1.2
Butter, 2.1% salt, 16.5% MC dwb	−10			3.95/3.8
Butter, ditto	−5			3.9/4.0
Butter, ditto	0			3.90/3.95
Butter, ditto	+10			4.07/4.02
Carrot	+22	—	68/20	65/15

continued. . .

Table 2.4 continued

Material	Tempera-ture (°C)	$\varepsilon'/\varepsilon''$ 30 MHz	1 GHz	2.5 GHz
Fish, cod	−25	5.5/1.4	4.0/0.7	—
Fish, cod	−5	13/12	8.5/4.5	—
Fish, cod	+10	83/250	66/40	—
Ham, cooked	−20	—	—	7/2
Ham, cooked	−10	—	—	10/5
Ham, cooked	−5	—	—	20/6
Ham, cooked	+0	—	—	42/21
Ham, cooked	+20	—	—	45/24
Honey 100%	+25	37/3 approx.	13.7/7	—
Honey 50%	+25	—	69/11	—
Lard	+82	—	2.6/0.14	2.5/0.15
Oil, corn	+25	—	2.6/0.17	2.53/0.14
Oil, corn	+49	—	2.7/0.17	2.6/0.14
Oil, corn	+82	—	2.7/0.15	2.6/0.16
Onion, white 92.9% MC, 0.97 g/cm^3	+23	—	66/14	63/15
Orange (Navel) 0.92 g/cm^3, 87.5% MC wwb	+23	—	73/15	70/15
Peach	+20	—	72/14	68/14
Pear	+23	—	68/13	64/14
Potato	+22	—	66/30	61/19
Strawberry, Duchesne 0.76 g/cc, 92.1% MC wwb	+23	—	73/15	71/13
Sugar, saturated solution	+25	45/5 approx.	24/6	—
Tallow	+82	—	2.6/0.13	2.5/0.14
Wheat flour, 14% MC dwb	+15	—	4.5/3.2	—
Wheat flour, 2% MC dwb	+15	—	2.7/0.67	—
Wheat, bulk, SG 0.79 10.6% MC wwb	+24	3.7/3.7	2.8/0.28	2.6/0.26
Yeast (raw)	c+15	—	46/18	40/14

Notes: 1. The dielectric properties are given in the format $\varepsilon'/\varepsilon''$.
2. The data are interpolated from adjacent frequencies to correspond approximately to the designated ISM frequency bands. Dipole resonance may, in a few cases, result in errors. The data are intended for preliminary guideline use, and confirmatory measurement is strongly recommended.

2.5 Penetration depth

As a wave progresses into a dielectric-heating workload, its amplitude diminishes owing to absorption of power as heat in the material. In the absence of reflected waves in the material, the field intensity and its associated power flux density fall exponentially with distance from the surface. Because the power absorbed in an elemental volume of material is proportional to the power flux density flowing through it, the power dissipation also falls exponentially from the surface. The rate of decay of the power dissipation is a function of both the relative permittivity ε' and the loss factor ε''. The 'penetration depth' D_p (metres) is defined as the depth into the material at which the power flux has fallen to $1/e$ (=0.368) of its surface value, and is readily shown (Metaxas and Meredith, 1993) to be given by:

$$D_p = \frac{\lambda_0}{2\pi\sqrt{(2\varepsilon')}} \cdot \frac{1}{\sqrt{\left[\left\{1+\left(\frac{\varepsilon''}{\varepsilon'}\right)^2\right\}^{0.5}-1\right]}} \tag{2.8}$$

Where $\varepsilon'' \le \varepsilon'$, eqn. 2.8 simplifies to eqn. 2.9, with an error up to 10%:

$$D_p \approx \frac{\lambda_o\sqrt{\varepsilon'}}{2\pi\varepsilon''} \tag{2.9}$$

The penetration depth is a very important parameter for a workload because it gives an immediate first-order indication of the heat distribution within it. In a semi-infinite slab of ideal material (i.e. having constant values of ε' and ε'' with temperature), and with a plane wave at normal incidence, the temperature rise θ_z at depth z is given by

$$\theta_z = \theta_0 \exp(-z/D_p) \tag{2.10}$$

An approximation of the heat distribution in a thick slab, heated consecutively from each side, can be made by superposition of two such exponential distributions having origins at the two faces; some thawing and tempering ovens use this principle.

Note that D_p does not mean that there is no heating at a depth exceeding the penetration depth. By integration of eqn. 2.10 it can be shown that the heat dissipated in the layer bounded by the surface and the plane at depth D_p is 63.2% of the total, the balance being dissipated in the material at depths greater than D_p.

Note also that penetration depth is not the same as skin depth. The penetration-depth effect arises essentially from the dielectric-loss factor ε''

which causes power dissipation in accordance with eqn. 2.7. Because of this power dissipation, the power-flux density falls as the wave propagates into the material. Skin depth (Section 4.5.6.1) refers to the surface 'skin' of a conducting wall in which the majority of the microwave current flows. The microwave energy is nearly fully reflected from the wall and only a very small proportion is absorbed. The currents are constrained to flow in a very thin skin by the incident magnetic field inducing currents to flow which, in turn, create a secondary magnetic field virtually equal and opposite to the excitation field within the conductor. However, this secondary magnetic field adds to the magnetic field of the incident wave grazing the surface of the conductor, and sets up a reflected wave from the surface.

Table 2.5 shows the normalised penetration depth as a function of the dielectric properties of materials ε' and ε'', calculated from eqn. 2.8. To obtain the penetration depth D_p (metres), multiply the normalised value by the free-space wavelength λ_0 (metres).

It is interesting to note, that at constant loss factor ε'', the penetration depth increases with the relative permittivity ε'. The reason is that, as ε' increases, the characteristic impedance Z_0 of the medium falls [as $1/(\sqrt{\varepsilon'})$], so that the electric-field stress E_i also falls where the power-flux density remains constant; see Section 1.6. Since the power dissipated per unit volume is proportional to E_i^2 (eqn. 2.7), it too will fall, so less power is abstracted from the wave as it progresses and the rate of decay of power-flux density will fall, correspondingly increasing D_p.

At the 27 MHz ISM frequency ($\lambda_0 = 11$ m) penetration depth is not usually a significant feature because the workload is small in comparison. Sometimes it is claimed that there is no penetration-depth effect because there is no immediately obvious power flux associated with a wavefront. Consider a conventional parallel-plate 27 MHz heating applicator forming a capacitor with a lossy dielectric. The current waveform will lead the voltage by nearly, but not quite, 90°, the shortfall being due to the power absorbed in the load which results in a small component of current in phase with the voltage. Consider now the Poynting vector (Section 1.6), where the electric field is a vector normal to the surface of the plates, and the magnetic field a vector circumferential to the plates and parallel to them. The component of electric field in phase with the magnetic field forms a Poynting vector in a direction radially inwards towards the centre of the plates, representing a net power flow into the lossy dielectric. The main component of the magnetic field, in phase quadrature with the electric field, also forms an instantaneous Poynting vector, but the associated power flow is then alternately into and away from the dielectric on consecutive half cycles, representing the ebb and flow of stored energy in the capacitor. There is therefore a wavefront approaching the applicator from the side, with a corresponding penetration-depth effect from the side towards the centre of the load.

Table 2.5   Normalised penetration depth

Loss factor	Relative permittivity									
	1	1.5	2	3	5	7	10	15	30	100
0.001	1.59E+02	1.95E+02	2.25E+02	2.76E+02	3.56E+02	4.21E+02	5.03E+02	6.16E+02	8.72E+02	1.59E+03
0.005	3.18E+01	3.90E+01	4.50E+01	5.51E+01	7.12E+01	8.42E+01	1.01E+02	1.23E+02	1.74E+02	3.18E+02
0.01	1.59E+01	1.95E+01	2.25E+01	2.76E+01	3.56E+01	4.21E+01	5.03E+01	6.16E+01	8.72E+01	1.59E+02
0.03	5.31E+00	6.50E+00	7.50E+00	9.19E+00	1.19E+01	1.40E+01	1.68E+01	2.05E+01	2.91E+01	5.31E+01
0.1	1.59E+00	1.95E+00	2.25E+00	2.76E+00	3.56E+00	4.21E+00	5.03E+00	6.16E+00	8.72E+00	1.59E+01
0.15	1.06E+00	1.30E+00	1.50E+00	1.84E+00	2.37E+00	2.81E+00	3.36E+00	4.11E+00	5.81E+00	1.06E+01
0.3	5.36E-01	6.53E-01	7.52E-01	9.20E-01	1.19E+00	1.40E+00	1.68E+00	2.05E+00	2.91E+00	5.31E+00
0.5	3.28E-01	3.95E-01	4.54E-01	5.53E-01	7.13E-01	8.43E-01	1.01E+00	1.23E+00	1.74E+00	3.18E+00
0.7	2.40E-01	2.86E-01	3.26E-01	3.96E-01	5.10E-01	6.02E-01	7.19E-01	8.81E-01	1.25E+00	2.27E+00
1	1.75E-01	2.05E-01	2.32E-01	2.79E-01	3.58E-01	4.22E-01	5.04E-01	6.17E-01	8.72E-01	1.59E+00
1.5	1.26E-01	1.43E-01	1.59E-01	1.89E-01	2.40E-01	2.82E-01	3.36E-01	4.11E-01	5.81E-01	1.06E+00
3	7.65E-02	8.26E-02	8.88E-02	1.01E-01	1.23E-01	1.43E-01	1.70E-01	2.06E-01	2.19E-01	5.31E-01
5	5.56E-02	5.83E-02	6.12E-02	6.69E-02	7.82E-02	8.89E-02	1.04E-01	1.25E-01	1.75E-01	3.18E-01
7	4.57E-02	4.73E-02	4.90E-02	5.24E-02	5.93E-02	6.61E-02	7.58E-02	9.03E-02	1.25E-01	2.28E-01
10	3.74E-02	3.83E-02	3.93E-02	4.13E-02	4.53E-02	4.93E-02	5.53E-02	6.47E-02	8.83E-02	1.59E-01
30	2.09E-02	2.11E-02	2.12E-02	2.16E-02	2.23E-02	2.31E-02	2.42E-02	2.61E-02	3.19E-02	5.36E-02
100	1.13E-02	1.13E-02	1.14E-02	1.15E-02	1.17E-02	1.18E-02	1.18E-02	1.21E-02	1.30E-02	1.30E-02

The penetration depth is obtained by multiplying the tabulated figure by the free-space wavelength.

2.6 Limits on power-dissipation density

In Section 2.3 the electrical parameters affecting the power-dissipation density are presented and it is clear from Table 2.1 that very high power-dissipation density is achievable with modest applied voltage stress, depending on dielectric properties.

In practice it is usually undesirable for many reasons to strive for the highest power density and shortest possible process time. Obviously, such an objective will result in at least one of the parameters of the process being close to its practical limit, e.g. voltage breakdown. Where dielectric heating is to replace a conventional process which takes many hours, its commercial advantage may be accomplished by a process time of, say, 15 min with very modest parameter values; there would be no significant advantage in saving a further 10 min by raising the parameter values to a point of low factor of safety. Management judgement is obviously relevant in such cases where there is no physical limitation within the proposed range of operating parameters.

Some specific limitations of performance are given is Sections 2.6.1–2.6.3.

2.6.1 Thermal limitations

Thermal limitations arise usually, but not exclusively, from nonuniform heating or an inhomogeneous workload, and are of three types:

(*a*) impressed temperature gradients which result in mechanical stress gradients in the workload and possible fracture;

(*b*) moisture gradients in drying, resulting in differential shrinkage and fracture, a particular problem in the rapid drying of ceramics. This effect occurs even with perfect heating uniformity. Process time is critical. Heat flows within the material not only from conventional heat conduction but also through the migration of hot water diffusing towards the surface. Surface heat-transfer conditions may be critical and rapid surface evaporation may need to be discouraged by maintaining a controlled high humidity. The mechanisms are collectively complicated and cannot reliably be quantified, and extensive experiment is essential for each material;

(*c*) internally generated steam pressure due to local boiling within the workload, causing mechanical stress leading to rupture, possibly with explosive violence; the bursting of eggs in a domestic microwave oven is a well known example. An unstable condition can arise with a poorly-mixed solution, e.g. sugar–syrup, where concentrated solution is at boiling temperature, in excess of other liquid at lesser concentration. Stirring, or just heat equilibration with time, may cause the less-concentrated liquid to boil on coming into contact with the concentrated solution, resulting in 'boiling over' which may occur after the microwave heating is finished. This is the probable cause of accidents in the home with domestic microwave ovens where violent boiling has suddenly and unexpectedly occurred after the workload has been removed from the oven. It is particularly hazardous in containment vessels with a narrow neck.

2.6.2 Voltage breakdown and arcing

In Sections. 2.3.1 and 2.3.2, the internal and external electric-field stresses (E_{int} and E_{ext}) are evaluated as a function of the power-dissipation density in the workload. Clearly, if either of these stresses reaches the critical level, even locally, voltage breakdown will occur. The resulting ionised gases form an arc which will then present a low-resistance electrical path, and considerable power will be dissipated locally resulting in burn damage to the workload and possibly the equipment (e.g. conveyor band). Usually, the electric-field stresses applied are considerably less than the breakdown levels, but where the required power-dissipation density is high, or the loss factor is low, or where operation is under partial vacuum, the conditions need careful scrutiny.

The field intensity required to maintain an arc is considerably less than that required to initiate it, and it is invariably necessary to switch off the microwave power to extinguish it. A time period of a few seconds (2–5 s. typical) must then be allowed for the ionised gases to disperse before reapplying the microwave power; if the workload is damaged in the episode, there is a high risk of breakdown recurring at the same location. High-speed detection and suppression of arcs, together with automatic restarting, are an important feature of advanced equipment.

For air at ambient atmospheric pressure and temperature (18°C = 288 °K), the voltage-breakdown stress is substantially constant from DC through the microwave frequency spectrum at about 30 kV/cm *in a uniform field*. The relevant voltage in the microwave context is the peak of the AC sine wave. In using this figure, care must be taken to allow for field-concentration effects, and any effects under abnormal or fault conditions which would erode the margin to breakdown.

Paschen's law (Joos, 1958) states that the breakdown voltage of a gas is proportional to the absolute pressure or density, i.e. the number of molecules of gas present. Since, at constant pressure, the density falls with rising temperature, it follows that the voltage-breakdown strength falls with rising absolute temperature. Thus at 200°C and atmospheric pressure, the voltage-breakdown stress of air is about $30 \times 288/473 = 18$ kV/cm. It is therefore very important to allow for the temperature of the surface of the workload and of the gases within the oven when estimating voltage-breakdown limits.

At low pressure, the voltage-breakdown stress E_b falls linearly in accordance with Paschen's law until the pressure falls to about 15 mbar, when $E_b = 450$ V/cm. It reaches a minimum around 1 mbar before rising again as the pressure is further reduced. This minimum voltage breakdown stress is frequency dependent and is about 190 V/cm at 2450 MHz, and 80 V/cm at 915 MHz. In most vacuum drying processes a drying temperature of 33°–46°C is usually low enough, equivalent, respectively, to 50–100 mbar, and at this pressure the voltage-breakdown stress is the same for all frequencies, in the range 1500–3000 V/cm. Figures 2.1 and 2.2 show the voltage-breakdown stress for air and water vapour, respectively.

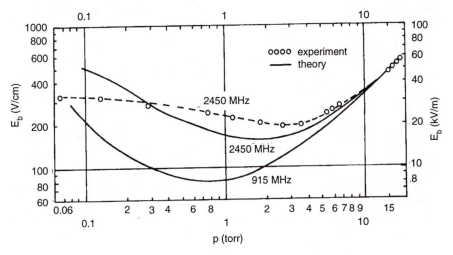

Figure 2.1   Voltage breakdown of air against vacuum pressure

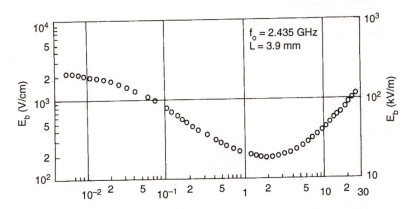

Figure 2.2   Voltage breakdown of water vapour against vacuum pressure

Because of the statistical nature of voltage breakdown, and the many factors which cause the voltage stress to rise substantially above the norm, it is necessary to operate with a wide margin between the average voltage stress and the breakdown voltage stress. Great care must be taken with detail design if the margin is less than 1:5.

Arcing may also result with workloads of large particulate form, where the particles are touching or in very close proximity (Figure 5.5). Analogously, consider an electrostatic model of two conducting spheres placed nearly touching each other in a uniform electric field, so that their centres are on a line parallel to the E-field vector. The displacement current 'captured' by the spheres will pass through the (small) gap between them over a small area so that the displacement-current density will be greatly increased, creating locally a high value of electric-field stress compared with the field previously existing. Especially if the particles have high relative permittivity, the electric-field stress in the space around the point of near contact will similarly be greatly intensified, and a correspondingly high stress occurs within the particle at the same point, resulting in intense local heating. There may be immediate voltage breakdown in the gap between the particles; but, if not, the intense local heating may so raise the temperature of the gas in the vicinity that its density falls with a corresponding fall in breakdown-voltage strength in accordance with Paschen's law, giving a delayed voltage breakdown.

This last phenomenon is especially likely when drying particulates and in heat treatment of foods, especially root vegetables and brassicas, where the high relative permittivity of the water present enhances this 'focusing' effect. Dielectric heating at all frequencies exhibits this effect, the only solution of which is to operate with a low power density.

2.6.3 Thermal runaway

Thermal runaway is a condition which arises when the power dissipation in a small elemental volume within a workpiece exceeds the rate of heat transmission to its surroundings, so that the rate of increase in enthalpy is greater than in its neighbours. The temperature increases at a faster rate than in the surroundings, until decomposition occurs. Thermal runaway invariably degenerates into arcing, and it is often difficult to decide which sequence of events has occurred.

In practice, the mechanisms involved in thermal runaway are very complicated and almost impossible to quantify. It is a statistical phemenon around which the operational conditions are set so that the probablity of a runaway incident is within acceptable bounds; conditions which can only be determined by experiment.

As runaway is a localised inequality in the heat-balance equation, it is easy to identify the parameters involved: the heat input, and the heat dissipation.

(i) *Heat input*
Attention is focused on the heating equation (eqn. 2.7), notably the E-field E and the loss factor ε''.

The E-field may be locally high due to field concentration near a boundary interface, or due to a standing wave or close proximity to a power-feed-point. The last two of these are field concentrations extending over a volume comparable with a sphere of diameter at least in the order of $0.1\lambda_d$, which is usually much larger than the volume of a runaway zone in its initial phase, certainly at 915 Mhz, and so these are not usually involved. It may also be increased locally within the workpiece by an abnormality in ε' in an inhomogeneous mixture. Metal particles, especially wire, are likely to create field concentration.

The loss factor ε'' is often a rising function of temperature, notably in plastics and rubber especially as high-temperature decomposition begins. The products of decomposition, especially carbon, may have a much higher value of ε''. As a result, the higher the temperature, the greater the power dissipation, causing the temperature to rise yet higher in a runaway.

(ii) Heat dissipation
The hot-spot element is cooled by heat diffusion to its surrounding material, being determined solely by the thermal diffusivity κ and the temperature gradient. The thermal conductivity K (see eqn. 3.10) may fall due to formation of voids with high temperature decomposition, reducing the heat transmission away from the hot-spot. Materials with low thermal conductivity and diffusivity are more prone to thermal runaway.

The above provides some guidelines for design to avoid thermal runaway:

(*a*)   operate with as low a mean power density as possible;
(*b*)   design the oven feed system to give as uniform an illumination as possible, specifically avoiding field concentrations;
(*c*)   ensure that the workload is uniformly spread in the oven;
(*d*)   take note of inhomogeneity of the workload (e.g. voids, lean and fat). If these cannot be avoided, the power density must be set accordingly; and
(*e*)   provide a detection system, e.g. smoke detector.

2.7 The engineering aspects of power-dissipation density in applicator design

In Section 2.6, the principal limits on power-dissipation density are given, and quantified where possible. Each application must be examined in detail for possible proximity to these limits, and indeed to any other limits peculiar to the application; such examination must be based on a simple theoretical assessment, followed essentially by experimental verification in a laboratory model. Usually the parameters affecting the limits are imperfectly known, and there can be a very wide range of values, for example in the voltage-breakdown level in wet materials

A major part of applicator design is therefore a survey of the limits on power-dissipation density imposed by voltage-breakdown and thermomechanical effects, with a particular emphasis on the likely range of variability of the parameters involved. Equally important is the estimate of power density required for the satisfactory accomplishment of the task, usually set by the process time required. Clearly, there is an optimum compromise.

Section 3.5 quantifies the heat balance for the process, i.e. the total amount of power to inject the desired enthalpy at the proposed mass flow of the workload. Section 2.3 relates the power-dissipation density in the workload to the applied internal and external electric field strengths. Obviously, the total power required and the power density are simply related via the exposure time and the volume of the workload V_w exposed in the heating applicator. Note that V_w is not the volume flow of the workload, but is linearly proportional to the exposure time. The greater V_w , the lower the power density, but the longer the exposure time, and the slower the rate-of-rise of temperature.

Clearly, V_w determines the dimensions of the oven whether a batch type or for a continuous process. However, where the thickness of the workload is a significant fraction of the penetration depth D_p (e.g. > 20%), due allowance must be made for the higher power density at the surface. Eqn. 3.7 quantifies this effect in defining the mean temperature rise T_m in a thick slab (thickness L) in terms of the surface temperature (T_s) and the penetration-depth ratio (L/D_p). As the local temperature rise is proportional to the local power density, eqn. 3.8 also gives the average and surface power densities. Figure 2.3 shows the increase in power density at the surface as the thickness increases relative to the penetration depth. Note that this curve assumes that no

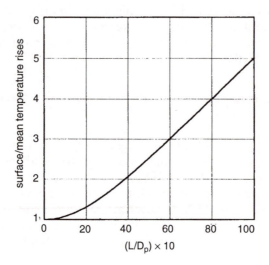

Figure 2.3 Surface/mean-temperature rise in thick slab heated from both sides sequentially against L/D_p

standing wave is present; presence of a standing wave may increase or reduce the surface power density, depending on its position.

In a multimode oven the distribution of power density in the workload varies considerably within the oven, even though the uniformity of heating overall is good, as discussed in Chapter 5. The electric-field intensity locally may rise substantially above the average value, particularly with a lightly loaded small oven with a low mode-spectrum density. The effect can only be evaluated by numerical solution of Maxwell's equations, or experimentally.

2.8 Processing at elevated pressure

Processing at pressure exceeding atmospheric is necessary where wet materials are to be treated at a temperature greater than boiling point at atmospheric pressure. Water usually is the liquid, but other volatiles are possible in chemical reactions.

Food is the most common workload requiring high process temperature, and the pressure required for a given temperature is determined by standard steam tables (e.g. Kaye and Laby, 1966), with an allowance, where appropriate, for any increase in temperature due to dissolved solids, e.g. sugars. These criteria are discussed in Section 3.5.5.1. Processes working at high pressure are typically sterilisation and high-speed cooking.

2.8.1 Pressure locks

Most pressure systems are continuous-flow, where the workload, in particulate form, enters and leaves the pressure vessel through mechanical pressure locks. These pressure locks may be continuously rotating star valves or alternating 'bins' in pairs. The rotary star valve comprises a circular drum with discrete pockets in its outer surface to hold the workload particles. It rotates within a stator having open ports diametrically and vertically opposite. The clearance between rotor and stator is fine to minimise direct gas leak. The workload enters from above, and is transferred to the lower port whence it falls onto a conveyor in the pressure vessel. The rotary star valve also forms a good microwave seal, but care must be taken to ensure there is no arcing between rotor and stator due to excessive wall-current intensity in its vicinity.

The alternating-bin pressure lock is suitable for larger particles. At each end of the oven it comprises a pair of storage bins each with sliding gate valves at their inlet and outlet ports. In operation, at the output end of the oven, one bin has its outlet valve closed and fills under pressure with its inlet valve open. Simultaneously, the other bin has its inlet valve closed and discharges workload to atmosphere through its open outlet valve. After a preset time delay, the roles are reversed giving thereby a substantially continuous-flow pressure lock The bins at the input end of the oven operate on a similar cycle. This pressure lock is more complicated than the star valve and introduces greater pressure surges into the process vessel. Horizontal gate valves are liable to jam due to workload particles being pressed into their seatings. Moreover, the bins and associated pipework are pressure vessels and must therefore conform to the same specifications as the main pressure vessel.

2.8.2 Microwave pressure windows

Pressure windows have two uses: first to prevent undesirable gases from entering the waveguide feed system, and secondly to provide a high-pressure lock in a vacuum or high pressure applicator. For the first window, the pressure differential is modest, in the region of a few millimetres water gauge, while for the second it may be 1 bar or more. The mechanical design reflects these differences; specifically, for the high-pressure window it is necessary to provide for a safe discharge of the stored pressure energy in the vessel should the window fail catastrophically.

It is not economically practical to operate the whole waveguide system, from oven to generator, at elevated pressure, because rectangular waveguide, without expensive reinforcement, distorts severely at the usual working pressure, greater than 2 bar. Moreover the output window of the generator (magnetron) is designed for working at atmospheric pressure. A pressure window is therefore required, and is usually placed at the wall of the pressure vessel.

The essential requirements of the pressure window are:

(i)   to form a perfect gas-tight seal;
(ii)   to transmit microwave power with minimum loss and reflection;
(iii)   to be of a material of very low loss factor ε'' to minimise microwave heating of the window;
(iv)   to be mechanically stable; some plastics materials creep at high temperature under stress, and some ceramics are prone to chip;
(v)   to operate with a high factor of safety between working and fracture mechanical stresses, taking into account fatigue effects from repeated pressure cycling;
(vi)   to be chemically inert so as not to react with gases within the vessel, and not to carbonise should an arc form;
(vii)   for food applications, to be a material acceptable for toxicity;
(viii) to be easy to clean, with smooth surfaces and no 'bug traps';
(ix)   to release the high-pressure gas within the vessel in a controlled, safe manner should the window fail; and
(x)   to be translucent, so that arc-detection equipment can be fitted on the atmospheric-pressure side of the window, looking through it.

In addition, the window must be protected from condensation and droplets from wash-down water and from other matter to avoid arcs. It is not safe to assume that the microwave power will evaporate water from the window surface before an arc can form. This requirement is readily achieved by heating the window conventionally; in most cases, such heaters would be provided for the whole vessel to minimise heat loss from the workload, and to ensure freedom from droplets on the walls. The temperature must be at least equal to the boiling temperature at the operating pressure.

2.8.2.1 Pressure windows: microwave design

Most pressure windows in industrial microwave equipment comprise a flat plate of dielectric material, mounted across a waveguide, or across the aperture of a horn. These are simple low-cost windows which can easily be replaced on site.

Such windows cause a reflection because they represent a section of waveguide with different propagation parameters to those of the main waveguide, and for thin windows form a parallel capacitive susceptance across the waveguide, but there is a more complicated transformation of admittance as the thickness increases.

The admittance of a window is readily calculated by applying the lossless-transmission-line equation (eqn. 4.11) to the dielectric-loaded section due to the window. The propagation constant β and the characteristic admittance Y_0 are adjusted to allow for the permittivity of the dielectric, as described in Section 1.5. The waveguide on the load side of the window is assumed reflectionless, and it is required to find the input admittance on the generator side of the window, and the corresponding VSWR. Table 2.6 shows the results of this computation at the principal frequencies, for a selection of commonly used window materials. There is a very important case of the window thickness being $\lambda_{dg}/2$ where λ_{dg} is the waveguide wavelength in the dielectric-loaded section of waveguide: this gives a perfect match, but is usually impractical because the thickness is excessive. However, for alumina, $\varepsilon' = 8.9$, the permittivity is sufficiently high to shorten the waveguide wavelength enough that a window satisfying this condition is only 20.57 mm thick at 2450 MHz.

These susceptances can readily be matched by a second reactance positioned to cancel that of the window. If a single window is to be used, it can be matched by an inductive or a capacitive iris by the procedure of Section 4.3.5. It is preferable to avoid placing the iris close to the window, because it will reduce the effective area and increase the power density. It is also good practice to place the iris on the generator side of the window because it may otherwise form a dirt trap in the applicator.

A double window is sometimes used as a protection against damage if a window fails catastrophically, releasing the pressure energy in the pressure vessel, as described in Section 2.8.2.2. If the windows are electrically identical, they can mutually match by placing them so that their corresponding surfaces are $\lambda_g'/4$ apart where λ_g' is the effective electrical length, comprising the waveguide wavelengths of a section of unfilled waveguide and of one of the windows.

Table 2.6   Admittance and VSWR of dielectric windows

a Frequency 915 MHz, waveguide size WG4 248 × 124 mm

t (mm)	Material	ε'	y	VSWR	Material	ε'	y	VSWR
1	Polyethylene	2.25	1.00+j0.013	0.987	Teflon	2.06	1.00+j0.012	0.988
2			1.001+j0.027	0.974			1.001+j0.024	0.976
4			1.005+j0.053	0.948			1.004+j0.049	0.952
6			1.010+j0.080	0.923			1.009+j0.073	0.930
8			1.019+j0.106	0.899			1.016+j0.097	0.908
10			1.029+j0.131	0.876			1.025+j0.12	0.886
15			1.066+j0.191	0.823			1.057+j0.17	0.837
20			1.117+j0.243	0.775			1.102+j0.223	0.792
25			1.182+j0.283	0.735			1.158+j0.262	0.753
1	Polycarbonate	2.70	1.00+j0.016	0.984	Alumina	8.90	1.001+j0.033	0.968
2			1.00+j0.032	0.969			1.006+j0.065	0.937
4			1.006+j0.063	0.939			1.023+j0.130	0.878
6			1.013+j0.094	0.910			1.051+j0.191	0.825
8			1.024+j0.124	0.882			1.092+j0.249	0.777
10			1.038+j0.154	0.856			1.145+j0.299	0.734
15			1.085+j0.223	0.796			1.328+j0.378	0.650
20			1.152+j0.280	0.744			1.558+j0.339	0.596
25			1.237+j0.319	0.701			1.737+j0.124	0.570
55			—	—			1.00+j0.002	0.998

continued. . .

Table 2.6a continued

t (mm)	Material	ε'	y	VSWR	Material	ε'	y	VSWR
1	Quartz	3.78	1.001+j0.020	0.980	Polypropylene	3.5	1.001+j0.019	0.981
2			1.002+j0.040	0.961			1.002+j0.038	0.962
4			1.009+j0.080	0.923			1.008+j0.076	0.927
6			1.020+j0.119	0.887			1.019+j0.114	0.892
8			1.036+j0.158	0.853			1.033+j0.150	0.860
10			1.057+j0.194	0.821			1.052+j0.185	0.829
15			1.129+j0.276	0.751			1.118+j0.265	0.761
20			1.231+j0.336	0.694			1.211+j0.325	0.705
25			1.359+j0.360	0.649			1.328+j0.354	0.660

Table 2.6b Frequency 2450 MHz, waveguide size WG9A 86.4 x 43.2 mm

t (mm)	Material	ε'	y	VSWR	Material	ε'	y	VSWR
1	Polyethylene	2.25	1.002+j0.044	0.957	PTFE	2.06	1.002+j0.040	0.961
2			1.010+j0.087	0.916			1.008+j0.080	0.923
4			1.039+j0.172	0.841			1.034+j0.157	0.854
6			1.088+j0.251	0.775			1.076+j0.230	0.792
8			1.158+j0.320	0.719			1.137+j0.294	0.739
10			1.249+j0.373	0.672			1.215+j0.345	0.693
15			1.547+j0.377	0.590			1.476+j0.182	0.611
20			1.790+j0.116	0.555			1.723+j0.182	0.571
25			1.717+j0.253	0.564			1.735−j0.157	0.569

continued...

Table 2.6b continued

t (mm)	Material	ε'	y	VSWR	Material	ε'	y	VSWR
1	Polycarbonate	2.70	1.003+j0.052	0.950	Alumina	8.90	1.012+j0.107	0.899
2			1.012+j0.103	0.902			1.047+j0.211	0.810
4			1.050+j0.202	0.816			1.169+j0.395	0.734
6			1.114+j0.293	0.743			1.457+j0.495	0.576
8			1.206+j0.369	0.682			1.785+j0.404	0.522
10			1.326+j0.420	0.633			1.989+j0.061	0.502
15			1.695+j0.346	0.555			1.392-j0.485	0.593
20			1.862-j0.068	0.536			1.004-j0.061	0.940
25			1.590-j0.405	0.571			1.242+j0.427	0.646

N.B. Alumina is a perfect match for t = 20.57 mm

t (mm)	Material	ε'	y	VSWR	Material	ε'	y	VSWR
1	Quartz	3.78	1.005+j0.040	0.961	Polypropylene	3.5	1.004+j0.063	0.939
2			1.008+j0.080	0.923			1.017+j0.124	0.883
4			1.034+j0.157	0.854			1.069+j0.243	0.783
6			1.076+j0.230	0.792			1.159+j0.349	0.702
8			1.137+j0.294	0.739			1.289+j0.428	0.637
10			1.215+j0.345	0.693			1.457+j0.462	0.588
15			1.476+j0.373	0.611			1.880+j0.196	0.524
20			1.723+j0.182	0.571			1.771-j0.343	0.536
25			1.735-j0.157	0.569			1.319-j0.439	0.626

2.8.2.2 Pressure windows: mechanical design

Where the differential pressure across the window is low (a few millimetres water gauge), for example for a window forming a vapour lock to prevent undesirable gases entering the waveguide feed system, the design of the window is straightforward. The mechanical forces acting on it are low and it can be a thin sheet of dielectric material: polypropylene is often used because of its low loss factor, reasonable mechanical stability and ability to work at 100°C. It has a smooth surface and is acceptable in the food industry, but is flammable and creates sooty deposits if subject to arcing.

Thin windows are usually sandwiched between a pair of plain waveguide flanges, cut to the outer dimensions of the flange and drilled to conform to the flange bolt holes. Adhesive aluminium foil is applied to the to the window on both sides, to form a 'picture frame' mating with the flange faces. The aluminium is bent over the outer edges so that there is a continuous electrical path from one side of the window to the other. In the final assembly, the window is clamped between the flanges with a copper–mineral-filled gasket on each side. This is a very simple window, cheap to make and easy to fit. There is a slight disadvantage that the transmission path includes a slot of depth equal to the flange face, which inserts a series reactance which slightly degrades the VSWR.

An alternative construction is to cut a rebate in one flange so that, when the two are brought together, a groove is formed into which the window fits. The groove is typically $0.015\lambda_0$ deep. The window is sealed with silicone rubber, and the flanges must be tightly clamped to ensure metal–metal contact. This is a more expensive construction than that above, but has an improved power-handling capability.

Where the window is a pressure lock to a pressure vessel, the mechanical forces on it are high, and careful analysis of stresses is necessary to determine the necessary thickness. Allowance must be made for fatigue fractures and creep, features common in ceramic and plastics materials respectively. There is no analytic solution to the stresses in thick flat rectangular plates, and either numerical or approximate methods must be used.

A simple and generally adequate approximation for the maximum stress in the plate is to consider it as a set of isolated simply supported beams under uniform load, spanning the narrow height of the waveguide. Simple consideration of bending moment and the stress s_{max} at the surface at the span centre leads to the approximate relation

$$s_{max} = \frac{3pb^2}{8t^2} \tag{2.11}$$

where t is the thickness of the window, b the waveguide height, and p the gas pressure. Eqn. 2.11 tends to overestimate the stress.

The maximum stress must be substantially less than the fracture tensile stress for the material of the window, with a factor of safety requiring careful consideration. Thick alumina plates, in particular, may fail due to stress concentration in microfissures in the surface, which grow with repeated pressure cycling until failure occurs (Entwistle, 1986).

Thick plates are mounted in grooves as described above for thin plates, but the tolerances on seating flatness must be such as to avoid stress concentrations in the plate from point loads. The gas seal is usually by a film of silicone-rubber sealant. However, if there is a tight specification on gas leakage, it may be necessary to consider metallising the edge of the window and brazing it into a metal frame for sandwiching between the flanges.

If the window fails, there is the possibility of an explosive release of stored pressure energy which may be hazardous if no provision is made for a controlled discharge. Such an arrangement comprises a second window spaced apart from the main window, with a reinforced waveguide between. An array of cutoff waveguides (i.e. nonpropagating) branches from this waveguide and discharges into an exhaust duct releasing the gases to atmosphere, or into a containment vessel.

2.8.2.3 Pressure windows: protection

Arcing in a pressure vessel frequently migrates to the window, where damage results if the microwave power is not switched off quickly. Optical sensors have proved the most reliable detectors for arcs, 'looking' through the window from the generator side if the window is a translucent material. If the window is opaque, the sensor must look at the applicator side, which may be difficult because the sensor is necessarily further away; a telescope lens may be necessary to achieve sufficient sensitivity. The integrity of the sensor can be tested regularly by a built-in test lamp.

2.9 Processing under vacuum

Microwave vacuum processing has three important industrial applications: in low-temperature evaporative drying, in freeze drying, and in the deposition of exotic materials on a substrate. In the first two cases, efficient heat transfer to the workload is the prime objective with no ionisation, while in the last a stable 'plasma' of ionised gas is required. For these processes the absolute pressure is usually in the range 1–200 mbar.

Evaporative drying under vacuum is a valuable process where rapid drying is required but the dry matter of the workload is damaged at the normal boiling point (100°C for water at atmospheric pressure). It has proved especially useful in the confectionery and drinks industries where the final drying of solid, friable foams (softening point c.60°C) is rapidly achieved. In some cases, the foam is controllably created by extension of the same process. Much experimental work was done in the 1970s on freeze drying by microwave under vacuum, but there was a problem in achieving high enough power density with the limited, low values of E-field possible at the vacuum pressure necessary; nonetheless, interest remains in this possibility, which could be rekindled by the advent of switch-mode power supplies (Section 10.3).

Electric-field stress is a critical parameter in vacuum processing, discussed in detail by Metaxas and Meredith (1983). Graphs of electric-field-breakdown stress as a function of pressure for air and water vapour at 15°C are shown in Figures 2.1 and 2.2 for the principal frequencies. It will be seen that the breakdown stress reaches a minimum about 1 mbar, with an E field of 90 V/cm at 915 MHz and 120 V/cm at 2450 MHz. for air, slightly higher for water vapour. At lower pressures, the breakdown stress rises, to the advantage of the freeze-drying process, but the maintenance of such a low vacuum pressure in an industrial process is a formidable problem. At higher pressures, 10–200 mbar, the pressure rises permitting usable field stress for vacuum drying at reduced temperature.

Typically, for air and water vapour at ambient temperature, evaporative drying is conducted in the pressure range 30–300 torr (40–300 mbar) where the breakdown stress is in the range 10^3–7×10^3 V/cm at both 915 and 2450 MHz. Freeze drying is conducted in the pressure range 0.1–1.0 torr; at 915 MHz. the E-field breakdown stress is 80–110 V/cm, and at 2450 MHz it is 110–300 V/cm.

As the temperature rises, the air density falls and the E-field breakdown stress falls too (Paschen's law):

$$E_{To} = \frac{288}{T_0 + 273} \times E_{amb} \qquad (2.12)$$

where T_0 is the operating temperature.

The relevant E-field breakdown strength is that given by eqn. 2.12 at the highest temperature present in the applicator vessel.

The safe working E-field stress is eroded by the usual factors of field concentrations near discontinuities, standing waves and local resonances. It is also reduced by amplitude modulation of the generator output power, this is a particular problem with conventional power supplies using rectification at power-line frequency, and especially so with saturating resonant power supplies (Section 10.2.3). Transient surges on the mains power supply may also momentarily raise the E-field significantly above the norm.

The development of switch-mode power supplies (Section 10.4 is a significant advance for vacuum processing in the substantial reduction in amplitude modulation in addition to their other advantages.

2.9.1 Applicators for vacuum processing

Most applicators can be adapted for vacuum processing. The mechanical-engineering standards required are high, however, because the vessel and associated equipment must conform to regulations for pressure vessels, and because of the need to avoid leaks which waste time and energy in vacuum pumping.

Circular cylindrical vessels are the most common in major industrial installations, operated as multimode ovens (Chapter 5). The design procedure for an evaporative dryer is:

(i)   Determine the allowed maximum workload temperature.
(ii) From steam tables determine the corresponding vacuum pressure.
(iii) Determine the limiting E-field stress at the principal microwave frequencies.
(iv) Make preliminary allowance for factor of safety on working E-field stress.
(v)   Determine the preferred operating microwave frequency by consideration of:
      • penetration depth, economics, total power required (heat balance)
      • power density in workload at the E field given by (iv).
(vi) From the power density calculate the volume of workload in the oven, and if continuous-flow, the conveyor-bed cross-section, dimensions and length.
(vii) From (vi), determine the practical dimensions of the oven.
(viii) Consider the microwave-vacuum-window design: its power density and corresponding E-field stress. How many windows are required?
(ix) Calculate the volume of the oven and, from the allowed pumping time at start-up, determine the size of the vacuum-pumping system. Allow a margin for leaks (typically +30%), and for air admitted through feed and discharge ports.
(x)   Consider automatic pressure control and its stability.
(xi) Consider the design of the vessel and associated equipment in conformity with applicable national standards for pressure vessels.
(xii) Reconsider the factor of safety (iv) above in a more detailed assessment of deratings due to the many possible factors.

The above is a repetitive procedure to which the optimum design compromise converges

2.9.2 Vacuum windows

Microwave vacuum windows follow the same design principles as those for pressure windows discussed in Section 2.8, and the matching techniques are identical. However, ionisation must be avoided at the window, and this means that the power density must be maintained at a low level so that the electric-field stress is well below the ionisation threshold. This is usually achieved by increasing the area of the window, using a horn tapered in both co-ordinates feeding from the atmospheric side of the window. The size of the horn aperture is determined by the mechanical strength of the window; for example, alumina plates 270 mm square and 20.5 mm thick, have been used at 2450 MHz. transmitting 6 kW into an oven at 50 mbar. A typical horn would have a taper of 30° included angle for the broad faces from the standard waveguide throat, with the narrow faces tapered to suit.

Other techniques for increasing the total power input include provision of more windows with a power-splitting system, and the use of a standing wave to reduce the E-field stress at the window surface. In the latter, an inductive iris is placed beyond the window, creating a standing wave adjusted so its minimum E-field is in the plane of the vacuum side of the window. This reactance is then matched by another on the atmospheric-pressure side. The electric-field intensity at the voltage minimum is proportional to $\sqrt{S}$ where S is the VSWR (< 1.0). To halve the electric-field intensity, it is necessary to create a VSWR of 0.25, which is a substantial mismatch representing 36% reflected power. With other reflections unavoidably present, this is a difficult system to optimise and should only be considered for small final adjustments.

Chapter 3

The thermodynamic aspects
of volumetric heating

3.1 Introduction

Early in the development of industrial-microwave-heating technology, it became clear that conventional heat transfer within the workload, and to its surroundings, could have a very important effect on the process, and equipment and process operation must be designed accordingly.

In addition to the elementary heat balance—that the energy, of all kinds, injected into the workload must equate with its change in thermal energy plus the energy transferred to its surroundings—no electroheat system can be evaluated and quantified properly without due attention to the heat and mass transfer in the workload during and following exposure to volumetric heating. Heat flow occurs within the workload through thermal conduction resulting from temperature gradients imparted during volumetric heating; and also from the surface to the surroundings via convection, radiation and conduction due to temperature difference between the surface and immediate environment.

Mass transfer, the migration of volatile substances such as water, both within the workload and to the surroundings, often has a first-order effect on heat flow because of its high value of total heat (enthalpy). Most commonly, the mass transfer is of water, which has high values of latent heat in both evaporation and thawing. Moreover, the thermal diffusivity is itself a function of the moisture content, and so varies with time and position in the workload volume.

Heat flow is a complicated subject, and attempts at quantifying it frequently founder on inadequate, or the absence of, data on basic physical properties. Nonetheless, it is helpful to the engineer to set out the principles as an aid to understanding the mechanisms involved.

3.2 Power and energy

It is important to distinguish between power and energy, terms which are often used loosely as though synonymous.

'Energy' (joules, or watt seconds) is a capacity to do work, and can be stored in many ways: as in a stretched spring, a given mass of fuel, a charged electrical capacitor, kinetic energy in a moving mass, or heat stored in a body, to give a few examples.

'Power' (watts, or joules per second) is the time rate-of-dissipation of energy. In a heating system, the dissipation of a given, *fixed amount of energy* in a passive nonvolatile workload results in a certain average temperature rise; in volumetric heating this may be accomplished very quickly. The time taken to transfer the energy does not affect the final average temperature, assuming no heat losses to the surroundings. However, if a fixed amount of continuous power is dissipated in a similar workload its temperature will rise linearly with time, in principle without limit but there are obvious practical limits.

3.3 Volumetric and conventional heating: a comparison

That a fixed continuous power dissipation into a dry, passive workload causes its average temperature to rise linearly with time is of fundamental importance, distinguishing it from conventional heating, in which the average temperature of the workload asymptotically reaches the oven temperature, and cannot rise above it.

In principle, with volumetric heating the average temperature of the workload continues to rise as long as power is applied, irrespective of the temperature of the oven walls or of the air inside the oven. The question sometimes asked, 'what is the temperature of the microwave oven?', has no fundamental relevance. The temperature and humidity inside the oven do, however, have an important secondary effect on the system performance because of surface heat transfer, causing surface heating or cooling depending on conditions. This subject is discussed in greater detail in Section 3.5.

Unlike conventional heating ovens, microwave ovens are very efficient in converting energy into heat in the workload. In a large industrial oven, the microwave efficiency, defined as the percentage of the applied microwave energy which is dissipated as heat in the workload, can be in the region of 95%, and the conversion of electrical power into microwave power can have an efficiency of 85%. Moreover, a conventional oven has to be heated to a temperature substantially in excess of the required temperature in the workload; a microwave oven, if heated at all, is normally heated to a temperature no greater than the required surface temperature of the workload. The radiation and convection heat losses from the microwave oven are therefore significantly less because of its lower temperature.

Further energy saving arises because a microwave oven has instantaneous control of power, which means that equilibrium conditions are rapidly re-established after a change, and start-up can be rapid. Very fast feedback control loops can be used to control process parameters accurately, leading to improved product quality.

3.4 Batch and continuous processes

Volumetric heating by microwave is either a batch or a continuous process. In a batch process a fixed mass of workload is placed inside the oven and a pre-determined amount of energy is injected as

$$E = \int_0^T p\,dt \tag{3.1}$$

where E = total energy injected into the workload, (Joules, or watt seconds)
 p = applied power to the workload, which may be a variable function with time (watts)
 T = total time of the process (seconds)
This energy E can then be used directly to calculate the thermal change in the workload as discussed in Section 3.4.

In a continuous-flow process the workload is fed through a microwave chamber, normally fed with a constant power whilst the workload travels through at a constant mass-flow rate. Again, the change in thermal state of the workload can be calculated directly by the methods of Section 3.4.

Very many factors determine whether a process should be continuous or batch, such as:

(*a*)   type of workload, e.g. blocks, discrete items, powders, liquids or particles in containers for batch processing; extruded solids, deposited powders or particles on a conveyor, or liquids flowing inside a microwave-transparent pipe for continuous processing;

(*b*)   ease of control of the workload parameters, e.g. batch weight, rate of mass-flow, or starting temperature, may be decisive in choosing the best procedure. Accurate control of continuous mass flow is often very difficult, precluding a continuous process;

(*c*)   existing up-stream and down-stream product handling and control systems;

(*d*)   available space for equipment and materials. Continuous-process plant is usually elongated in floor plan, whereas batch systems have a generally square layout with space allocated nearby for product storage, both that awaiting treatment and treated material;

(*e*)   manpower requirements—a continuous-flow system usually requires fewer operators;

(*f*)   automatic monitoring and control of process parameters is usually easier with a continuous process;

(g)   start-up and shut-down wastage of product may be unacceptable in a continuous-flow process;

(h)   the duty-cycle of actual heating time to total process time, i.e. heating time plus loading/unloading time, may be unacceptable in a batch process; the peak microwave heating capacity becomes greater than for the corresponding continuous process, because the mean power is reduced by the duty cycle, resulting in a higher capital cost of the generator equipment; and

(i)   Management control may be easier with one process than the other, depending on local conditions.

3.5 Heat balance

By definition, in an ideal volumetric-heating system, energy is dissipated uniformly throughout a workload. Assuming, in the first instance, that the workload is dry and passive and remains in the same state throughout the process, the rate of rise of temperature $d\theta/dt$ (degrees Celsius per second) is related to the power dissipation in the workload P (watts), the mass of the workload M (grams) and its mass specific heat s (joules per gram per degree Celsius) simply as

$$P = M\frac{d\theta}{dt}s \qquad\qquad (3.2)$$

This relation is fundamental to all heating processes, irrespective of the nature of the power source. As presented above it corresponds to a batch process heating a fixed mass of workload. For a continuous-flow process the temperature rise (T degrees Celsius) is fixed and the mass flow throughput dM/dt (grams per second) is related to the power as

$$P = \frac{dM}{dt}\Delta\theta s \qquad\qquad (3.3)$$

Specific heat s is often quoted in units of calorie per gram per degree Celsius. Conversion to joules per gram per degree Celsius involves division by 4.18, the mechanical equivalent of heat. These parameters are sometimes given in kilograms, kilojoules and kilowatts, and care must be taken to use consistent units in calculation.

3.5.1 Heat balance: stability

It will be appreciated from the form of the energy balance relations given above that the temperature rise in the workload is related linearly to the other parameters. It follows that small variations in these parameters (mass or mass-flow, specific heat or power) will have a pro-rata effect on temperature rise. Moreover, the energy balance involves temperature rise, so that a change in input temperature of the workload will have a corresponding, additive, change to the output temperature.

Accurate control of the heating process principally requires control of the above physical parameters of the workload and the effective energy (or power and time) delivered to it. Residual temperature variations are then due to aberrations in the accuracy of control; however, by measuring the parameters it is possible to improve stability of final temperature by 'feed-forward' techniques, and by closed-loop control of generator power to set the output temperature constant.

In practice there are many factors which complicate this ideal view of the heating and control system. It is assumed that the volumetric heating is uniform throughout the workload; for several reasons discussed in later chapters this is only approximate, and the heating is nearly always more intense near the surface. The specific heat is variable with temperature in most materials, although this is not usually a significant effect in comparison with others. However, the microwave (or RF) loss factor ε'', which determines the local energy dissipation, varies considerably with temperature, almost always rising as the temperature increases. This means that as the temperature rises the heat dissipation rises too, so the rate of rise of temperature in that region increases. This effect, in very extreme cases, can lead to a thermal 'runaway', discussed in greater detail in Section 2.6.

3.5.2 Sensible and specific heat

Sensible heat is the term used to describe heat in a body which has solely raised its temperature in accordance with eqn. 3.2 or 3.3. For a batch process eqn. 3.2 gives directly the power required to secure a given average temperature rise (i.e. sensible heat) in a workload of chosen mass: the amount of heat input required is directly proportional to the specific heat s, which is a property of the material of the workload. Table 3.1 gives some typical values of specific heat and Table 3.2 gives average rate of rise of temperature as a function of dissipated power density (kilowatts per kilogram) and specific heat.

Table 3.1   Specific heats of typical materials

Material	Temperature	Specific heat	
	(°C)	(kJ kg^{-1} K^{-1})	(kcal kg^{-1} K^{-1})
Apples	−20	3.46	0.83
Carrots	−20	2.27	0.54
Fish	−20	1.99	0.47
Beef	−20	2.40	0.57

continued. . .

Table 3.1 continued

Material	Temperature	Specific heat	
	(°C)	(kJ kg^{-1} K^{-1})	(kcal kg^{-1} K^{-1})
Lamb (loin)	−20	1.65	0.39
Carrots	+15	3.77	0.90
Fish (fresh)	+15	3.35	0.80
Blackberries	+15	3.68	0.88
Brussels sprouts	+15	3.68	0.88
Cabbage	+15	3.93	0.94
Ham (shoulder)	+15	2.51	0.60
Peas	+15	3.30	0.79
Pork (fresh)	+15	2.09	0.50
Potatoes	+15	3.35	0.80
Tomatoes	+15	3.97	0.95
Aluminium	+15	1.00	0.24
Brass	0	0.38	0.09
Steel, carbon	+75	0.48	0.11
Steel, stainless	+75	0.51	0.12
Granite	+20	0.86 typical	0.12 typical
Cellulose	+20	1.4	0.33
Neoprene	+20	2.1 typical	0.5 typical
Nylon	+20	2.0 typical	0.5 typical
Polystyrene	+20	1.2	0.29
Polyethylene	+20	2.3	0.55
Polyvinylchloride (PVC)	+20	1.04	0.25
Polytetrafluoroethylene (PTFE or Teflon®)	+20	1.34	0.32
Water, distilled	+15	4.185	1.00
Ice, pure	−21 to −1	2.0 to 2.1	0.478 to 0.502
Clay (dry)	+15	1.0	0.24
Coal (bituminous)	+15	1.4	0.33
Concrete	+15	0.71	0.17
Soil (dry)	+15	1.2 typical	0.3 typical
Fibreboard (hard)	+15	2.1	0.5
Glass (Pyrex)	+15	0.84	0.2
Glass (quartz)	+15	1.17	0.28
Leather (dry)	+15	1.9	0.45
Limestone (dry)	+15	0.92	0.22
Porcelain	+15	0.92	0.22
Slate	+15	0.75	0.18
Brick (Fletton)	+15	0.79	0.19
Firebrick	+15	0.84	0.20
Oak	+15	2.4	0.57
Pine	+15	1.9	0.53
Cork	+15	1.8	0.43

Using eqn. 3.2, Table 3.2 is calculated to show the rate of rise of temperature (degrees Celsius per minute) for various values of specific heat and heating power density (kilowatts per kilogram) for a batch process. For a continuous-flow process Table 3.3 can be used.

Table 3.2   Rate of rise of temperature for given power density

Specific heat	Power density (kW/kg)							
	0.3	0.6	1.0	3.0	6.0	10.0	30.0	60.0
(kJ/kg)	(degC/ min)	(degC/ min)	(degC/ min)	(deg/C min)	(deg/C min)	(deg/C min)	(deg/C min)	(deg/C min)
0.4	45	90	150	450	900			
0.6	30	60	100	300	600			
0.8	22	45	75	225	450	750		
1.0	18	36	60	180	360	600		
1.2	15	30	50	150	300	500		
1.4	13	26	43	129	257	429		
1.6	11	22	37	112	225	375		
1.8	10	20	33	100	200	333		
2.0	9	18	30	90	180	300	900	
2.2	8	16	27	82	164	273	818	
2.4	7	15	25	75	150	250	750	
2.6	7	14	23	69	138	273	692	
2.8	6	13	21	64	128	214	643	
3.0	6	12	20	60	120	200	600	
3.2	6	11	19	56	112	187	562	
3.4	5	11	18	53	106	176	529	
3.6	5	10	17	50	100	170	500	1000
3.8	5	9	16	47	95	158	473	947
4.0	5	9	15	45	90	150	450	900
4.2	4	9	14	43	86	143	428	857

Using this Table the power required to accomplish a simple (i.e. sensible-heat only) heating process can quickly be established. For example, a batch process having a workload with specific heat 2.4 kJ/kg (or 2.4/4.18 = 0.57 kcal/kg) and a required rate of rise temperature of 75°C/min requires a power density of 3.0 kW/kg. If the workload weighs 8 kg the power required to be dissipated is $3 \times 8 = 24$ kW, and if the required temperature rise is 90°C the time required is 90/75 = 1.2 min.

Table 3.3 shows the power requirement for a continuous-flow process in which the workload mass flow is shown in kilograms per minute for a range of values of specific heat, and the specific power requirement is given in kilowatts per degree Celsius. For example, a workload with a specific heat of 0.53 kcal/kg/°C, and a mass flow of 2.5 kg/min has a specific power, reading from the Table, of 9.17E-2 kW/°C. To achieve 120°C temperature rise would therefore require $9.17 \times 10^{-2} \times 120 = 11$ kW.

Note that the figures given by Tables 3.2 and 3.3 are for ideal conditions. In practice, more power must be generated for the following reasons which are discussed in detail later:

(a)   transmission loss from generator to oven;
(b)   reflected power from oven back towards the generator;
(c)   losses within the oven itself;
(d)   Derating factor arising from nonuniform heating to account for power wasted in heating hot spots whilst awaiting cold spots to reach the desired temperature; and
(e)   Surface heat transfer at the workload, which may be positive or negative depending on operating conditions.

3.5.3 Heat capacity of mixtures

The workload usually consists of several substances mixed together, though the mixing may well be far from perfect; extreme situations are:

(a)   mixtures such as rubber compounds comprising a polymer mixed intimately with powdered carbon and other substances;
(b)   food such as ready-to-serve meals where the components of the meal are essentially separated, but the assembly has an overall heat capacity in the context of heating efficiently to a desired final temperature.

To calculate sensible heat it is necessary to estimate the effective specific heat of the mixture from the known specific-heat values of the ingredients forming the mixture. Note that these mixtures are not chemical compounds and may have specific heat values which cannot be simply calculated from the substances which have reacted chemically to form them.

The calculation procedure is based on considering the sensible-heat energy added to each component of a mixture for a chosen temperature rise, say 1°C, knowing the proportions of each component present. For example, consider a wet clay, in which 1 kg of mixture contains 60% dry clay (specific heat 0.24 kcal/kg/°C) and 40% water (specific heat 1.00 kcal/kg/°C) by weight. For 1°C temperature rise the sensible heat added to each component is:

> Clay, weight 0.600 kg: $0.6 \times 0.24 = 0.144$ kcal
> Water, weight 0.400 kg: $0.4 \times 1.00 = 0.400$ kcal
> Total = 0.544 kcal

for a total weight of 1.00 kg. of mixture with 1°C temperature rise. Thus the specific heat of the mixture of dry clay and water in the proportions given is 0.544 kcal/kg/°C, or $4.2 \times 0.544 = 2.28$ kJ/kg/°C.

Table 3.3   *Process power requirements as function of mass flow and specific heat*

Specific heat		Mass flow (kg/min)					
		1.00E+00	1.26E+00	1.59E+00	2.00E+00	2.52E+00	3.17E+00
(kJ/kg degC)	(kcal/kg degC)	(kW/ degC)	(kW/ degC)	(kW/ degC)	(kW/ degC)	(kW/ degC)	(kW/ degC)
0.4	0.1	6.67E-03	8.40E-03	1.06E-02	1.33E-02	1.68E-02	2.11E-02
0.6	0.14	1.00E-02	1.26E-02	1.59E-02	2.00E-02	2.52E-02	3.17E-02
0.8	0.19	1.33E-02	1.68E-02	2.12E-02	2.67E-02	3.36E-02	4.23E-02
1	0.24	1.67E-02	2.10E-02	2.65E-02	3.33E-02	4.20E-02	5.28E-02
1.2	0.29	2.00E-02	2.52E-02	3.18E-02	4.00E-02	5.04E-02	6.34E-02
1.4	0.33	2.33E-02	2.94E-02	3.17E-02	4.67E-02	5.88E-02	7.40E-02
1.6	0.38	2.67E-02	3.36E-02	4.24E-02	5.33E-02	6.72E-02	8.45E-02
1.8	0.43	3.00E-02	3.78E-02	4.77E-02	6.00E-02	7.56E-02	9.51E-02
2	0.48	3.33E-02	4.20E-02	5.30E-02	6.67E-02	8.40E-02	1.06E-01
2.2	0.52	3.67E-02	4.62E-02	5.83E-02	7.33E-02	9.24E-02	1.16E-01
2.4	0.57	4.00E-02	5.04E-02	6.36E-02	8.00E-02	1.01E-01	1.27E-01
2.6	0.62	4.33E-02	5.46E-02	6.89E-02	8.67E-02	1.09E-01	1.37E-01
2.8	0.67	4.67E-02	5.88E-02	7.42E-02	9.33E-02	1.18E-01	1.48E-01
3	0.71	5.00E-02	6.30E-02	7.95E-02	1.00E-01	1.26E-01	1.59E-01
3.2	0.76	5.33E-02	6.72E-02	8.48E-02	1.07E-01	1.34E-01	1.69E-01
3.4	0.81	5.67E-02	7.14E-02	9.01E-02	1.13E-01	1.43E-01	1.80E-01
3.6	0.86	6.00E-02	7.56E-02	9.54E-02	1.20E-01	1.51E-01	1.90E-01
3.8	0.9	6.33E-02	7.98E-02	1.01E-01	1.27E-01	1.60E-01	2.01E-01
4.2	1	7.00E-02	8.82E-02	1.11E-01	1.40E-01	1.76E-01	2.22E-01

Specific heat		Mass flow (kg/min)				
		4.00E+00	5.04E+00	6.35E+00	8.00E+00	1.01E+01
(kJ/kg degC)	(kcal/kg degC)	(kW/ degC)	(kW/ degC)	(kW/ degC)	(kW/ degC)	(kW/ degC)
0.4	0.1	2.67E-02	3.36E-02	4.23E-02	5.33E-02	6.72E-02
0.6	0.14	4.00E-02	5.04E-02	6.35E-02	8.00E-02	1.01E-01
0.8	0.19	5.33E-02	6.72E-02	8.47E-02	1.07E-01	1.34E-01
1	0.24	6.67E-02	8.40E-02	1.06E-01	1.33E-01	1.68E-01
1.2	0.29	8.00E-02	1.01E-01	1.27E-01	1.60E-01	2.02E-01
1.4	0.33	9.33E-02	1.18E-01	1.48E-01	1.87E-01	2.35E-01
1.6	0.38	1.07E-01	1.34E-01	1.69E-01	2.13E-01	2.69E-01
1.8	0.43	1.20E-01	1.51E-01	1.91E-01	2.40E-01	3.02E-01
2	0.48	1.33E-01	1.68E-01	2.12E-01	2.67E-01	3.36E-01
2.2	0.52	1.47E-01	1.85E-01	2.33E-01	2.93-01	3.70E-01
2.4	0.57	1.60E-01	2.02E-01	2.54E-01	3.20E-01	4.03E-01

continued. . .

Table 3.3 continued

Specific heat		Mass flow (kg/min)				
		4.00E+00	5.04E+00	6.35E+00	8.00E+00	1.01E+01
(kJ/kg degC)	(kcal/kg degC)	(kW/ degC)	(kW/ degC)	(kW/ degC)	(kW/ degC)	(kW/ degC)
2.6	0.62	1.73E-01	2.18E-01	2.75E-01	3.47E-01	4.73E-01
2.8	0.67	1.87E-01	2.35E-01	2.96E-01	3.73E-01	4.70E-01
3	0.71	2.00E-01	2.52E-01	3.18E-01	4.00E-01	5.04E-01
3.2	0.76	2.13E-01	2.69E-01	3.39E-01	4.27E-01	5.38E-01
3.4	0.81	2.27E-01	2.86E-01	3.60E-01	4.53E-01	5.71E-01
3.6	0.86	2.40E-01	3.02E-01	3.81E-01	4.80E-01	6.05E-01
3.8	0.9	2.53E-01	3.19E-01	4.02E-01	5.07E-01	6.38E-01
4.2	1	2.80E-01	3.53E-01	4.45E-01	5.60E-01	7.06E-01

3.5.4 Latent heat

When a substance changes its state from solid to liquid, or liquid to gas, a quantity of heat energy must be added which does not change the temperature. This heat energy is called latent heat, and can be considered as heat energy necessary to break molecular bonds holding the substance together as a solid, or freeing the molecules from the mutual attractions of a liquid. The process of change from one state to the other is precisely reversible, heat being supplied to change from solid to liquid and being released from liquid to solid, and similarly from liquid to gas. The two quantities latent Heat of fusion (solid to liquid) and latent heat of evaporation (liquid to gas) are not the same in magnitude, and both represent substantial quantities of heat.

Many important applications of industrial microwave heating involve latent heat, e.g. thawing and tempering (i.e. partial thawing), and drying. For water the latent heat of fusion L_f is 80 kcal/kg and latent heat of evaporation L_e is 540 kcal/kg at 0°C and 100°C, respectively. Kaye and Laby (1986) give data for Le for other boiling points. L_f does not change significantly and for most purposes can be taken as constant at the above figures, unless extreme temperatures are involved, e.g. in freeze drying. These latent-heat figures represent considerable heat energy; the heat required to melt 1 kg of ice at 0°C to water at 0°C is the same as that required to heat 1 kg of water through a temperature rise of 80°C, and nearly seven times as much energy is required to boil away 1 kg of water, starting at 100°C.

Clearly drying and thawing processes are very energy-intensive and often the latent heat is the predominant factor in the calculation of total energy required.

3.5.5 Drying techniques

The simplest and quickest drying method for many materials is to use microwave power to boil the water present to dryness, or to the desired final moisture content (forced drying). Unfortunately, it is also the most expensive in both capital cost and running cost because of the very large amount of power required, especially if the initial moisture content is high. Only where the intrinsic value of the product is very high and the initial moisture content low is microwave (or RF) forced drying justified. There may also be problems of low porosity in some materials where the steam generated within the workload cannot escape freely so that high pressure is developed internally, causing physical rupture; the rate of drying then must be slowed.

Drying can be performed more economically by combining microwave (or RF) with conventional drying methods (microwave-accelerated drying), where hot air, sometimes in combination with infrared radiant heat, is used to provide most of the latent heat of evaporation, and is used particularly in the early stages of drying where it is most effective, reserving the microwave power for final drying. In particular microwave and RF are especially efficacious when applied after the well known 'falling-rate' stage of conventional drying is reached (Osborne and Turner, 1966). This is at the point in the drying cycle where the surface of the workload is substantially dry, but there remains a wet core, enveloped by dry material at the surface which has a very poor thermal conductivity. Transmission of heat to this inner core is then extremely slow by conventional techniques, but the dry skin is virtually transparent to microwave or RF, allowing the central wet core to be heated easily; 30%–50% reduction in drying time is often achieved, together with other substantial advantages such as reduction in floor space occupied by the dryer, or enhanced throughput, and improvements in product quality and control.

3.5.5.1 Microwave forced drying

In this Section we examine the heat balance of a microwave forced-drying system, in which all the energy required is provided from a microwave source. The workload is assumed to start from an initial temperature θ_1 degrees Celsius, with a moisture content m_1 per cent, and is required to be dried to a final moisture content m_2 percent, where the moisture contents are the percentage water, by weight, of the dry matter (known as the 'dry-weight basis').

The energy required is then the sensible heat initially to raise the temperature of the workload to 100°C, plus the energy required to evaporate the quantity of water implied by the reduction of moisture content from m_1 to m_2. The sensible heat comprises that for the dry matter, plus that for the total mass of water present initially at the moisture content m_1. The heat balance equation can then be written as

$$E = \frac{4.2}{60}\left\{ s_d\,(T_b - T_0) + \frac{m_1}{100}s_1\,(T_b - T_0) + \frac{(m_1 - m_2)}{100}L \right\} \qquad (3.5)$$

where
  E = the total energy required (kW/min/kg)
  s_d = the specific heat of the dry matter (kcal/degC/kg)
  s_l = the specific heat of the liquid, for water, 1.00 kcal/degC/kg
  m_1 = initial moisture content % (dry-weight basis)
  m_2 = final moisture content % (dry-weight basis)
  L = latent heat of evaporation of liquid (kcal/kg)
  T_0 = initial temperature C (deg C)
  Tb = boiling temperature of liquid (deg C)

This equation gives the total heat input required to dry the workload through the moisture content specified $(m_1 - m_2)$, which in microwave forced drying is totally from the microwave source without any heat transfer to or from the surroundings. Table 3.4 gives the energy required for total drying $(m_2 = 0)$ for a range of starting moisture contents and initial temperatures for some typical values of specific heat for the dry matter.

Eqn. 3. 5 is valid where the liquid is pure, without dissolved solids which lower the vapour pressure of the solution by Raoult's law (Hicks, 1971) and hence raise the boiling temperature of the solution according to its concentration. As the drying proceeds in these cases the required sensible heat rises as the boiling temperature of the saturated solution remaining increases. The boiling temperature is then a variable and must be treated as such in eqn. 3.5. This effect is well known in the confectionery industry where the boiling temperature of sugar–syrup solutions is well documented

The above treatment of drying assumes that the water is held in the dry-matter as totally free water, i.e. it is present as liquid and not in any way attached to the dry matter by molecular forces. While this is true for some substances, there are many where water is retained in the dry matter as water of crystallisation, or is chemically *adsorbed* in the dry matter, a physical surface effect rather than a chemical combination. Two important features arise: first the amount of heat energy required to release the molecules of water is greater than the normal latent heat of free water, and secondly the microwave absorption rate of the bound water is much less than for free water. The loss factor ε'' of a wet material with all the water adsorbed or chemically combined is usually less than 10% of its value if the same quantity of water is 'free'. It is therefore more difficult to remove water other than free water by microwave or RF treatment.

3.5.5.2 Microwave-assisted drying

Conventional drying, in which a workload is placed inside a hot-air oven in either a batch or continuous process, normally follows three distinct phases:

Table 3.4 Energy required for forced drying at atmospheric pressure based on eqn. 3.5

m	s	Temperature (deg C)					
		0	20	40	60	80	100
(%dwb)		(kW/min)	(kW/min)	(kW/min)	(kW/min)	(kW/min)	(kW/min)
0	0.3	2.09	1.67	1.26	0.84	0.42	0.00
5	0.3	4.32	3.83	3.34	2.86	2.37	1.88
10	0.3	6.55	5.99	5.43	4.88	4.32	3.76
15	0.3	8.78	8.15	7.52	6.90	6.27	5.64
20	0.3	11.01	10.31	9.61	8.92	8.22	7.52
40	0.3	19.93	18.95	17.97	17.00	16.02	15.04
100	0.3	46.68	44.86	43.06	41.24	39.42	37.61
120	0.3	55.59	53.50	51.41	49.32	47.22	45.13
160	0.3	73.43	70.78	68.13	65.48	62.83	60.18
200	0.3	91.26	88.05	84.84	81.64	78.43	75.22
0	0.4	2.79	2.23	1.67	1.12	0.56	0.00
5	0.4	5.02	4.39	3.76	3.14	2.51	1.88
10	0.4	7.25	6.55	5.85	5.16	4.46	3.76
15	0.4	9.48	8.71	7.94	7.18	6.41	5.64
20	0.4	11.71	10.87	10.03	9.20	8.36	7.52
40	0.4	20.62	19.51	18.39	17.28	16.16	15.04
100	0.4	47.37	45.42	43.47	41.52	39.56	37.61
120	0.4	56.29	54.06	51.83	49.60	47.36	45.13
160	0.4	74.12	71.33	68.55	65.76	62.97	60.18
200	0.4	91.96	88.61	85.26	81.92	78.57	75.22
0	0.5	3.49	2.79	2.09	1.39	0.70	0.00
5	0.5	5.72	4.95	4.18	3.41	2.65	1.88
10	0.5	7.95	7.11	6.27	5.43	4.60	3.76
15	0.5	10.17	9.27	8.36	7.45	6.55	5.64
20	0.5	12.40	11.43	10.45	9.47	8.50	7.52
40	0.5	21.32	20.07	18.81	17.55	16.30	15.04
100	0.5	48.07	45.98	43.89	41.79	39.70	37.61
120	0.5	56.99	54.62	52.25	49.87	47.5	45.13
160	0.5	74.82	71.89	68.96	66.03	63.11	60.18
200	0.5	92.65	89.17	85.68	82.19	78.71	75.22
0	0.6	4.18	3.35	2.51	1.67	0.84	0.00
5	0.6	6.41	5.51	4.60	3.69	2.79	1.88
10	0.6	8.64	7.67	6.69	5.71	4.74	3.76
15	0.6	10.87	9.83	8.78	7.73	6.69	5.64
20	0.6	13.10	11.99	10.87	9.75	8.64	7.52
40	0.6	22.02	20.62	19.23	17.83	16.44	15.04
100	0.6	48.77	46.54	44.31	42.07	39.84	37.61
120	0.6	57.68	55.17	52.66	50.15	47.64	45.13
160	0.6	75.52	72.45	69.38	66.31	63.25	60.18
200	0.6	93.35	89.73	86.10	82.47	78.85	75.22

m = moisture content s = specific heat

(i)   a warm-up phase where the wet material rises in temperature asymptotically to the temperature of the oven. In practice the relevant temperature of the oven is the wet-bulb temperature because the surface of the product is usually wet at this stage. Since drying occurs at the surface, the water inside the workload must migrate to the surface for the process to proceed; the migration rate rises with temperature and so this early phase of the process is important. Where the workload is thick, the rate of heat flow to the centre will be very slow, limited by the thermal diffusivity of the material in a conventional process;

(ii)   a constant-rate phase where the rate of evaporation remains in equilibrium between the rate of flow of water to the surface, and the rate of evaporation to the hot-air stream flowing across it. This phase is maintained until the rate of flow of water from within cannot sustain the rate of drying from the surface, and so the surface begins to dry. A large moisture gradient develops from the surface to the wet material at the centre, where the moisture content may still be substantially equal to its starting value; and

(iii) a falling-rate phase where the surface temperature rises to the dry-bulb temperature of the oven, and water migrates towards the surface but evaporates below it at a rather indistinct boundary where there is little air movement. Latent heat for evaporation must flow through the dry material to the wet boundary, which is a very slow process, especially as there is almost always a limit imposed on the maximum surface temperature. This slow phase continues until final dryness is achieved, and the time taken to remove the last 10% of moisture may well compare with, or exceed, that for the total of phases (i) and (ii).

Quantifying and predicting the performance of such a drying cycle is notoriously difficult, and most drying processes and ovens are designed around experience amassed previously. The difficulties arise largely through an imperfect understanding of the mechanisms in detail, the uncertainties of basic physical constants of the material (e.g. thermal diffusivity, porosity, dynamics of mass transfer of water), inhomogeneity of the workload material, and nonuniformity of temperature and airflow within the oven; all these combine to make mathematical modelling imprecise. Control of the air temperature (dry-bulb) and the humidity is critical, and the requirements for an optimised process are different in the three main stages of drying.

Microwave and RF heating have a significant role in reducing the time of drying, but many of the same difficulties given above remain, so that system performance remains uncertain, even when it is proposed to apply volumetric heating to an existing conventional oven of known performance. The need for experimental evaluation on a model of realistic size (e.g. not less than one-quarter scale) is paramount, and it must be appreciated that the properties of the workload are as important as those of the oven in determining the overall performance.

Phases (i) and (ii) are two obvious stages which can be accelerated by volumetric heating; but it must be emphasised that the objective is, in general, not to provide latent heat by this method because of the high power this implies. In phase (i) volumetric heating can quickly raise the average temperature of the workload to its optimum (a compromise between the needs of product quality and high temperature for drying). The heat energy required is simply sensible heat and is readily calculated, but its effect on the drying rate in phase (ii) requires experimental determination. The temperature reached by volumetric heating in phase (i) will be higher than the wet-bulb temperature of the air stream in phase (i), so that some cooling will occur; further injections of volumetric heating may be beneficial in phase (ii) to restore the workload temperature at its optimum.

Phase (iii) is well known for the effective application of volumetric heating: microwave or RF energy is absorbed directly by the remaining water, and is not significantly impeded by the envelope of dry, thermally insulating material at the surface of the workload. The residual water at this point can readily be evaporated at some ten times the rate in the conventional falling-rate stage, with obvious advantages in size of plant or of throughput because the transition from phase (ii) to phase (iii) can be transferred to the end of an existing oven. This technique has been adopted widely in the biscuit industry, where RF heating has been applied as an 'add-on' unit to the end of a conventional continuous oven. Here, volumetric-heating power is used as the energy source so there is a need for care in equating the economics of enhanced performance with the capital and running costs. Moreover, it is found that there are other advantages gained beyond increased throughput, in product quality (in biscuits a reduction of the incidence of cracking, 'checking'), and the 'smoothing' of variations of moisture content always present at the exit from a conventional drying oven.

In the early 1970s an experimental drying plant (Meredith) was built for resin-bonded friction plates, where accelerated drying was required, but with minimum polymerisation of the resin, which was to be cured at a later stage in production. Microwave energy was used for phase (i), and hot air in combination with infrared radiation from quartz IR lamps for phase (ii), followed by further microwave heating for phase (iii); the dryer was a continuous-flow conveyorised configuration. The use of infrared radiation in phase (ii) is interesting because it was recognised that evaporative cooling from the surface would take place during the constant-rate stage. Since infrared heating is essentially a surface-heating process, it was considered the ideal to counterbalance the surface cooling. A drying time of some 40 min was achieved compared with around 2 h by conventional drying, and the proportion of resin prematurely cured was significantly less, resulting in a final product of superior density. About 75% of the total energy required for drying was provided by hot air and infrared radiation, the balance from microwave heating.

In the early 1980s RF-assisted drying was developed, named 'air radio-frequency-assisted' drying (ARFA dryer) (Swift and Jones 1983; Metaxas 1996). This uses the same thermodynamic principles, though the boundaries between the drying phases become blurred, and is reported to give substantial increases (about ×2) in drying rate with only 10% contribution by RF heating to the total energy required on selected types of workload. The most suitable are porous but with small pores, and with low-loss dry matter. Certain building materials (e.g. plasterboard) have been found appropriate examples. Again, any application for this technique must be scrutinised very carefully experimentally before a decision is made to employ it in production.

3.5.6 Tempering and thawing

One of the most widely used industrial microwave processes is the thawing of frozen foodstuffs, either partially (tempering) or fully. Tempering is most common, being used extensively in the meat-products industry where the meat is received in frozen blocks, at –15 to –25°C and typical weight 25 kg. It has been found that the subsequent process of slicing, dicing or mincing gives best results if the meat is not fully thawed because a residue of ice gives firmness to the meat allowing cleaner cutting; a final temperature in the range –2 to –4°C being optimum. Fish and poultry are also tempered by microwave.

Complete thawing is used in the final preparation of butter, where it is stored frozen at –12 to –18°C, and is required to have a temperature of +2 to +5°C for final blending and packing into retail packs. The bulk frozen blocks are usually rectangular, 267 mm square, 406 mm long, and weigh 25 kg. Thawing is also used for frozen vegetables and fruit, which may be frozen into blocks, or individually quick frozen (IQF) and loose, the latter being placed in boxes for treatment.

Performed conventionally, all the above processes require between one and four days to accomplish, because of the size of the blocks and their bulk storage volume which makes heat transfer extremely slow. Even after this period the uniformity of temperature and ice content is often inadequate, resulting in production difficulties and frequent damage to downstream machinery. Such long process times result in many serious problems; for example, of inflexibility of production throughput, of large expensive areas devoted to tempering, of poor hygiene and poor quality control.

Microwave processing reduces tempering and thawing to about 15 min, and may be either batch or continuous. Throughput of some existing installations exceeds 20 t/h. All the above problems are eliminated or reduced to negligible proportions, with outstanding cost savings.

3.5.6.1 Heat balance in tempering and thawing

As previously the total heat for the process is the sum of the sensible heat to raise the temperature from its initial value, and the latent heat required to melt the water present. However, two factors arise which complicate direct calculation of the energy required:

(i) The water present nearly always has a mixture of dissolved salts, all of which, by differing amounts, depress the freezing point of the solution.
(ii) In tempering, only part of the ice is thawed, and although the percentage of ice thawed is probably the most significant parameter in specifying the process, it is extremely difficult to measure, and so temperature is used to specify the end point. When more than about half the ice has thawed, the rate of change of temperature with enthalpy is almost zero until all the ice has thawed. This effect, combined with nonuniformity of heating arising fundamentally from limitations on penetration depth, and additionally in practice from the inhomogeneous nature of the workload, makes precise calculation of the energy required unreliable. Performance prediction is based on experience with existing installations and empirical data.

The presence of dissolved salt in the free water has a great effect on the shape of the graph of enthalpy plotted against temperature. In refrigerating pure (distilled) water, as the temperature is reduced from normal ambient temperature towards freezing point, normal sensible heat is removed in accordance with the specific heat and temperature change. At 0°C the temperature remains fixed until all the water has frozen, and then continues to fall at a rate again determined by the specific heat of ice. The enthalpy-temperature graph comprises three straight lines, representing the sensible heat of water (above 0°C); the latent heat, a vertical line at 0°C; and the sensible heat of ice (below 0°C). If a dilute salt solution is frozen in the same manner, freezing begins at 0°C as for distilled water, but the liquid being frozen initially is the surplus pure water in excess of that which forms a saturated salt solution at a temperature just lower than 0°C. As the temperature is lowered further more pure water freezes, and the remaining liquid becomes more concentrated, until eventually a critical temperature is reached at which all the remaining saturated salt solution freezes. For NaCl this temperature is –21.5°C: between 0°C and –21.5°C the enthalpy change comprises both latent and sensible heat, the former being progressively removed in a linear manner except at these two critical temperatures. Moreover the effective specific heat is itself a function of temperature because there is a mixture of ice (specific heat 0.4 cal/g/deg C) and water (specific heat 1.0 cal/g/deg C), in proportions depending on temperature.

It will therefore be appreciated that where, as in food applications, the free water present in the workload includes a variety of dissolved salts in uncontrolled quantities, and the total quantity of water is in itself a variable, estimation of enthalpy can only be by measurement, or previous experience.

Figure 3.1 shows a set of curves of the specific enthalpy (kilojoules per kilogram) required to temper or thaw a variety of foodstuffs; note that the curves generally follow the shape described above but the transitions from ice to water do not have the sharp changes of distilled water, because of the salts present. Enthalpy is a relative quantity, arbitrarily given a zero value at a convenient point, e.g. –20°C. To find the change in enthalpy required between two chosen temperatures, simply read the corresponding enthalpy values at those temperatures and subtract one from the other. The resulting specific enthalpy (kilojoules per kilogram or kilowatt seconds per kilogram) permits direct calculation of the power required to process a given throughput. (e.g. kilograms per hour). In the temperature range –1 to –3°C, the enthalpy changes very quickly for most foods, and therein is the difficulty in precise throughput prediction. Because the workload is often very inhomogeneous, and of irregular shape and size (even though of precise weight), the uniformity of heating is impaired beyond the normal microwave effects due to penetration depth. Normal microwave penetration-depth limitation results in an energy-dissipation gradient through the thickness of the workload, so different levels in the product are at different points on the enthalpy diagram.

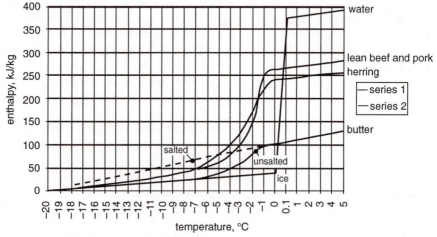

Figure 3.1   Enthalpy of frozen foods

A histogram of temperature distribution immediately after treatment may show 90% of readings within ±2 deg C with a very few (<1%) hot spots over +30°C. Since the product is invariably to be thoroughly cooked within less than one hour, this is not generally regarded as significant by users.

3.6 Surface heat transfer

Heat flow through the surface of a workload subject to volumetric heating can have a very important effect on the overall temperature or enthalpy distribution. Moreover, surface heat transfer continues after volumetric heating has ceased, and where the treatment is a process involving temperature and time, particular attention must be given to it. All the conventional forms of heat transfer must be considered:

- convection
- radiation (infrared)
- thermal conduction
- steam condensation, or evaporative cooling

3.6.1 Convection

Convection is the transfer of heat from the surface by air movement across the surface, and may be either *natural*, in which the air flow is induced by thermal expansion of the air due to its being in contact with the hot surface, or *forced* where the airflow is created by a fan. The amount of heat transfer is obviously proportional to the temperature difference between the air and the surface, and increases with the air velocity. However, many other factors are involved, such as the surface roughness, the orientation of the surface, proximity of other hardware, and humidity. In addition, the rate of heat flow obviously equals the flow of heat to the surface from within, which is proportional to the temperature gradient normal to the surface within the workload, and the thermal diffusivity. Because the magnitudes of most of these parameters are generally not known precisely, it is very necessary to confirm data by experiment, on a realistic scale, where conventional heat transfer is perceived as having an important effect on overall performance.

For natural convection, the typical rate of heat flow to or from an exposed, dry surface is in the range 2–5 $W/m^2/deg$ C, assuming that the thermal diffusivity of the workload is sufficiently high to sustain this magnitude of heat flux for the internal temperature gradient existing just below the surface. If this last condition is not satisfied, the surface temperature will tend toward the air temperature at the surface. This rate of heat flow at the surface is given as a guide to assess whether or not it would be a significant factor in the heat balance of the workload, and then to make a decision whether to include a controlled hot-air facility in the machine design.

Since, in practice, the objective is usually to avoid surface heat transfer, the temperature of the surrounding air is controlled to be, as closely as possible, equal to the internal temperature of the workload. The internal temperature of the workload is, of course, rising during volumetric heating and so in principle there should be an increase in air temperature to match. Arranging for the air temperature to rise at the same rate as the internal temperature of the workload would be extremely difficult, in either a batch process (linear rise in temperature with time) or continuous flow (linear temperature gradient along the length of the machine). The question 'how closely must the air temperature match the internal temperature?', and the above data, coupled essentially with experimental verification, lead to optimum design which in practice becomes a compromise between the process needs and cost.

For forced convection, a heated air stream is directed over the surface of the workload, substantially at the same temperature as the workload, most often in a contraflow direction to the travel of the workload in a continuous-flow process. This is the technique usually adopted to minimise surface heat loss for dry workloads. Figure 3.2 shows curves of heat transfer from a turbulent-flow airstream to a smooth flat surface, in a flow of confined headroom.

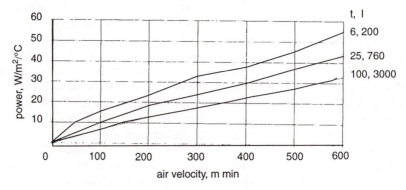

Figure 3.2   Heat transfer from air to a smooth surface - confined flow

t = tunnel height, mm
l = tunnel length, mm
(Adapted from Osborne and Turner (1967)

The amount of power necessary to heat the air is usually substantial and for economy a recirculation procedure is often used, where hot exhaust air is mixed with fresh air, typically with the volume flow of recirculated air three to six times that for the air intake. Forced convection wastes power, not only in heating the air, but also in the circulation fan, and its use must be optimised. Hot air must not be allowed to escape from the entry/exit ports of the oven, which means careful design of the air-duct system and fan such that neutral pressures are obtained at these zones. Lagging of hot surfaces is clearly important, not only to avoid waste but also to prevent discomfort to operators and a general rise in ambient temperature. Heating may be by any conventional system—electric, steam heat exchanger, direct or indirect gas or oil firing. The power required for direct heating an air stream at constant pressure is given in Table 3.5.

Table 3.5 Power for heating air

Volume flow (m³/min)	Volume flow (ft³/min)	Mass flow (kg/s)	Temperature (deg C)							
			60 (kW)	80 (kW)	100 (kW)	120 (kW)	140 (kW)	160 (kW)	180 (kW)	200 (kW)
1	35.30	0.02	1.23	1.64	2.05	2.46	2.87	3.28	3.69	4.10
2	70.60	0.04	2.46	3.28	4.10	4.92	5.74	6.56	7.38	8.20
4	141	0.08	4.92	6.56	8.20	9.84	11.48	13.12	14.76	16.40
8	282	0.16	9.84	13.12	16.40	19.68	22.96	26.24	29.52	32.80
16	565	0.32	19.68	26.24	32.80	39.37	45.93	52.49	59.05	65.61
32	1130	0.65	39.37	52.49	65.61	78.73	91.85	104.98	118.10	131.22

3.7 Heat conduction within the workload

For the many reasons discussed in this and other chapters, the uniformity of volumetric heating is often imperfect, so that temperature gradients exist through the bulk of the workload, with a complicated distribution. Unlike conventional heating via the surface, where diffusion is the prime mechanism of heat transfer within the workload, in volumetric heating diffusion has only a secondary role in equalising the temperatures of hot and cold spots. It occurs both during the volumetric heating process, and, importantly, afterwards; it is sometimes called 'soaking'.

Heat flow is often difficult to quantify because of uncertainty of the values of the many parameters involved, not least the actual temperature distribution following volumetric heating; but useful guideline estimates are possible using the methods of Carslaw and Jaeger (1959). The most important criterion is usually the time required to reach a specified minimum temperature following volumetric heating, but often the peak instantaneous temperature reached anywhere within the workload is important too, because it may represent a point of potential damage. In many processes, time and temperature in combination are the important criteria, for example in chemical reactions or in killing bacteria.

3.7.1 Heat flow in a thick slab

A feel for the temperature/time profile at points within a workload following or during volumetric heating is given by considering a slab of material of thickness L and of infinite extent in other dimensions. Carslaw and Jaeger (1959) examine heatflow in such a slab for various conditions, providing analytic solutions which are readily evaluated. Of particular interest is the temperature profile in the slab following volumetric heating, illuminating first from one face and then the other in quick succession. It is assumed that the volumetric heating is compensated to cancel standing-wave effects (Section 6.2.4). In such a case, the temperature distribution immediately following heating is the sum of two exponentials extending from the two faces, with exponential decay of the form $e^{-x/D}$ where D is the penetration depth.

There are two special cases for the subsequent soaking:

(*a*) with the surfaces maintained at a constant temperature equal to that attained during heating; and

(*b* with the surfaces thermally insulated so that there is zero heat flow at the surface.

Generally, case (*a*) is the broad objective in machine design, with the workload passing from the microwave-heating oven into a dwell tunnel heated to the required temperature by a hot-air stream. Case (*b*) is rarely adopted as a working practice, but otherwise represents an extreme condition where the surface heat transfer implied in case (*a*) does not occur.

In both cases the extremes of temperature are at the surface (maximum) and the centre plane (minimum), occurring at the time immediately following volumetric heating, and it is assumed that the time taken for the volumetric heating is very much less than the thermal time constant of the heat-soaking process. If the latter is not satisfied, Carslaw and Jaeger develop a further analysis for simultaneous volumetric heating with significant diffusion.

3.7.1.1 Heat distribution after volumetric heating

The simplest heat distribution immediately following volumetric heating is the classic exponential decay of temperature from the surface inwards, with a penetration depth D. Many heating ovens provide sequential heating from above and then below a thick slab (e.g. frozen meat), and so there is a second exponential extending from the opposite face, giving a saucer-shaped heat distribution. In practice, there may be superimposed a sinusoidal standing-wave distribution due to reflections, and while this can be included in the analysis, it is omitted here because the reflections are compensated in a well designed applicator.

Figure 3.3 shows a thick slab of thickness L which has been sequentially illuminated from both sides in a time assumed short compared with the thermal time constant of its material. T_s is the surface temperature and is the maximum value, whilst the coolest point is at the midplane of the slab.

It is readily shown that the temperature T_x at the plane at depth x is given by

$$T_x = T_S \frac{e^{-x/D} + e^{-(L-x)/D}}{1 + e^{-L/D}} \tag{3.6}$$

and the average temperature in the slab T_m is

$$T_m = \frac{1}{L} \int_0^L T_x \, dx = \frac{1}{L} \int_0^L T_s \frac{e^{-x/D} + e^{-(L-x)/D}}{1 + e^{-L/D}} dx \tag{3.7}$$

whence
$$T_m = \frac{2 T_S}{1 + e^{-L/D}} \frac{D}{L} \left(1 - e^{-L/D}\right) \tag{3.8}$$

Also, by putting $x = L/2$ in eqn. 3.6, the centre-plane temperature T_C (i.e. minimum) is given as

$$T_C = T_S \cdot \frac{2e^{-L/2D}}{1 + e^{-L/2D}} \tag{3.9}$$

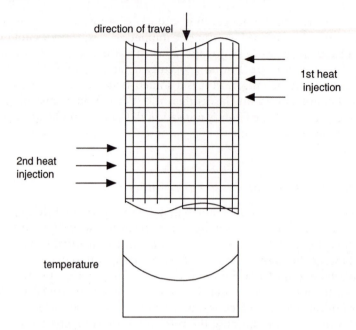

Figure 3.3   Temperature distribution in a thick slab following volumetric heating sequentially from both sides

These equations give the relationship between the surface, centre and mean temperatures in the slab as a function of the ratio of penetration depth to slab thickness. The mean temperature corresponds to the temperature rise calculated from the heat balance equation. It is assumed that the dielectric properties of the slab remain constant during heating, and where latent heat is involved the temperatures must be interpreted as energy input, and the heat balance evaluated accordingly.

3.7.1.2 Transient heat flow

Following volumetric heating, there are temperature gradients present within the workload which cause heat to flow from hot areas to cool, with the whole mass eventually reaching a constant equilibrium temperature. The rate at which heat flows is a function of the thermal conductivity K (cal/cm/deg C/s) of the material, and inversely to its specific heat s (cal/g/deg C) and density ρ (g/cm^3). These are embodied in the 'thermal diffusivity' κ for the material, where

$$\kappa = K/\rho s \ (\text{cm}^2/\text{s}) \tag{3.10}$$

Some values of thermal diffusivity for typical substances are given in Table 3.6.

Table 3.6   Thermal constants of materials

Material	Density (kg/m^3)	Specific heat (J/kg/deg C)	Thermal conductivity (W/m/deg C)	Thermal diffusivity (m^2/s)
Aluminium	2700	861	217	9.3×10^{-5}
Copper	8940	382	389	1.1×10^{-4}
Brass 70:30	8500	376	104	3.3×10^{-5}
Mild steel	7850	493	389	1.1×10^{-5}
Stainless steel				
18/8 Cr:Ni	7920	510	13.7	3.4×10^{-6}
80/20 Cr:Ni	8400	510	15.8	3.6×10^{-6}
Granite	2600	820	2.5	1.17×10^{-6}
Wood				
Spruce parallel to grain	410	1250	0.23	4.5×10^{-7}
Spruce perpendicular to grain	410	1220	0.12	2.4×10^{-7}
Water				
0°C	1000	4217	0.56	1.33×10^{-7}
10°C	1000	4192	0.58	1.38×10^{-7}
60°C	1000	4184	0.65	1.56×10^{-7}
90°C	1000	4205	0.66	1.57×10^{-7}
Beef				
−3°C	830	3000	1.15	4.6×10^{-7}
−16.5°C	830	1672	1.45	1.0×10^{-6}
Lamb				
−4°C	830	1651	1.15	4.6×10^{-7}
−14°C	830	1651	1.4	1.02×10^{-6}
Cod				
−10°C	830	1670	1.6	1.15×10^{-6}
−20°C	830	1670	1.7	1.2×10^{-6}
Beans				
−13°C	800	1670	0.93	7×10^{-7}
Strawberry				
−20°C	520	2300	1.3	0.67×10^{-6}
−10°C	520	3400	1.2	0.67×10^{-6}
0°C	570	4000	0.33	0.14×10^{-6}
+10°C	570	4000	0.34	0.15×10^{-6}
+20°C	580	4000	0.35	0.15×10^{-6}
Neoprene	1230	2090	0.19	7.4×10^{-8}
Nylon	1140	1980	0.25	1.1×10^{-7}
Polyethylene	920	2300	0.31	1.5×10^{-7}
PVC	1350	1045	0.15	1.0×10^{-7}

continued. . .

Table 3.6 continued

Material	Density	Specific heat	Thermal conductivity	Thermal diffusivity
	(kg/m^3)	$(J/kg/deg\ C)$	$(W/m/deg\ C)$	(m^2/s)
Cellulose	1505	1400	0.23	1.1×10^{-7}
(compressed cotton)				
Silicone rubber	1200	1045	0.15	1.2×10^{-7}
PTFE (Teflon®)	2200	1045	0.25	1.1×10^{-7}
Rubber (natural crepe)	1200	1600 av	0.19	1.1×10^{-7}
Air	1.29	1010	0.0242	1.86×10^{-5}

In the heat flow equations discussed below, the transient term includes the thermal diffusivity in the form

$$e^{-\kappa x^2/t} \tag{3.11}$$

where x is distance (centimetres) and t time (seconds). It will be seen that the heat-flow rate is independent of the form of the temperature distribution within the workload and in general the equilbrium time is not affected by the form of the volumetric heating.

3.7.1.3 Heat flow with surfaces maintained at fixed temperature during soaking

It is frequently required to provide an extended high-temperature environment following volumetric heating, achieved by holding the workload in a conventionally heated oven so that the surfaces of the workload are maintained at substantially the same temperature as reached during heating. The energy-input distribution, and corresponding temperature distribution, assuming only sensible heat is involved, are then defined by eqn. 3.6. Carslaw and Jaeger (1959) (para 3.4 eqn. 5, p. 101) give an equation describing the subsequent temperature-time characteristic of a preheated slab with initial temperature distribution $f(x)$ and one face held at fixed temperature, the other thermally insulated. With both surfaces maintained at the same temperature, and with a symmetrical initial temperature distribution, it follows that the subsequent temperature distribution must remain symmetrical, so that the rate of change of temperature with position (x) is zero at the midplane, and so there is no heat flow across the midplane. Thus Carslaw and Jaeger's model is equivalent to half the thickness of the slab considered in Figure 3.4, and their eqn. 5 (para 3.4) is eqn. 3.12 below, where the slab thickness is $2L$ (i.e. $\pm L$, and $x=0$ is the midplane):

$$T(x,t) = T(L,0) + \frac{2}{L}\sum_{n=0}^{\infty}\ \exp\left\{-\kappa(2n+1)^2\pi^2 t/4L^2\right\}\cos\frac{(2n+1)\pi x}{2L}\left\{\begin{array}{l}\frac{2L(-1)^{n+1}T(L,0)}{(2n+1)\pi}\ + \\[2mm] \int_0^L f(x')\cos\frac{(2n+1)\pi x'}{2L}dx'\end{array}\right\}$$

$$(3.12)$$

In this equation $T(x,t)$ is the temperature at a distance x from the midplane of the slab at time t. $T(L,0)$ is the surface temperature which remains constant at all times by definition of the boundary conditions, which is seen to be satisfied by putting $x=L$ in the first cosine term of the summation. The infinite series represents the transient temperature changes for $t>0$, and converges rapidly. $f(x')$ is the initial temperature given by eqn. 3.6; making this substitution and performing the integration, an expression is obtained for the temperature–time characteristic at any position within the slab.

The main interest is in the temperature transient at the midplane for which $x=0$, and Figure 3.4 shows a set of curves of the centre-plane temperature as a function of $\log(\kappa t/L^2)$, for a selection of values of relative penetration depth D/L. These curves show that there is a delay before the temperature starts to rise in the centre of the slab which is not much affected by the penetration depth; but, for example, the time taken to reach, say, 95% of the final temperature rise is of the order of one-tenth of that for conventional heating alone, where volumetric heating with $L/D = 0.4$ is applied.

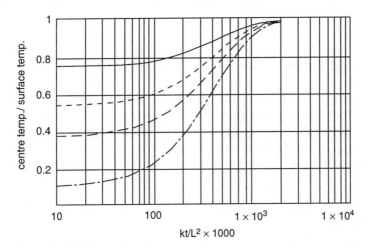

Figure 3.4 *Heat soaking after volumetric heating with exposed surfaces at constant temperature*

Normalised temperature as a function of time with penetration-depth fraction L/D as parameter L = slab thickness ——— $L/D = 0.3$
----$L/D = 1.0$ -----$L/D = 1.5$ — ·—$L/D = 3.0$

3.7.1.4 Heat flow with surfaces insulated during soaking

An alternative extreme to holding the surfaces at constant temperature during soaking is to insulate them thermally so that there is no external flow of heat. The final equilibrium temperature is now the mean temperature corresponding to the total energy injected during volumetric heating; the surface temperatures will fall, whilst the centre temperature will rise, both asymptotically to the mean.

Carslaw and Jaeger consider this case also in their eqn. 6 (para 3.4, p. 101) for a slab of thickness $2L$, which, after substituting the initial temperature distribution given by eqn. 3.6 gives

$$T(x,t) = T_s \frac{D_p}{L} \frac{1-e^{-2L/D_p}}{1+e^{-2L/D_p}} \left\{ 1 + \sum_{n=1}^{\infty} \frac{\exp\left(-kn^2p^2t/l^2\right)}{1+\left(\frac{n\pi D_p}{L}\right)^2} \cos\left(\frac{n\pi x}{L}\right) \right\} \tag{3.13}$$

In this expression the term outside the main bracket, with the unity inside the bracket, represents the mean temperature; whilst the summation term corresponds to the transient temperature change when $t > 0$, tending to zero for $t = \infty$. Inserting values of $x = 0$ and $x = L$ gives, respectively, the surface and centre temperatures as functions of time. Some typical values are plotted in Figure 3.5.

3.8 Power derating due to nonuniform heating

If a process requirement allows insufficient time for thermal equilibrium to be established after volumetric heating but before the next process stage, more energy has to be injected into the workload than is required to heat to the desired *average* final temperature. This is because the minimum temperature at a cold spot has to reach the specified figure, so that the temperature of the remainder of the workload is higher, exceeding the desired figure and thereby representing a waste of energy.

The effect is readily illustrated and quantified for a thick slab heated alternately from both sides so that there is a temperature distribution through the slab corresponding to the sum of two exponential patterns, one from each face, as discussed in Section 3.7.1.1, where the penetration depth is D and the thickness of the slab is L. For this condition, the midplane temperature T_c is the lowest and is given by eqn. 3.9, while the mean temperature T_m is given by eqn. 3.8 and corresponds to the total energy

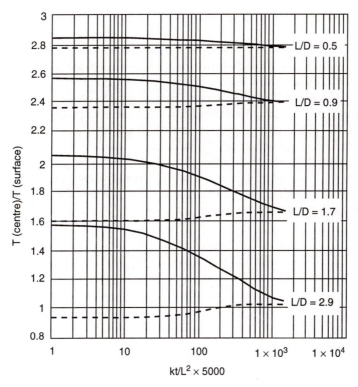

Figure 3.5 *Heat soaking after volumetric heating with exposed surfaces thermally insulated*

Normalised temperature as a function of time with penetration-depth fraction L/D as parameter L = slab thickness

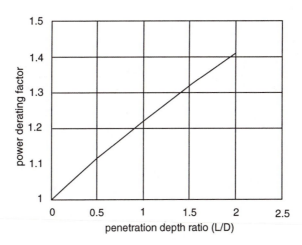

Figure 3.6 *Power-derating factor against penetration depth*

injected during volumetric heating. Dividing eqn. 3.8 by eqn. 3.9 gives T_m/T_c as a function of L/D. The factor T_m/T_c is a derating factor representing the amount by which the total energy input must be increased in order to ensure that the coldest point reaches the minimum required temperature.

Figure 3.6 is a computed curve based on eqns. 3.8 and 3.9, showing the magnitude of the derating factor as a function of D/L for the above configuration. It shows, for example, that where the thickness of the slab is equal to the penetration depth, the required energy input is 22% more than that calculated from a simple heat-balance consideration.

The temperature distribution is often more complicated than that assumed above, and each case must be carefully assessed, but the above approach illustrates a loss of performance which is frequently overlooked in determining the power requirement for an installation.

Chapter 4

Microwave transmission: theory and practice

4.1 Introduction

In this Chapter we first consider basic concepts of microwave power transmission, introducing the essential parameters commonly used, familiarity with which is vital to a proper understanding of the technology. Transmission in coaxial, parallel-plate and open-wire lines in the elementary transverse electromagnetic (TEM) mode is then discussed, before waveguides are introduced. There follows a discussion on waveguide practice, with descriptions and characteristics of the components commonly used in an industrial microwave-heating installation.

At microwave frequencies, where the wavelength is comparable with the dimensions of the equipment, the traditional electrical engineer's view of power transmission along a pair of conductors, using discrete components such as inductors and resistors, breaks down: the physical separation between the conductors is a significant part of a wavelength, so at a distant point there is a phase difference between components of field due to the conductors. Energy is radiated wastefully into space. Closed waveguides are used for efficient transmission, and their analysis is derived directly from Maxwell's equations. Fortunately, a waveguide behaves like a transmission line at lower frequencies in many of its important aspects, and so the electrical engineer's concepts of forward and reflected waves used in the analysis of long open parallel lines can be applied.

A rigorous analysis of transmission line and waveguide theory is well presented in many textbooks on microwave theory (e.g. Slater, 1942; Marcuvitz, 1986; Ramo *et al.*, 1965; Stratton, 1941), and it is unnecessary to reproduce it here. Instead, we give a sound and realistic presentation of these topics sufficient for the nonspecialist to understand and be aware of the salient features important in industrial heating practice.

4.2 Impedance matching

Impedance matching is an everyday topic of electrical engineers, and particularly of those involved in the design of dielectric-heating equipment. It is essentially a process of arranging that a generator with a certain available power dissipates all that power into a load. In fact, it is a process familiar to everyone, being universal in all situations where power is to be transferred optimally to a load, e.g. in acoustics, mechanical engineering, hydraulics; it is by no means unique to electrical engineering. And yet, because it is so commonplace, few stop to consider its implications. Were it not that nature provided us with a voice mechanism which is impedance-matched to sound waves in air, and an eardrum which similarly is matched to air, we would not be able to speak and hear.

An obvious mechanical example of impedance matching is the gear train between the engine and wheels of a car. The power output of the engine is the torque developed at the crankshaft multiplied by the rotation speed. If the engine speed is very low because it is heavily loaded, the torque may also be low, giving a low power output. Similarly, if the speed is very high because it is lightly loaded, the torque may again be low, so that the power output is again less than peak. In between these extremes there are a speed and a torque where the engine develops its maximum power output. The speed and the torque are analogous to voltage and current from an electrical generator: the quotient voltage/current is the internal resistance of the generator. The engine has an optimum value of speed/torque for maximum power output, i.e. a definitive output impedance. The load varies widely from a high torque at low speed at the wheels climbing an incline to a high speed and relatively low torque cruising on a level road. A variable gearbox between the engine and wheels is adjusted so that the engine runs substantially at its preferred speed irrespective of the final load. The gearbox *matches* the output impedance of the generator (engine) to the load. It is the electrical equivalent of an adjustable matching transformer with a variable turns ratio. At microwave frequencies, wound transformers do not exist, but other structures are used which have equivalent electrical properties and are often called matching transformers.

4.2.1 Generator output impedance: optimum matching of load

Let us now quantify impedance matching, and consider some misconceptions which often arise. Figure 4.1 shows a battery with internal EMF V volts, and internal resistance R_{INT} ohms connected to a load R_{load} ohms. Ohm's law gives the current I as

$$I = \frac{V}{R_{INT} + R_{load}} \tag{4.1}$$

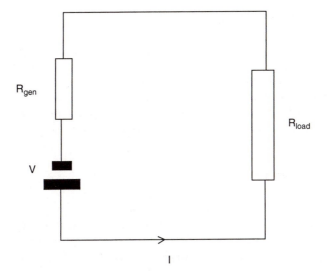

Figure 4.1   Basic generator and load circuit

The power dissipated in the load resistor is $P_{load} = I^2 R_{load}$. Substituting for I_{load} and differentiating dP_{load}/dR_{load} to find the maximum condition in the usual way, it is found that maximum power is transferred to the load when $R_{load} = R_{gen}$.

We have taken the simplest case of a DC battery. For an AC generator which has internal reactance as well as resistance, a more detailed analysis shows that the load resistance still must equal that of the generator, but that the load must have a reactance equal and opposite to that of the generator (e.g. Slater, 1942). The reactances cancel, leaving the load current determined solely by the resistance of the circuit.

Now consider a load terminating a very long or infinite transmission line. Looking back into the line from the load terminals, the impedance of the line is its characteristic impedance. For an ideal parallel-wire line such as that in Figure 4.2, the characteristic impedance is $Z_0 = \sqrt{(L/C)}$ where L is the inductance, and C the capacitance, both per unit length of line. Note that although L and C represent pure lossless reactances, the characteristic impedance is purely resistive. Thus to transfer maximum power to the load from the line, the load must have a resistance $R_{load} = \sqrt{(L/C)} = Z_0$.

Since there are no resistive losses in the line by definition, there is no power loss therein.

Now consider again the circuit of Figure 4.1. Maximum power is transferred to the load when $R_{gen} = R_{load}$. Since the same current flows through both, the power dissipated in the generator is the same as that in the load, representing only 50% efficiency. While in low-power signal generators this is a desirable objective, in high-power equipment it would be totally unacceptable, first as a waste of energy, and secondly as a cooling problem in removing heat internally dissipated in the generator. In high-power practice, the internal dissipative resistance of the generator is low compared with the Z_0 of the transmission line it feeds, and the design technique is focused on achieving high efficiency by coupling the generator optimally to the transmission line. However, the generator has a resonant circuit (L, C) where the inductance and capacitance determine the operating frequency as a product (i.e. $f = 1/2\pi\sqrt{LC}$), and the nondissipative source impedance is equal to $\sqrt{(L/C)}$. The latter is matched to the transmission line.

4.3 Forward and reflected waves

Consider a parallel-wire transmission line, long compared with the wavelength corresponding to the frequency of the generator feeding it, terminated in a resistance load R_{load}, as shown in Figure 4.1. The line is erected in free space with relative dielectric constant $\varepsilon' = 1$.

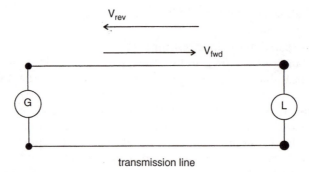

Figure 4.2   *Generator connected to load via transmission line: forward and reflected waves*

The frequency of the generator is f hertz, for which the wavelength λ (metres) is given by eqn. 1.23, in which c is the velocity of propagation of energy along the line. For of a two-wire line, essentially with a 'go' and 'return' circuit, the velocity is that of a plane wave in space as discussed in Chapter 1.

The generator impresses at the input to the line a sinusoidal voltage V_F which travels at velocity c towards the load.

The transmission line has a characteristic impedance Z_0, and this is the impedance 'seen' initially by the wavefront as it progresses along the line from the generator. Clearly, if there is a discontinuity in the line at any point, the impedance will locally change because the capacitance and/or inductance per unit length will change at that point. Moreover, it is obvious that energy can propagate with equal facility in either direction along the line. Since the condition for maximum power transfer is now broken at the discontinuity, less power will be transmitted onward from it. In general, the balance is of three components: power absorbed at the discontinuity, power reflected back towards the generator and, of course, power transmitted onward towards the terminating load. If the discontinuity comprises a simple localised change in capacitance or inductance, there will be no power absorbed at the discontinuity, since perfect capacitors and inductors absorb no power. Such a discontinuity would be a conducting iris or projection into the transmission line. However, if the discontinuity includes a resistance component, power will be dissipated at the discontinuity. As shown in Section 4.3.2 *et seq.*, an unwanted reflection can be cancelled by using a second reflecting device as a matching element. In high-power transmission lines great care must be taken in design of matching devices, since even a small effective resistance, while absorbing a relatively small proportion of the transmitted power, may in fact dissipate sufficient to create a local overheating problem.

4.3.1 Reflection and transmission coefficients, scattering coefficients

In Section 4.3, the concept of forward and reflected waves was introduced, and in this section they will be quantified.

Eqns. 1.18–1.21 show how the field components of a travelling wave vary sinusoidally in time (ωt), and in space ($\beta z = (2\pi/\lambda)z$, typically as $V_f \sin(\omega t - \beta z)$, representing a forward travelling wave of amplitude V_f. As in elementary electrical engineering, such an electric field or voltage, can be represented as a vector, the length of which is proportional to the voltage, rotating about an origin at a speed of one revolution in one complete cycle of the sine wave (i.e. $\omega t = 2\pi$, whence $t = 2\pi/\omega = 1/f$, where f is the frequency). Following the usual practice, it is more convenient to make the vector stationary by allowing its space origin to rotate at the speed f hertz (revolutions per second)

The effect of the term βz is to cause the vector to be in a different phase angle βz radians between two points a distance z apart along the line. If the vector is determined at a point on the transmission line, and the point of observation moves, the vector rotates through an angle βz, where z is the distance moved. The direction of rotation is determined by a sign convention that if the observation point moves *towards the terminating load*, the vector rotates in an *anticlockwise* direction. Conversely, the vector rotates clockwise if the observation point moves towards the generator.

For a wave travelling from the load towards the generator (i.e. a reflected wave), the representation is $V_{ref} \sin(\omega t + \beta z)$, where V_{ref} is its amplitude.

Again following electrical practice, it is more convenient to express the sine term in its exponential form $\exp\{j(\omega t \pm \beta z)\}$ (e.g. Slater, 1942; Ramo *et al.*, 1965), where the negative/positive signs represent the forward and reflected waves. Dividing the forward by the reflected waves, we have

$$\frac{V_F}{V_{Ref}} \frac{\exp\{j(\omega t - \beta z)\}}{\exp\{j(\omega t + \beta z)\}} = \frac{V_F}{V_{Ref}} \exp(-j2\beta z) = \rho \exp(-j2\beta z) \quad (4.2)$$

where $\rho \exp(-j2\beta z)$ is the reflection coefficient.

Note that the reflection coefficient is a vector, having two components: an amplitude, which can have a value $0 < \rho < 1$, and a phase angle $-2\beta z$. As the observation point moves along the transmission line, the phase angle of the reflection coefficient changes accordingly. Substituting $\beta = 2\pi/\lambda$ and $z = \lambda/2$, the phase angle becomes 2π, showing that as the observation point moves through a half wavelength the reflection-coefficient vector rotates a complete revolution. At all points a half wavelength apart, the reflection coefficient has the same magnitude and phase angle. It must be emphasised that, in the above, the transmission line is assumed lossless. For most practical purposes, this is true for a plain waveguide feeding power from a generator to a load. For a travelling-wave applicator where power is being dissipated in a workload, it is certainly not a valid assumption because there is attenuation of both forward and reflected waves within the applicator; this is discussed in detail in Section 6.2.

Figure 4.3*a* shows a vector diagram illustrating how the forward and reflected waves add vectorially to give a resultant whose amplitude relative to the forward wave varies as a function of both the amplitude and the phase angle of the reflection coefficient. By dividing the amplitudes by the forward-wave amplitude, the reflection coefficient represents the relative, or 'normalised', amplitude of the reflected wave, and the resultant is similarly normalised as shown in Figure 4.3*b*.

Note that the resultant of adding the forward and reflected waves, although a vector, is not in itself a propagating wave. It is simply the field existing at the observation point on the line.

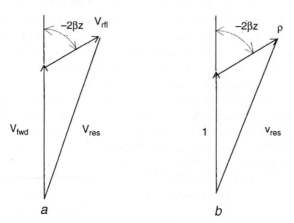

Figure 4.3   *Vector diagrams showing forward and reflected waves*

　　　a   With resultant
　　　b   Normalised to show reflection coefficient

Some very important transmission line properties are illustrated in Figure 4.3:

(i)   At the position on the line where $z = 0$ or $n\lambda/2$ (n is any integer), the forward and reflected waves are in phase with each other and so add arithmetically. Intuitively and by analogy with vector diagrams at lower frequencies, this suggests that there is no reactive component present at that position, and so the impedance on the line is purely resistive. The line could be terminated at this point with a pure resistance load; if it were infinite, i.e. an open circuit, no power would be absorbed and so the reflected wave would equal the forward wave in amplitude. Since they are in phase, the voltage at the open-circuit point is twice that of the forward wave.

(ii)   At the position on the line where $z = \lambda(1+2n)/4$ (i.e. a quarter-wavelength with half-wave increments therefrom), the forward- and reflected-wave vectors are in phase-opposition so that the resultant is their arithmetic difference. Again, if no power is absorbed the reflected wave must equal the forward, and the resultant is zero. This would be the result of placing a short-circuit across the line at that point.

(iii)   If the observation point is located a quarter-wavelength from that of case (ii) the situation is that of case (i), where the impedance is infinite. This illustrates an extreme case of the very important property of a transmission-line impedance transformation. It means that a quarter-wavelength section of line short-circuited at one end appears as an open circuit at the other; this is a property widely used in 'choking' systems and in providing all-metal supports where an insulator would have to be used at lower frequencies.

(iv)   At other locations for the observation point, the impedance is complex, and is discussed in Sections 4.3.3–4.3.5.

Since *power* is proportional to (voltage)^2 it follows that the reflected power is $|\rho|^2 P_F$ where P_F is the forward power, and $|\rho|$ means the amplitude of ρ. If the reflected wave is caused by a nonabsorbing (i.e. lossless) obstacle, the power transmitted beyond it is $P_F(1 - |\rho|^2)$, and is often written $|\tau|^2 P_F$ where τ is the transmission coefficient. Clearly,

$$|\tau|^2 = 1 - |\rho|^2 \tag{4.3}$$

The concept of reflection and transmission coefficients can be extended to embrace structures which have several output terminals or ports (e.g. junctions between transmission lines, directional couplers), and 'scattering coefficients' can be used to define precisely their electrical characteristics. Like reflection coefficients they have amplitude and phase, and they define, at specified reference planes at each output, the amplitude and phase of each component of output signal due to unit input to each of the other ports in turn. They are most conveniently arranged as a matrix, called a 'scattering matrix' (Ramo *et al.*, 1965).

4.3.2 Standing waves and standing-wave ratio

It will be seen from Figure 4.3*b* that, as the observer moves along the line, the unit vector representing the forward wave remains fixed but the vector representing the reflected wave rotates so that the locus of its tip is a circle. The vector sum representing the field intensity at the observer's position varies from a maximum of $(1 + |\rho|)$ to a minimum of $(1 - |\rho|)$. The positions of these two conditions are $\lambda/4$ apart. In general, the resultant can readily be calculated trigonometrically from the vector diagram as

$$V_{res} = \sqrt{\left\{ 1 + |\rho|^2 + 2|\rho|\cos(-2\beta z) \right\}} \tag{4.4}$$

If the reflection is small ($\rho < 0.2$), the resultant varies approximately sinusoidally with position. However if $\rho \to 1.0$ it can be seen by inspection of the Figure 4.4A that the amplitude of the resultant varies quite slowly with position around its maximum value, but very rapidly near the zero value.

This quasisinusoidal variation of the resultant electric field or voltage with position is called a standing-wave pattern. It is stationary in position with time, and changes only if the reflection coefficient changes in amplitude and/or phase angle.

The voltage standing-wave ratio (VSWR) S is defined as the voltage minimum amplitude divided by the voltage maximum or vice versa:

$$S = \frac{1 - |\rho|}{1 + |\rho|} \quad \text{or} \quad S = \frac{1 + |\rho|}{1 - |\rho|} \tag{4.5}$$

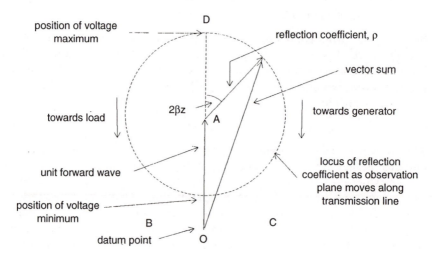

Figure 4.4A   *Vector diagram of forward and reflected waves as observer moves along transmission line*

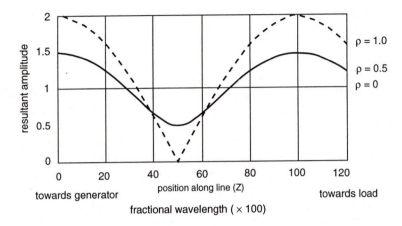

Figure 4.4B   Generation of standing-wave pattern: resultant field amplitude as a function of position along transmission line

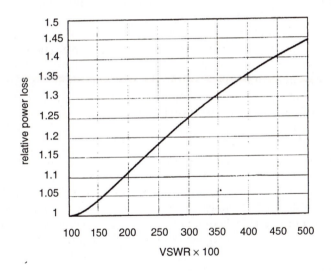

Figure 4.4C   Increase in power loss in a transmission line due to a standing wave (S > 1 definition of VSWR)

Both definitions are in regular use, but it is obvious which is used in a particular case because for the first $S < 1$ and the second $S > 1$.

The associated magnetic field also has a standing-wave pattern of the same ratio, but its minima are not, in general, located coincidentally at the voltage minima, because the load may have a reactive as well as a resistive component. Indeed, if the load is purely reactive, i.e. $|\rho| = 1.00$, the maxima of the electric-field pattern fall coincidentally with the minima of the magnetic-field standing-wave pattern.

The VSWR is a very important parameter because it is relatively easy to measure by a probe mounted on a moveable carriage, inserted into the transmission line to sample the E-field intensity. The carriage is adjusted in position to locate a voltage minimum, and the intensity measured. It is then moved to a voltage maximum, and the intensity is again measured. The quotient of the two readings is the VSWR. The reflection coefficient can then be calculated from eqn. 4.5. By inspection of Figures 4.4A–C it is clear that the position of the VSWR pattern is determined by the phase of the reflection coefficient; by measuring the position of a voltage minimum the phase can be determined, giving a complete determination of ρ.

Before the advent of network analysers in the 1980s, standing-wave detectors using this principle were commonplace instruments in all microwave laboratories. They are very tedious to use because measurements have to be taken at several spot frequencies over a required frequency band, and then plotted by hand on a Smith chart, as discussed in Section 4.3.4.

VSWR, reflection and transmission coefficients are all related as shown above. Another related parameter often used is the 'return loss', expressed in decibels. This is simply the voltage-reflection coefficient translated into a power-reflection coefficient and expressed in decibels.

$$\text{return loss} = 10 \log_{10} |\rho|^2 = 20 \log_{10} |\rho| \qquad (4.6)$$

Table 4.1 quantifies the these parameters as a function of VSWR. In Table 4.1 the reflected power is $|\rho|^2$. The normalised resistance r is the value of a resistance termination (relative to the characteristic impedance of the line) which would give the VSWR shown. If r is less than 1.00, there is a voltage minimum of the standing wave at the termination, and if r is greater than 1.00 then the line terminates in a voltage maximum. See Section 4.3.3 for details.

From the foregoing it is evident that a standing wave should be avoided because it represents power reflected away from the terminating load. Moreover, the standing wave of magnetic field has an associated standing wave of current flow in the walls of the waveguide. Since the power dissipated in the walls is proportional to $I^2 R$, the heating is similarly patterned. The wall current pattern is the same form as given by eqn. 4.4, and it is found by squaring this equation and integrating over a whole wavelength that the total power loss $I^2 R$ is increased by the standing wave by the factor $(1 + |\rho|^2)$ where ρ is the corresponding reflection coefficient. Figure 4.4C shows the increase in power loss in the line as a function of VSWR ($S < 1$); although in a low-loss

Table 4.1   *Reflection and transmission coefficient, reflected power, return loss and normalised resistance as functions of VSWR*

VSWR <1	VSWR >1	ρ	$\|\rho\|^2$	τ	Return loss (dB)	r	1/r
1	1.00	0.00	0.00	1.00	$-\infty$	1	1.00
0.9	1.11	0.05	0.0025	1.00	−25.58	0.9	1.11
0.8	1.25	0.11	0.0121	0.99	−19.08	0.8	1.25
0.7	1.43	0.18	0.0324	0.98	−15.07	0.7	1.43
0.6	1.67	0.25	0.0625	0.97	−12.04	0.6	1.67
0.5	2.00	0.33	0.109	0.94	−9.54	0.5	2.00
0.4	2.50	0.43	0.185	0.90	−7.36	0.4	2.50
0.3	3.33	0.54	0.292	0.84	−5.38	0.3	3.33
0.2	5.00	0.67	0.449	0.75	−3.52	0.2	5.00
0.1	10.00	0.82	0.672	0.57	−1.74	0.1	10.00
0.03	33.33	0.94	0.884	0.34	−0.521	0.03	33.33
0.01	100.00	0.98	0.960	0.20	−0.174	0.01	100.00
0	∞	1.00	1.00	0.00	0	0	∞

line the increase in loss may not be significant in terms of available power, it may represent a large increase in temperature, particularly near a current maximum.

4.3.3 Impedance and admittance

Following standard AC theory and Ohm's law, impedance is simply the quotient V/I in circuits where these parameters are readily measurable, or E/H in microwave systems where VSWR and reflection coefficients are more easily measured.

Although in low-frequency AC circuits the actual impedance values are used, at microwave frequencies they are not so readily measured, and it is more practical and equally valid to use 'normalised' values which are the actual impedances divided by the characteristic impedance of the transmission line. Note that the characteristic impedance of a lossy transmission line is complex, which has relevance in the calculation of the characteristics of certain classes of heating applicators; see Sections 6.2, 7.8 and 7.9, and eqn. 4.12. Impedance measurements made via VSWR measurements inherently give normalised values.

The impedance of a terminating load is in general complex, i.e. it has both resistive and reactive components, the latter due to associated inductance or capacitance. The complex load impedance ($z_L = r_L + jx_L$) is directly related to its reflection coefficient $\rho_L \exp(j\phi)$ (Ramo *et al.*, 1965) as

$$\rho_L \exp(j\phi) = \frac{z_L - 1}{z_L + 1} \qquad (4.7)$$

whence
$$z_L = r_L + jx_L = \frac{1 + \rho_L e^{j\phi}}{1 - \rho_L e^{j\phi}} \qquad (4.8)$$

Clearly, if z_L is a pure resistance ($x_L = 0$), ρ is also a real number with $\phi = 0$. Comparison with eqn. 4.5 shows that in this case $r_L = s$. For other cases Z_L is complex.

Again following AC theory, normalised admittance y_L is simply the reciprocal of normalised impedance z_L. Engineers use admittance and impedance with equal facility, one is often less cumbersome to manipulate algebraically than the other. In general admittance is more appropriate for parallel circuits and components shunted across the line, while impedance is easier to manipulate for series circuits (Slater, 1942; Ramo *et al.*, 1965).

4.3.4 Impedance transformation: the Smith chart

In the foregoing, it is shown how, as an observing plane moves along a mismatched transmission line, the phase of the reflection coefficient changes, giving rise to a standing-wave pattern. A specific case is cited where the locally high resistive impedance at a voltage maximum changes to a correspondingly low resistance value at the adjacent voltage minimum, at a distance of $\lambda/4$ away. The high impedance is *transformed* to a low impedance, the quarter-wavelength of line behaving as an ideal transformer with a turns ratio $N = S$. Also, the whole pattern repeats at regular intervals along the line, spaced apart by a distance $\lambda/2$.

This is a special case of a general property of impedance transformation of a transmission line. If two points are chosen spaced apart along a transmission line, the impedance at the first is transformed to another value at the second dependent on the phase change of the reflection coefficient between the two locations. Any phase angles are possible and so, in general, the resulting impedances are complex (i.e having both resistive and reactive parts).

Note that in Figure 4.4A and 4.4B, showing the generation of a standing wave, the resultant field intensity, close to a voltage minimum when $\rho \to 1$, changes both in amplitude and phase extremely rapidly with the position of the observation plane. However, near a voltage maximum the corresponding rates of change are very slow. Inspection of eqn. 4.8 reveals that, near the voltage minimum, the impedance changes very rapidly too, compared with near the maximum. These properties of impedance transformation are shown to be represented by the transmission-line equation (e.g. Ramo *et al.*, 1965; Slater, 1942). For a *lossless* transmission line of length L and characteristic impedance Z_0 it is

$$Z_i = Z_0 \frac{Z_L \cos\beta L + jZ_0 \sin\beta L}{Z_0 \cos\beta L + jZ_L \sin\beta L} \qquad (4.9)$$

where Z_i is the impedance looking into the line from the generator, Z_L is the load impedance and $\beta = 2\pi/\lambda$ is the phase constant.

In normalised values (i.e. dividing by Z_0),

$$z_i = \frac{z_l \cos \beta L + j \sin \beta L}{\cos \beta L + j z_l \sin \beta L} \tag{4.10}$$

The corresponding admittance transmission-line equation is

$$y_i = \frac{y_l \cos \beta L + j \sin \beta L}{\cos \beta L + j y_l \sin \beta L} \tag{4.11}$$

Until recently, eqns. 4.9–4.11 were cumbersome to manipulate and their graphical solution was often simpler, certainly in that it shows more clearly the relationships between the parameters, which is especially helpful when evaluating an impedance-matching problem. Although the advent of personal computers has made their solution very easy, including their complex form eqns. 4.12 below for lossy transmission lines, the graphical solution embodied in the 'Smith chart' remains very important. Its derivation and proof are well presented in the standard textbooks (Slater, 1942; Ramo *et al.* 1965) and are not given here. Instead there is a non-rigorous description of the chart intended to highlight its important features.

The Smith chart is essentially a development of the forward/reflected-wave vector diagram of Figure 4.3a, and comprises three families of geometrically related circles representing values of VSWR (or reflection coefficient), normalised resistance (or conductance) and normalised reactance (or susceptance). It can be based on either impedance or admittance representation; because most of the microwave applications involve shunt elements, the following description is of an admittance Smith chart.

Referring to Figure 4.4a, as the largest value of reflection coefficient ρ is unity for a passive circuit, the boundary of the chart represented by the vector diagram is a circle of unit radius centred on the tip A of the unit-amplitude forward-wave vector. Any value of ρ will therefore fall within this boundary. Moreover, movement along the transmission line results in the vector ρ describing a circular locus about A, the distance moved being the angle $4\pi z/\lambda$ as shown in Figure 4.4b, remembering that $\beta = 2\pi/\lambda$. **Clockwise rotation corresponds, by sign convention, to moving the observer's reference plane towards the generator.**

The first family of circles is therefore a concentric set centred on A, of radii equal to a range of values of $|\rho|$, or, more commonly, VSWR related to $|\rho|$ via eqn. 4.5. Note that for a reflectionless load ($|\rho| = 0$), VSWR = 1.0 and the centre of the diagram A is the point representing the load. The objective in 'impedance matching' is to aim the load point at A, or within a predetermined circle centred on A representing the worst acceptable value of VSWR. Usually, this has to be done over a range of frequencies, a procedure known as broadband matching.

The second family of circles is a set with convenient radii, having their centres on the forward wave vector OA and passing through the point O, so that they are tangential to each other at O. These are derived from eqn. 4.11 and are circles of constant conductance (or resistance for an impedance diagram). Note that the circle for $g = 1$ passes through the match point A, and similarly for $r = 1$. Also note that the circle $g = 0$ is the boundary of the Smith chart.

The third family of circles comprises two sets of arcs which are mutual mirror images. One set has its centres on the axis OB, at right angles to OA, each circle passing through O so that the set is mutually tangential at O. The other set has its centres on the axis OC, forming a mirror image of the first on the opposite side of the axis OA. The arcs originate at O and terminate at the boundary of the Smith chart. These arcs are loci of constant susceptance (or reactance for an impedance diagram). Note that these arcs are orthogonal to the conductance circles at all intersections. For the pair of circles of infinite diameter, the arcs are the axis OAD corresponding to the reflected wave being in-phase or antiphase with the forward wave. We have already observed that this condition is for a purely conductive (resistive) load, and so this axis OAD is the axis of zero susceptance (reactance).

Consider an open circuit transmission line: its resistance at the load terminal is infinite and therefore the conductance is zero. Since it is an open-

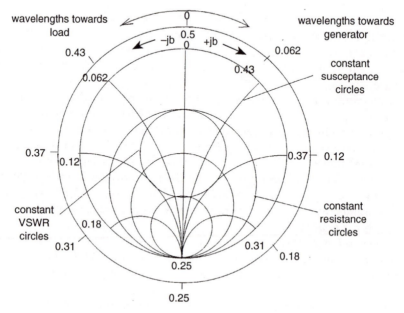

Figure 4.5   Simplified Smith chart

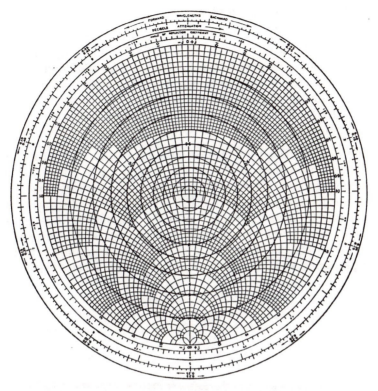

Figure 4.6   Full Smith chart

circuit termination, the vector diagram in Figure 4.4a applies with 2βz = 0. The open circuit is represented at the reference plane of the load terminal by the point D on the Smith chart. If we now move the reference plane towards the generator, the admittance locus will follow the g = 0 circle, moving clockwise. As open-circuited transmission lines less than $\lambda/4$ long appear capacitive, the susceptance circles on the right of the chart correspond to capacitive susceptances (+jb), and on the left to inductive susceptances (−jb).

For an impedance chart, the point D corresponds to a short circuit. Moving towards the generator a short distance, less than $\lambda/4$, the line 'looks' inductive, and so for an impedance Chart the reactance circles on the right of the chart are inductive (+jx) and those on the left capacitive (−jx).

A skeleton Smith chart is shown in Figure 4.5, showing a few of the principal circles described above for clarity. Figure 4.6 is the chart normally used; it simply has more circles than in Figure 4.5, allowing greater accuracy of plotting.

4.3.5 Impedance-matching procedure

Using the Smith chart, it is a straightforward procedure to match a load to its feed waveguide. In many cases the matching can only be optimised under one set of operating conditions for the load, and an additional, adjustable, matching device must be used to 'match out' the residual reflection for other conditions. Thus a coarse fixed matching element is supported by an adjustable device. The procedure for establishing the coarse matching element is as follows:

(i)   Determine the load impedance (admittance) as 'seen' at its point of connection to the transmission-line feeder, e.g its waveguide flange. This may be determined using a network analyser, or by using a 'slotted line' or standing-wave detector. The latter illustrates the procedure best.

(ii) Two measurements are taken with the standing wave detector, the VSWR, and the distance d_{min} from the flange to the nearest voltage minimum. At the voltage minimum the admittance is a pure conductance, so at that point we can plot on the chart (Figure 4.7) its

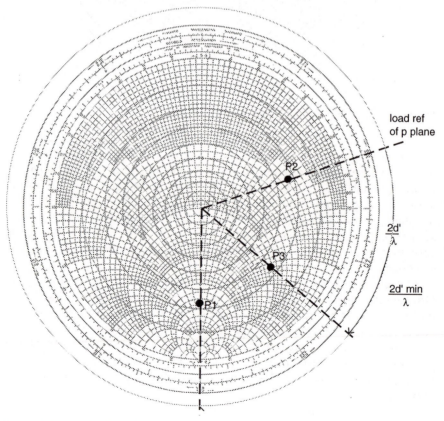

Figure 4.7   Smith chart showing matching procedure (admittance diagram)

admittance as the point P_1 where the measured VSWR intersects the $b = 0$ axis. Since we are plotting a voltage minimum, the conductance must be greater than 1.00, so the plotted point lies on the lower half of the axis.

(iii) To plot the admittance at the reference plane of the flange, we simply measure the angle $d_{min}/\lambda g$ from the 6 o'clock position around the chart in an anticlockwise direction (i.e. towards the load), and replot the admittance at P_2, at the same radius from the centre of the chart as P_1.

(iv) Next we move clockwise from P_2, at constant radius, until we intersect at P_3 the $(1 + jb)$ circle, which passes through the centre of the chart. At this point, we have the correct conductance for matching but there is a susceptance b remaining. The distance from the flange is $d' = \lambda g\,(\Delta\phi)$, where $\Delta\phi$ is the angular shift from P_2 to P_3 measured on the outer scale of the chart.

(v) At P_3 we can insert a lossless passive susceptance (typically an iris or post) of magnitude $-b$ to cancel the susceptance of the load at this point so that the combined admittance becomes $\{1 + j(+b - b)\} = 1 + j0$. The load is now matched to the waveguide.

Note that it would be possible to select a second intersect with the $(1 + jb)$ circle for P_3 by moving further towards the generator. The susceptance would then be of opposite sign and would require an inductive instead of a capacitive matching element, or vice versa. The disadvantage would be that the matching element is now further from the load and so the combination becomes more frequency sensitive (narrower band). However, the matching element may be more realisable, and capable of higher power handling. Also note that the two possible positions for the matching element are *not a quarter wavelength apart*. This distance is only approximately $\lambda/4$ where the initial mismatch is small. Because of the convergence of the susceptance and conductance lines at the lower half of the chart, it is difficult to match a large mismatch with a multiscrew tuner: there is considerable interdependence between the screws.

4.3.6 Impedance transformation in lossy transmission lines

When the transmission line is lossy, the amplitudes of the forward and reflected waves are attenuated with distance in their respective direction of travel. In simple waveguides used for power transmission, the losses are very small and can be ignored, allowing the transformation (eqns. 4.9–4.11) and the Smith chart to be used. However, in many applicators (e.g. the longitudinal waveguide, and parallel-plate line) the presence of the workload causes the transmission-line losses to be significant. Under these conditions, attenuation must be taken into account, resulting in eqns. 4.9–4.11 changing to a more complicated form involving the hyperbolic functions. Moreover, the propagation constant ($\gamma = \alpha + j\beta$) becomes complex, since both α and β have finite nonzero values.

The modified transmission line equation for lossy transmission is, on a normalised impedance basis (Slater, 1942),

$$z_i = \frac{\sinh \gamma L + z_t \cosh \gamma L}{\cosh \gamma L + z_t \sinh \gamma L} \tag{4.12}$$

The similarity to eqn. 4.10 is immediately apparent, and an equivalent admittance equation can readily be written, similar to eqn. 4.11. Note that the lossy line has a complex characteristic impedance.

Evaluation of eqn. 4.12 was previously tedious, but is straightforward using a computer program such as 'Mathcad Plus 6'® which is able to process arrays of complex numbers.

4.4 Twin open-wire, coaxial and parallel-plate transmission lines

In this Section and Sections 4.5 and 4.6, we consider the properties of transmission lines, beginning with the most familiar before discussing waveguides. To the 'conventional' engineer, the most important difference between hollow waveguides and open-wire, coaxial and parallel-plate transmission lines is that DC power can be transmitted in the latter, which is obviously not possible in an empty closed conducting tube forming a waveguide. 'Go' and 'return' paths exist in the open wires and their ability to transmit DC sets them apart as a special-case solution to Maxwell's equations (Lamont 1942; Ramo *et al.*, 1965; Slater, 1942).

The essential feature of these transmission lines is that, in the fundamental mode, the energy propagates with the associated electric and magnetic fields transverse to the axis of the line: the axial components of both the magnetic and electric fields are zero. This fundamental mode is called the transverse electromagnetic (TEM) mode. As an example, two straight parallel conductors carrying a direct or low frequency alternating current to a terminating resistor have a voltage between them. Both the resulting electric- and magnetic-field patterns comprise only components lying in the transverse plane to the axis, as shown in elementary electrical theory. There is no theoretical upper limit to the frequency for the TEM mode. In that there is no axial component of either the E or H fields, there is a direct similarity to a plane wave (Section 1.5). The propagation velocity of the TEM wave is the same as that for a plane wave in the dielectric medium (Stratton, 1941; Ramo *et al.*, 1965)

The characteristic impedance of a TEM transmission line is $\sqrt{(L/C)}$ ohms, (Ramo *et al.*, 1965) where L and C are the inductance and capacitance, respectively, per unit length of the line. These values are known for the above three TEM configurations, giving values of Z_0 as shown in Table 4.2. These data are for air dielectric; where the lines are filled with dielectric $\varepsilon_d = \varepsilon' - j\varepsilon''$ the value of ε_0 must be multiplied by ε_d.

Table 4.2   Characteristics of TEM transmission lines

Line type	L	C	Z_0
	(H/m)	(F/m)	(Ω)
Parallel plate spacing a, width b	$\mu_0 a/b$	$\varepsilon_0 b/a$	$\dfrac{a}{b}\sqrt{\dfrac{\mu_0'}{\varepsilon_0}}$
Coaxial line inner diameter a, outer bore b	$\dfrac{\mu_0 \log_e(b/a)}{2\pi}$	$\dfrac{2\pi\varepsilon_0}{\log_e(b/a)}$	$\dfrac{1}{2\pi}\left\{\sqrt{(\mu_0/\varepsilon_0)}\right\}\log_e(b/a)$
Parallel conductors, conductor radius r, spacing d	$\dfrac{\mu_0}{\pi}\log_e(d/r)$	$\dfrac{\pi\varepsilon_0}{\log_e(d/r)}$	$\dfrac{1}{\pi}\left(\log_e(d/r)\right)\sqrt{\dfrac{\mu_0}{\varepsilon_0}}$

Note that the velocity of propagation is $v = 1/\sqrt{(LC)}$, and in each of the above three cases reduces to $1/\sqrt{(\mu_0\varepsilon_0)}$, which is the velocity of the plane wave.

Of the above TEM-mode lines, the parallel-plate line is of importance as the basis of heating ovens with wide conveyor bands, especially with metal bands, as discussed in more detail in Section 6.2.3. The coaxial line often forms the output section of magnetrons. It is sometimes a convenient flexible line for low-power use (<1 kW), but care must be taken to ensure that overheating does not occur due to the attenuation of the line. Coaxial lines have high attenuation compared with waveguides.

Parallel conductor lines are not used at microwave frequency because they are inevitably long compared with the wavelength, and radiate heavily. However, they are used in RF applications at 27 MHz.

4.4.1 Coaxial-line TEM-mode data

The electric and magnetic fields E and H are very simple for the TEM mode in a coaxial line with concentric conductors (Marcuvitz, 1986; Slater, 1942; Ramo *et al.*, 1965). The electric field E is a set of radial vectors normal to the surfaces of the inner and outer conductors. The magnetic field vector H is a family of concentric circles centred on the geometric centre of the conductors. Using cylindrical polar co-ordinates (r,ϕ,z) these fields are represented by

$$E_r = \frac{V}{r \ln \frac{b}{a}} \tag{4.13}$$

$$H_\phi = \frac{I}{2\pi r} \tag{4.14}$$

where V = voltage between the inner and outer conductors
   I = current flowing
   a = radius of outer conductor
   b = radius of inner conductor
All the other field components (E_ϕ, E_z and H_r, H_z) are zero.
   The maximum electric-field intensity occurs at the surface of the inner conductor, and is

$$E_{max} = \frac{|V|}{b \ln \frac{b}{a}} \tag{4.15}$$

For constant outer radius and E_{max} the power flow is a maximum when $a/b = 1.65$ (Marcuvitz, 1986). The transmitted power is

$$P = 2\pi E_m^2 \, b^2 \log_e \left(\frac{b}{a}\right) \sqrt{\frac{\varepsilon_0}{\mu_0}} \qquad \text{watts} \tag{4.16}$$

The attenuation of the TEM mode, for inner and outer conductors of the same characteristic or skin resistance R (Marcuvitz, 1986; Maxwell, 1947; Metaxas and Meredith, 1993) is:

$$\alpha = 8.686 \, R \left(\frac{1}{a} + \frac{1}{b}\right) \frac{1}{Z_0 \ln(a/b)} \qquad \text{decibels/metre} \tag{4.17}$$

$$\text{where} \quad R = 34.4 \sqrt{\frac{1}{\sigma \lambda_0}} \qquad \text{ohms} \tag{4.18}$$

It is informative to consider the characteristics of a typical coaxial line. Consider two lines with outer diameter ($2a$) 76 mm and inner diameter ($2b$) of $76/1.65 = 46$ mm for maximum power flow, and also of 38 mm (a common size), as listed in Table 4.3. These data are for a matched line. The attenuation data are approximate and the magnitude may be substantially higher (~50%) than quoted due to surface roughness, contamination and work hardening.

Table 4.3   Characteristics of some coaxial lines

Size 2a/2b mm	76/46	76/38	76/46	76/38
Material	Al	Al	Cu	Cu
Z_0 (Ω)	30	41	30	41
P_{max} (MW)	433	150	433	150
Voltage at 100 kW (V)	1732	2024	1732	2024
Current at100 kW (A)	58	49	58	49
Attenuation (dB/m)	0.0247	0.020	0.015	0.012
Power dissipation at				
100 kW transmitted (W/m)	570	461	346	277
Cutoff wavelength, H_{11} mode (m)	0.192	0.179	0.192	0.179
Cutoff wavelength, E_{01} mode (m)	0.150	0.190	0.150	0.190

Conductivity values : Copper = 4.10^7 mho/m
Aluminium = $1.5 . 10^7$ mho/m
Voltage breakdown strength of air, ambient conditions = 3000 kV/m
Wavelength = 0.33 m

The data in Table 4.3 illustrate the virtues and disadvantages of coaxial lines. Their main advantage is compact size compared with waveguides. In principle, they are capable of handling extremely high power on a voltage-breakdown basis; however, this cannot be realised under CW operation because of the high power dissipation occurring, even at 100 kW, which creates a severe cooling problem. In practice, the mechanical support for the inner line creates difficulties; dielectric discs or 'spiders' attract dust electrostatically which eventually causes surface tracking and breakdown. Quarter-wave stub supports create field concentrations and are inconvenient and expensive. Bends are particularly difficult to realise.

In addition, the currents are high, and this creates difficulties in design of flanges, especially with the inner line where the current density is highest. An indifferent contact results in severe I^2R heating and eventual oxidation, then to micro-arcing which may precipitate voltage breakdown.

Coaxial lines are also capable of supporting waveguide modes for which the cutoff wavelengths are well known (Marcuvitz, 1986). The last two rows of Table 4.3 show the cutoff wavelengths for the lowest order E and H modes, i.e. E_{01} and H_{11}, calculated from approximate data given by Marcuvitz. If the excitation wavelength exceeds the cutoff wavelength (see Section 4.5), that mode cannot be supported. If a waveguide mode can propagate, it will contaminate the TEM desired mode and so it is important to ensure that the dimensions are chosen so that the lowest-frequency waveguide mode is at least 25% higher than the operating frequency. The coaxial lines shown in Table 4.3 have dimensions suitable for use at 896 or 915 MHz ($\lambda = 0.33$ m), but not at 2450 MHz ($\lambda = 0.122$ m).

For the above reasons, coaxial lines should only be considered for short lengths where their compact form gives a clear advantage, or in low-power (<10 kW) applications where the above problems are deemed negligible.

4.5 Rectangular waveguides

In almost every industrial microwave-heating installation, waveguides have an essential role in conveying power from the generator to the load, and in many cases the heating chamber itself is based on waveguide technology. It is therefore essential for the engineer, as equipment designer or user, to have a sound understanding of the principal features of waveguide engineering. It is not necessary to be able to derive the field equations from Maxwell's equations; more to have a grasp of the results, e.g. the forms of E- and H-field distribution, the wall-current patterns, waveguide wavelength and cutoff wavelength, attenuation and power handling.

For a detailed derivation of the 'waveguide equations' the reader is referred to the standard literature (e.g. Slater, 1942; Ramo *et al.*, 1965; Baden-Fuller, 1979). Only the salient features necessary for a good understanding are presented here. First, Maxwell's equations are derived in rectangular co-ordinates *Ox, Oy, Oz*, the subscripts *x,y,z* to E and H representing, respectively, the components of the fields in the directions of those axes. Also, the time variation of the fields is assumed to be sinusoidal, with an angular frequency $\omega = 2\pi f$. After some mathematical manipulation, two important differential equations (eqns. 4.19 and 4.20) emerge, either of which represents a class of wave propagating in the waveguide along its longitudinal axis *Oz*:

$$\frac{\partial^2 H_z}{\partial x^2} + \frac{\partial^2 H_z}{\partial y^2} + k^2 H_z = 0 \tag{4.19}$$

$$\frac{\partial^2 E_z}{\partial x^2} + \frac{\partial^2 E_z}{\partial y^2} + k^2 E_z = 0 \tag{4.20}$$

$$\text{where } k^2 = \omega^2 \mu_0 \varepsilon_0 + \gamma^2 \tag{4.21}$$

$$\text{and } \gamma = \alpha + j\beta \tag{4.22}$$

γ is the complex propagation constant
α is the attenuation constant (nepers per metre) and
β is the phase constant (radians per metre), defined by eqn. 1.21.

Propagating waves based on eqn. 4.19 are designated H modes or TE (transverse-electric) modes: they have an axial component of magnetic field H_z and no axial electric field ($E_z = 0$). Only the transverse electric fields E_x and E_y are possible for solutions based on eqn. 4.19, hence the label TE mode.

Similar reasoning applies to eqn. 4.20, which essentially has an axial component E_z to the solution, with $H_z = 0$. Its modes are called E modes or TM modes.

A third solution is also possible in which $k = 0$, allowing both E_z and H_z to be zero. Eqns. 4.21 and 4.22 then degenerate, using eqn. 1.22 and putting $\alpha = 0$, to $f\lambda = \sqrt{(\mu_0\varepsilon_0)}$. This is a TEM wave, being the solution for the class of transmission lines considered in Section 4.4.

4.5.1 H or TE modes in rectangular waveguide

To obtain the field equations, which show how the E and H fields are distributed in a waveguide, we have first to consider the boundary conditions. These are the values of the E and H at the surface of the waveguide walls, and assuming walls of perfect conductivity it is obvious that the E-field intensity tangential to the surface must be zero; this first boundary condition is often written $E_{tan} = 0$. Considering now the magnetic field, there can be no net AC magnetic field within the wall of perfect conductivity, and because the magnetic field must form a closed loop the magnetic-field component *normal* to the wall surface must be zero, i.e. $H_{norm} = 0$.

Applying these boundary conditions to the rectangular waveguide shown in Figure 4.8, it is clear that

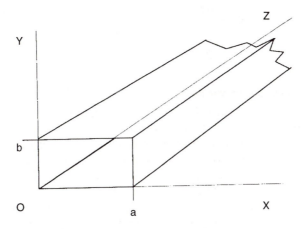

Figure 4.8   Rectangular-waveguide coordinate system

$$E_x = 0 \text{ at } y = 0 \text{ and } y = b$$
$$H_x = 0 \text{ at } x = 0 \text{ and } x = a$$
$$H_y = 0 \text{ at } y = 0 \text{ and } y = b$$

Functions of sinusoidal form may be expected to fit these boundary conditions, for example E_y = constant (sin $m\pi x/a$), where m is an integer, clearly satisfies the first condition. Sinusoidal functions also satisfy eqn. 4.19 and the Maxwell equations, and the complete set of field equations for H or TE waves in a rectangular waveguide of dimensions $a \times b$ $(a > b)$ is:

$$E_x = \frac{j\mu_0\omega k_y}{k^2} H_0 \cos\left(k_x x\right)\sin\left(k_y y\right)\exp(j\omega t - \gamma z) \tag{4.23}$$

$$E_y = -\frac{j\mu_0\omega k_x}{k^2} H_0 \sin\left(k_x x\right)\cos\left(k_y y\right)\exp(j\omega t - \gamma z) \tag{4.24}$$

$$Ez = 0 \tag{4.25}$$

$$H_x = \frac{\gamma k_x}{k^2} H_0 \sin\left(k_x x\right)\cos\left(k_y y\right)\exp(j\omega t - \gamma z) \tag{4.26}$$

$$H_y = \frac{\gamma k_y}{k^2} H_0 \cos\left(k_x x\right)\sin\left(k_y y\right)\exp(j\omega t - \gamma z) \tag{4.27}$$

$$H_z = H_0 \cos\left(k_x x\right)\cos\left(k_y y\right)\exp(j\omega t - \gamma z) \tag{4.28}$$

where $$k_x = \frac{m\pi x}{a} \qquad k_y = \frac{n\pi y}{b} \tag{4.29}$$

and $$k^2 = k_x^2 + k_y^2 \tag{4.30}$$

and k is defined by eqn. 4.21.

The six equations, eqns. 4.23–4.27 inclusive, give the field intensities of E and H in their three co-ordinate directions as functions of position within the waveguide, for values of m, n which are positive integers greater than or equal to zero. (Note that only one can be zero to avoid a trivial solution of all-zero fields.) The values m,n are called the mode numbers, and the H or TE modes are labelled as $H_{m,n}$ or $TE_{m,n}$. These integers m,n show the number of half cycles of sinusoidal variation of intensity existing within the waveguide: m along the Ox axis, n along the Oy. Clearly the simplest mode is the $TE_{1,0}$, having field patterns illustrated in Figure 4.9.

Other features to note in the field equations are:

(i)   the amplitude factor H_0 to each equation;
(ii) the other coefficients before the sine/cosine terms which in combination are impedance functions for the E fields, and dimensionless numbers for the H fields. Note also that the electric fields are in phase quadrature with the magnetic field, as indicated by the j operator in eqns. 4.23 and 4.24; and
(iii) the exponential terms which show sinusoidal time variation $e^{j\omega t}$, and propagation $e^{-\gamma z}$.

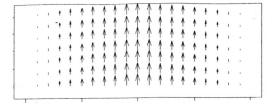

E-field: transverse section across waveguide

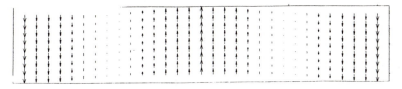

E-field: longitudinal view along axis of propagation

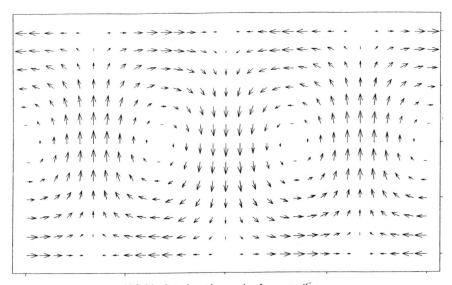

H-field: plan-view along axis of propagation

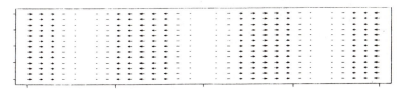

H-field: side view along axis of propagation

Figure 4.9   Electric- and magnetic-field patterns, rectangular waveguide, TE_{10} mode

4.5.2 Cutoff and waveguide wavelengths

Clearly, if γ is real there is exponential decay along the waveguide axis. For free propagation to occur, γ must be imaginary, and, from eqns. 4.21, 4.29 and 4.30, this can occur only if

$$\omega^2 \mu_0 \varepsilon_0 = 4\pi^2 f^2 \mu_0 \varepsilon_0 \geq \left(\frac{m\pi}{a}\right)^2 + \left(\frac{n\pi}{b}\right)^2 \tag{4.31}$$

Also, from eqns. 1.22 and 1.23,

$$f\lambda = \sqrt{\frac{1}{\mu_0 \varepsilon_0}}$$

Substituting in eqn. 4.30,

$$\left(\frac{2}{\lambda}\right)^2 \geq \left(\frac{m}{a}\right)^2 + \left(\frac{n}{b}\right)^2 \tag{4.32}$$

This last equation gives the condition for free propagation to occur: that the excitation wavelength must be less than the critical value λ_c which gives equality in eqn. 4.32, to give eqn. 4.33:

$$\left(\frac{2}{\lambda_c}\right)^2 = \left(\frac{m}{a}\right)^2 + \left(\frac{n}{b}\right)^2 \tag{4.33}$$

λ_c is called the cutoff wavelength, and clearly is dependent on the mode number.

The lowest-order mode in a rectangular waveguide of dimensions $a \times b$ with $a > b$ is for $m = 1$, $n = 0$ i.e the H_{10} or TE_{10} mode. Substituting these values in eqn. 4.33, the cutoff wavelength for the TE_{10} mode in a rectangular waveguide is

$$\lambda_c = 2a$$

Thus, for a rectangular waveguide to propagate in the TE_{10} mode without attenuation (other than that due to finite conductivity of the walls), the width must be at least half the free-space wavelength.

Reverting now to eqns. 4.21 and 4.22, putting $\alpha = 0$, using eqns. 1.22 and 1.23 and writing λ_g as a wavelength in the waveguide which satisfies eqns. 4.21 and 4.22, and rearranging,

$$\beta^2 = \left(\frac{2\pi}{\lambda_g}\right)^2 = \mu_0 \varepsilon_0 \omega^2 - k^2 = \left(\frac{2\pi}{\lambda_0}\right)^2 - \left(\frac{m\pi}{a}\right)^2 - \left(\frac{n\pi}{b}\right)^2 \tag{4.34}$$

Substituting eqn. 4.33, putting $m = 1$ and $n = 0$ for the TE_{10} mode and simplifying,

$$\frac{1}{\lambda_g^2} = \frac{1}{\lambda_0^2} - \frac{-1}{\lambda_c^2} \qquad (4.35)$$

Eqn. 4.35 shows that the wavelength in the waveguide λ_g is greater than the free-space excitation wavelength λ_0, and indeed is infinite if $\lambda_0 = \lambda_c$. As the waveguide size increases, $\lambda_g \to \lambda_0$. Values of λ_0 λ_c and λ_b are given in Table 4.4 for waveguide sizes and frequencies commonly used in industrial heating equipment.

Table 4.4   Frequency, free-space wavelength and waveguide wavelength

f_0	λ_0	λ_g	f_0	λ_0	λ_g
(MHz)	(cm)	(cm)	(MHz)	(cm)	(cm)
(a) f = 2450±50 MHz, waveguide size WG 9A (86 × 43 mm)			(b) f = 900±25 MHz, waveguide size WG 4A (248 × 124 mm)		
2400	12.49	18.1	875	34.26	47.38
2410	12.44	17.94	880	34.07	46.87
2420	12.39	17.79	885	33.87	46.37
2430	12.34	17.64	890	33.68	45.89
2440	12.29	17.49	895	33.5	45.42
2450	12.24	17.35	900	33.31	44.96
2460	12.19	17.21	905	33.13	44.51
2470	12.14	17.07	910	32.94	44.07
2480	12.09	16.94	915	32.76	43.64
2490	12.04	16.8	920	32.59	43.22
2500	11.99	16.67	925	32.41	42.81

4.5.3 Field equations: the fundamental TE_{10} or H_{10} mode

The TE_{10} mode in a rectangular waveguide is of such importance that its field equations require a separate statement. They are essentially eqns. 4.23– 4.30 rewritten with $m = 1$ and $n = 0$, and simplify to:

$$E_x = 0 \qquad (4.36)$$

$$E_y = -j\frac{\mu_0 \omega a}{\pi} H_0 \sin\left(\frac{\pi x}{a}\right) \exp(j\omega t - \gamma z) \qquad (4.37)$$

$$E_z = 0 \qquad (4.38)$$

$$H_x = \frac{\gamma a}{\pi} H_0 \sin\left(\frac{\pi x}{a}\right) \exp(j\omega t - \gamma z) \qquad (4.39)$$

$$H_y = 0 \tag{4.40}$$

$$H_z = H_0 \cos\left(\frac{\pi x}{a}\right) \exp(j\omega t - \gamma z) \tag{4.41}$$

Figures 4.9 shows the field patterns of the E and H fields, and Figure 4.10 shows the wall currents.

Some important features arise from the above:

(i)   The E-field has one component only, E_y, which has its maximum value in the centre of the waveguide, polarised normal to the broad faces. Its intensity is constant in the direction O_y.

(ii) The wall currents at the centre of the broad faces are purely longitudinal; a longitudinal thin slot in the broad faces on the centreline therefore does not interrupt the current path, and negligible microwave energy leaks from such a slot. This effect forms the basis of a simple class of heating chambers for thin webs where the web passes through such slots and couples to the high value of electric field resulting from (i) above.

4.5.4 Characteristic impedance

Analogously with plane waves, the characteristic impedance can be defined as E_x/H_y. Using the field equations (eqns. 4.23 and 4.26) and other relations above, it is readily shown that, for the TE_{10} mode, the characteristic-wave impedance Z_{0H} is

$$Z_{0H} = \frac{\lambda_g}{\lambda_0}\sqrt{\frac{\mu_0}{\varepsilon_0}} = \frac{\lambda_g}{\lambda_0} Z_0 \tag{4.42}$$

Note that this impedance tends to infinity near cutoff. It is also independent of the height of the waveguide b, which is anomalous. Intuitively, two waveguides of the same width a but different heights b must have different characteristic impedances. This problem is overcome by using other definitions than E_x/H_y. (Ramo *et al.*, 1965; Schelkunoff, 1943). One is based on calculating a voltage V between opposing faces of the waveguide, and for the TE_{01} mode between the broad faces at their centres, and similarly calculating a current I flowing axially in the broad faces. The characteristic impedance of the waveguide is taken as V/I, thus:

$$V = \int_0^b E_y dy = j\frac{2ab}{\lambda_0}\left(\sqrt{\frac{\mu_0}{\varepsilon_0}}\right)H_0 \tag{4.43}$$

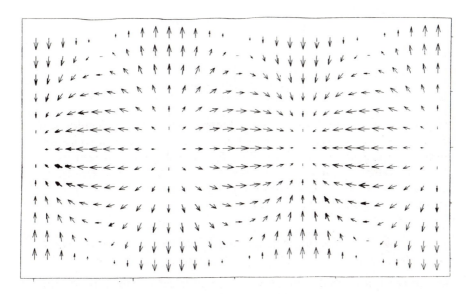

⇨ direction of propagation
broad face wall currents

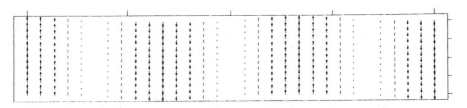

⇨ direction of propagation
side-wall currents

Figure 4.10   Wall-current patterns, rectangular waveguide, TE$_{01}$ mode

$$I = \int_0^a H_x\, dx = aH_x = j2\frac{a^2}{\lambda_g}H_0 \qquad (4.44)$$

$$Z_{VI} = \frac{V}{I} = \frac{b}{a}\frac{\lambda_g}{\lambda_0}\sqrt{\frac{\eta_0}{\varepsilon_0}} = \frac{b}{a}Z_{0H} \qquad (4.45)$$

This last impedance involves both dimensions of the waveguide and is therefore a more realistic representation. Other definitions of characteristic impedance are based on power flow P, and voltage or current as above defined, evaluating the relations $Z = V^2/P$ or $Z = P/I^2$. These give results of the same form as eqn. 4.45 but multiplied or divided, respectively, by 2. All are valid, but it is important to be consistent.

4.5.5 Power flow, electric- and magnetic-field intensity, wall-current density

The power transmitted through a waveguide is calculated using Poynting's vector, evaluating $P = \int E \times H\, dA$ over the waveguide cross-section. For the TE_{01} mode this involves just E_y and H_x as given by eqns. 4.37 and 4.39, and then using eqn. 4.37 again with $x = a/2$ to relate the E-field, E_{ymax} at the centre of the waveguide, to H_0. The result of this procedure gives the peak intensity of electric field at the centre of the waveguide as (Metaxas and Meredith, 1993).

$$E_{ymax} = \sqrt{\left|\left\{\frac{4P}{ab}\frac{\lambda_g}{\lambda_0}\left(\frac{\mu_0}{\varepsilon_0}\right)^{0.5}\right\}\right|} \quad \text{volts per metre peak} \qquad (4.46)$$

Note that this is for a perfectly matched waveguide (VSWR = 1.00).

The transverse and longitudinal magnetic fields H_x and H_z are readily derived from eqns. 4.37, 4.39, 4.41 and 4.46, and give maximum values as in eqns. 4.47 and 4.48:

$$H_{xmax} = \sqrt{\left|\left\{\frac{2P\lambda_0}{ab\lambda_g}\left(\frac{\varepsilon_0}{\eta_0}\right)^{0.5}\right\}\right|} \quad \text{amperes per metre RMS} \qquad (4.47)$$

$$H_{zmax} = \sqrt{\left|\left\{\frac{2P\lambda_0\lambda_g}{a^3 b}\left(\frac{\varepsilon_0}{\mu_0}\right)^{0.5}\right\}\right|} \quad \text{amperes per metre RMS} \qquad (4.48)$$

It is instructive to consider some values of these parameters under typical conditions in an industrial plant, as in Table 4.5

Table 4.5   Typical conditions in an industrial plant

Waveguide	Internal dimensions	f	P	E_y max	H_x max	H_z max
	(mm)	(MHz)	(kW)	(V/cm peak)	(A/cm RMS)	(A/cm RMS)
WG4	248 × 124	900	75	690	0.99	1.72
WG9A	86 × 43	2450	25	1200	1.71	3.20

At the power levels quoted, the voltage stresses are modest compared with the voltage-breakdown strength of air at ambient conditions, and the magnetic-field stresses are low. Since the wall-current densities are equal to the grazing magnetic field stresses, the resulting currents are low and can be easily managed with simple flanges and standard waveguide construction methods. Also, since these wall currents result in I^2R heating, the power loss and resulting heating are small. Note that, because the maximum value of H_z is at the side walls of the waveguide (and $H_z > H_x$), these will heat more than the broad faces. Using eqn. 4.18 with aluminium of conductivity 1.5×10^7 mho/m, the skin resistivity R is 0.0155 Ω at 900 MHz and 0.0245 Ω at 2450 MHz, resulting in peak heat dissipation of 0.046 W/cm^2 and 0.25 W/cm^2, respectively, under the conditions in Table 4.5. While natural cooling would suffice at 900 MHz, some forced cooling may be necessary for the 2450 MHz case.

4.5.6 Attenuation

In an air-filled waveguide, attenuation is entirely due to I^2R heating of the walls of the waveguide. This remains true even if the relative humidity of the air is high; water in the gaseous state is very low-loss. Calculation of attenuation involves evaluating the power loss along a unit length of waveguide, by integrating $\int I^2 R \, dA$ over the internal surface, and then relating this to the power transmitted. For the TE$_{10}$ mode in rectangular waveguide, the attenuation is (Marcuvitz, 1986).

$$\alpha = R \sqrt{\frac{\varepsilon_0}{\mu_0}} \left[\frac{1 + \dfrac{2b}{a}\left(\dfrac{\lambda_0}{2a}\right)^2}{\sqrt{\left\{1 - \left(\dfrac{\lambda_0}{2a}\right)^2\right\}}} \right] \text{ nepers per metre} \qquad (4.49)$$

Values of attenuation and power dissipation are given in Table 4.6 for typical materials. The effective surface resistance R has been increased 50% over the theoretical figure to allow for surface conditions, and its variation with frequency is included via eqn. 4.18. The attenuation in decibels per metre is 8.686 times that in neper per metre.

Note that as the frequency rises the attenuation falls; inspection of eqn. 4.49 shows that as cut-off is approached the denominator tends to zero, so the attenuation approaches infinity. Physically, the wall currents tend to infinity as cutoff is approached, with consequent high loss.

Choice of material for the waveguide is very important; although the fraction of power lost due to attenuation may be acceptably small, it may represent a serious temperature rise in the waveguide. Most industrial equipment uses aluminium waveguide for power transmission, as it has low attenuation and is relatively low-cost. The other materials listed are only used where aluminium is unacceptable, e.g. due to high temperature or corrosion risk. Stainless steel is popular in the food and pharmaceutical industries, but the attenuation is very high. Cadmium-plated mild steel has been used where high temperatures are necessary, as in rubber-vulcanising equipment. Copper and brass are usually confined to laboratory equipment at 2450 MHz.

It must be emphasised that the above attenuation values are approximate. Attenuation is always significantly higher than theoretical due to surface roughness, crystalline structure/microfissures, surface-contamination corrosion and occlusions. Although a 50% allowance is made in Table 4.6, actual values often exceed those given, especially in equipment which has been in service in hostile conditions.

4.5.6.1 Skin depth

When a high-frequency current flows in a conductor, the magnetic field associated with it induces an EMF in the conductor, which causes a secondary current to flow in a direction opposing the principal current. This well known effect (Stratton, 1941; Marcuvitz, 1986; Davies, 1990; Metaxas 1996) causes the resultant current to be confined to the surface in a thin skin. In more detail, there are layers of current at increasing depth into the conductor, flowing parallel to the surface, but heavily attenuated with depth. In each layer, the current flows in opposite direction to its neighbour.

Skin depth δ_s is defined as the depth into the conductor from the surface at which the current density is $1/e$ ($= 0.368$) of its value at the surface, and is given by

$$\delta_s = \sqrt{\frac{2}{\sigma \omega \mu_0}} \qquad \text{metres} \qquad (4.50)$$

It will be seen that skin depth reduces with increasing frequency, and at microwave frequencies is extremely small. It is a significant effect at

Table 4.6 Attenuation of selected rectangular waveguides, TE_{01} mode

Parameter	WG size (mm)	f (MHz)	Cu	Al	Stainless steel 18% CR, 8% Ni	Brass 70% Cu	Cd plate
σ (mho/m)	n/a	n/a	5.5×10^7	3.0×10^7	1.4×10^6	1.45×10^7	1×10^7
WG4, attenuation dB/m	248 × 124	850	4.70E-3	6.31E-3	2.93E-2	9.12E-3	1.09E-2
		900	4.39E-3	5.94E-3	2.75E-2	8.55E-3	1.03E-2
		950	4.20E-3	5.72E-3	2.63E-2	8.16E-3	9.82E-3
WG4 power loss per metre at 75 kW transmitted, 900 MHz	248 × 124	900	76	103	476	148	178
WG9A, attenuation dB/m	86 × 43	2400	2.35E-2	3.18E-2	1.47E-1	4.57E-2	5.54E-2
		2450	2.28E-2	3.09E-2	1.43E-1	4.46E-2	5.36E-2
		2500	2.23E-2	3.02E-2	1.40E-1	4.35E-2	5.24E-2
WG9A, power loss, watts per metre at 25 kW transmitted, 2450 MHz	86 × 43	2450	132	178	837	258	310

frequencies from 1 kHz upwards, and results in the need to use stranded, separately insulated, conductors in some high-frequency circuits, so called 'Litz' wire.

Wall losses in a structure are given by

$$P_s = \frac{1}{2} R \left| H_t \right|^2 \qquad \text{watts per square metre} \qquad (4.51)$$

where $R = 1/(\delta_s \sigma)$ is the equivalent resistance.

Table 4.7 gives δ_s and R_s for typical materials at representative frequencies. Some of the materials listed may be used as a plated surface (Cd, Cr, Sn, Zn) or galvanised surface (Zn), or applied to a substrate by metal spraying (Al). These techniques have been used effectively on mild-steel structures, but care must be taken to ensure a sound coverage of plated material. Note the very thin skin depth at microwave frequencies, in the region of a few microns, comparable with or even less than the surface roughness of the metal. The elongated path length followed by the microwave surface currents due to the rough surface contributes to losses exceeding the theoretical value, usually by about 30–50%. Experience has shown that polishing the surface in an attempt to reduce the surface roughness gives limited reduction in losses, probably because the surface becomes work hardened, which tends to raise the resistivity.

4.5.7 TM or E modes in rectangular waveguide

Whereas TE modes have, uniquely, an axial component of magnetic field H_z and are a solution of the wave equation (eqn. 4.19), the TM modes are a solution of eqn. 4.20, and are characterised by a sole axial component E_z of E-field. They are subject to the same boundary conditions as the TE modes, and have cutoff wavelengths λ_c for each mode in the same manner. The waveguide wavelength λ_g is given by eqn. 4.35. The same nomenclature is used for mode-number designation, i.e. $TM_{m,n}$ where m,n are positive integers greater than unity. The lowest-order mode is the TM_{11}, the modes with $m = 0$ or $n = 0$ being nonexistent.

The general field equations for the TM_{mn} modes are

$$E_x = -\frac{\gamma}{k^2} \frac{m\pi}{a} E_0 \cos\frac{m\pi x}{a} \sin\frac{n\pi y}{b} \quad \exp(j\omega t - \gamma z) \qquad (4.52)$$

$$E_y = -\frac{\gamma}{k^2} \frac{n\pi}{b} E_0 \sin\frac{m\pi x}{a} \cos\frac{n\pi y}{b} \quad \exp(j\omega t - \gamma z) \qquad (4.53)$$

$$E_z = E_0 \sin\frac{m\pi x}{a} \sin\frac{n\pi y}{b} \quad \exp(j\omega t - \gamma z) \qquad (4.54)$$

Table 4.7 Skin depth and equivalent resistance for selected materials and frequencies

Material	Conductivity	Resistivity	Skin depth at			Equivalent resistance at		
			25 kHz	915 MHz	2450 MHz	25 kHz	915 MHz	2450 MHz
	(mho/m)	(Ω m)	(m)	(m)	(m)	(Ω)	(Ω)	(Ω)
Aluminium 100%	3.43+07	2.92E-08	0.000543513	2.84E-06	1.74E-06	5.36409E-05	0.010262112	0.01679228
Brass 70% Cu	1.45E+07	6.90E-08	0.000835935	4.37E-06	2.67E-06	8.2501E-05	0.015783373	0.025826927
Brass 90% Cu	2.41E+07	4.15E-08	0.000648408	3.39E-06	2.07E-06	6.39933E-05	0.012242646	0.020033103
Cadmium	1.11E+07	9.05E-08	0.000957581	5.01E-06	3.06E-06	9.45066E-05	0.018080177	0.029585274
Chromium	3.20E+07	3.13E-08	0.000563146	2.94E-06	1.80E-06	5.55786E-05	0.010632818	0.017398881
Copper 100% OFHC	5.81E+07	1.72E-08	0.000417466	2.18E-06	1.33E-06	4.1201E-05	0.007882211	0.012897958
Gold	4.52E+07	2.21E-08	0.000473209	2.47E-06	1.51E-06	4.67024E-05	0.0089347	0.014620185
Lead	4.07E +06	2.46E-07	0.00157879	8.25E-06	5.04E-06	0.000155816	0.029809273	0.048778034
Mild steel (μ' = 100)	6.25E+06	1.60E-07	0.000127326	6.66E-06	4.07E-07	0.001256618	0.024040518	0.393384033
Nickel	7.94E+06	1.26E-07	0.001129905	5.91E-06	3.61E-06	0.000111514	0.021333843	0.03490937
Nickel/chromium 80/20 alloy	9.26E+03	1.08E-04	0.033080236	1.73E-04	1.06E-04	0.003264789	0.624590976	1.022041699
Phosphor bronze	2.09E+07	4.78E-08	0.000695938	3.64E-06	2.22E-06	6.86843E-05	0.013140073	0.021501595
Platinum	1.02E+06	9.81E-07	0.003152762	1.65E-05	1.01E-05	0.000311156	0.059527594	0.097407241
Silver	6.10E+07	1.63-08	0.00040756	2.13E-06	1.30E-06	4.02234 -05	0.007695182	0.012591916
Stainless steel (19.1% Cr, 8.1% Ni, 0.6%W)	1.44E+06	6.95E-07	0.002653683	1.39E-05	8.48E-06	0.0002619	0.050104442	0.081987782
Tin	7.58E+06	1.32E-07	0.001156495	6.05E-06	3.69E-06	0.000114138	0.021835885	0.035730879
Titanium	1.86E+06	5.38E-07	0.00233479	1.22E-05	7.46E-06	0.000230428	0.044083388	0.072135305
Zinc	1.74E+07	5.74E-08	0.000762628	3.99E-06	2.44E-06	7.52661E-05	0.014399247	0.023562029

Permeability = 1.0

$$H_x = \frac{j\omega\varepsilon_0}{k^2}\frac{n\pi}{b}E_0\sin\frac{m\pi x}{a}\cos\frac{n\pi y}{b}\;\exp(j\omega t - \gamma z) \tag{4.55}$$

$$H_y = -\frac{j\omega\varepsilon_0}{k^2}\frac{m\pi}{a}E_0\cos\frac{m\pi x}{a}\sin\frac{n\pi y}{b}\;\exp(j\omega t - \gamma z) \tag{4.56}$$

$$H_z = 0 \tag{4.57}$$

The cutoff wavelengths $\lambda_{c(m,n)}$ of the TM_{mn} modes are

$$\lambda_{c(m,n)} = \frac{2\sqrt{(ab)}}{\sqrt{\left(m^2\dfrac{b}{a} + n^2\dfrac{a}{b}\right)}} \tag{4.58}$$

Substituting the waveguide sizes in eqn. 4.58 for the commonly used waveguides at 900 MHz and 2450 MHz, the cutoff wavelengths for the TM_{11} modes are:

WG4: 22.18 cm (1351 MHz);   WG9A: 7.69 cm (3897 MHz)

These are some 50% higher in frequency than the respective operating frequencies, and so the TM_{11} modes will not be excited, unless there is a substantial deformation of the waveguide or a large slab of dielectric material is present.

Marcuvitz (1986) gives relations for the attenuation, maximum value of E field and its location, and the power flow of the TM modes.

4.6 Circular waveguides

Circular waveguides are mostly used in specialised heating ovens; as transmission lines, they have the serious disadvantage that, in the fundamental mode analogous to the TE_{10} in rectangular waveguides, the plane of polarisation may twist causing power loss and mismatch. In the rectangular waveguide, this cannot happen because the orthogonal mode, i.e the TE_{01}, is normally cut off for a waveguide with an aspect ratio around 2:1. Nonetheless, specialised uses of circular waveguides are of sufficient importance to justify the inclusion of an introduction.

Circular geometry dictates the manipulation of Maxwell's equations in circular cylindrical co-ordinates (r, θ, z). The derivation of the field equations is given by Marcuvitz (1986) and follows the same pattern as for rectangular waveguides, arriving at a pair of wave equations like eqns. 4.19–4.22, expressed in r, θ, z co-ordinates. These are readily developed into Bessel's equations, the solutions of which are expressed in Bessel functions and their derivatives. These functions have tabulated values (Jahnke and Emde, 1933), and are treated in the same way as trigonometric functions; many computer programs include Bessel functions (e.g. Mathcad 6.0 Plus® or higher).

Like rectangular waveguides, circular waveguides can support both TM and TE waves, for each of which there is a family of modes, each with its individual cutoff frequency λ_c. Knowing λ_c for a given mode, the value of λ_g can be calculated using eqn. 4.34.

The general field equations for a lossless circular waveguide of radius a are (Marcuvitz 1986):

The $TE_{m,n}$ or $H_{m,n}$ modes:

$$E_r = \frac{j\omega\mu_0 n}{k_{nl}^2 r} H_0 J_n\left(\frac{p'_{nl}r}{a}\right)\sin n\theta \exp j(\omega t - \beta z) \tag{4.59}$$

$$E_\theta = \frac{j\omega\mu_0}{k_{nl}} H_0 J'_n\left(\frac{p'_{nl}r}{a}\right)\cos n\theta \exp j(\omega t - \beta z) \tag{4.60}$$

$$E_z = 0 \tag{4.61}$$

$$H_r = -\frac{j\beta}{k_{nl}} H_0 J'_n\left(\frac{p'_{nl}r}{a}\right)\cos n\theta \exp j(\omega t - \beta z) \tag{4.62}$$

$$H_\theta = \frac{j\beta}{k_{ln}^2 r} H_0 J_n\left(\frac{p'_{nl}r}{a}\right)\sin n\theta \exp j(\omega t - \beta z) \tag{4.63}$$

$$H_z = J_n\left(\frac{p'_{nl}r}{a}\right)H_0 \cos n\theta \exp j(\omega t - \beta z) \tag{4.64}$$

The TM_{ln} or E_{ln} modes:

$$E_r = -\frac{j\beta}{k_{nl}} E_0 J'_n\left(\frac{p_{nl}r}{a}\right)\cos n\theta \exp j(\omega t - \beta z) \tag{4.65}$$

$$E_\theta = \frac{j\beta}{k_{nl}^2} E_0 J_n\left(\frac{p_{nl}r}{a}\right)\sin n\theta \exp j(\omega t - \beta z) \tag{4.66}$$

$$E_z = E_0 J_n\left(\frac{p_{nl}r}{a}\right)\cos n\theta \exp j(\omega t - \beta z) \tag{4.67}$$

$$H_z = 0 \tag{4.68}$$

$$H_r = -\frac{j\omega\varepsilon_0 n}{k_{nl}^2} r E_0 J_n\left(\frac{p_{nl}r}{a}\right)\sin n\theta \exp j(\omega t - \beta z) \tag{4.69}$$

$$H_\theta = -\frac{j\omega\varepsilon_0}{k_{nl}} E_0 J'_n\left(\frac{p_{nl}r}{a}\right)\cos n\theta \exp j(\omega t - \beta z) \tag{4.70}$$

Notes on eqns. 4.59–4.70:

(i) p_{nl} is the lth nonvanishing root of the nth-order Bessel function (TM modes);

(ii) p'_{nl} is the lth nonvanishing root of the derivative of the nth-order Bessel function (TE modes);

(iii) n is an integer in the range $0 < n < \infty$, and represents the number of complete cycles of field variation circumferentially;

(iv) l is an integer in the range $0 < l < \infty$, and represents the number of cycles of field variation diametrically;

(v)   these eigenvalues (l,n) must not be confused with those for the rectangular waveguide which are specified for Cartesian co-ordinates. For example, the lowest-order TE_{10} mode in rectangular waveguides has an obvious similarity to the TE_{11} lowest-order mode in circular waveguides;

(vi)   The roots p_{nl} and p'_{nl} of the lower-order modes are shown in Table 4.8, and graphs of some Bessel functions $Jn(x)$ and their derivatives $d/dx\{J_n(x)\} = J'_n(x)$ are presented in many textbooks [e.g. Metaxas and Meredith (1983)];

(vii)   Table 4.8 also shows the cutoff wavelength for several low-order modes.

Marcuvitz gives data for power flow and attenuation in circular waveguides, together with an extension to waveguide modes in coaxial lines.

Table 4.8   Roots and cutoff wavelengths of several lower-order modes

Mode	Root	p'_{nl} or p_{nl}	$\lambda_c = 2\pi a/p'_{nl}$ (TE modes) or $2\pi a/p_{nl}$ (TM modes)
TE_{11}	1st of J'_1	1.841	$3.413a$
TE_{21}	1st of J'_2	3.054	$2.057a$
TE_{01}	1st of J'_0	3.832	$1.640a$
TE_{31}	1st of J'_3	4.20	$1.50a$
TE_{41}	1st of $J'4$	5.30	$1.18_6\,a$
TE_{12}	2nd of $J'1$	5.33	$1.17_9\,a$
TM_{01}	1st of J_0	2.405	$2.613a$
TM_{11}	1st of J_1	3.832	$1.640a$
TM_{21}	2nd of J_1	5.135	$1.223a$
TM_{02}	2nd of J_0	5.520	$1.138a$
TM_{31}	1st of J_3	6.379	$0.985a$

4.6.1 Cutoff circular-waveguide chokes

Circular waveguides, of bore size less than the cutoff diameter for the lowest-order mode at the operating frequency, have a high attenuation rate (decibels per metre); they have an important role in forming choke tubes through which dielectric drive shafts and probes can be introduced safely into a 'hot' microwave zone.

The critical (cutoff) diameter, for the lowest-order mode (TE_{11}) is, from Table 4.8, $\lambda_{0d}/4.413$, where λ_{0d} is the excitation wavelength in the dielectric material of the inserted shaft (i.e. $\lambda_{0d} = \lambda_0/\sqrt{\varepsilon}$) . The bore of a choke must be substantially less than this diameter, and is chosen as a compromise between the mechanical requirement of the size of the shaft or rod to be inserted and the length of the choke to achieve the required attenuation for safety, or RFI leakage. The diameter a for the choke having been chosen, its cutoff wavelength is 4.413 $a \sqrt{\varepsilon}$.

Using the procedure outlined in Section 4.5.2 the phase constant β becomes imaginary, giving a real, positive value to γ, signifying attenuation. The attenuation rate is then

$$\mathrm{Im}(\beta) = 2\pi \times 8.686 \sqrt{\left|\left\{\left(\frac{\varepsilon'}{\lambda_0}\right)^2 - \left(\frac{1}{\lambda_c}\right)^2\right\}\right|} \quad \text{decibels per metre} \qquad (4.71)$$

Figures 4.11 and 4.12 are curves of attenuation rate (at 2450 MHz and 915 MHz, respectively) for cutoff circular waveguides filled with the dielectrics shown. For waveguides partially filled with dielectric material, the attenuation is greater, tending to the air-filled case depending on the proportion of filling.

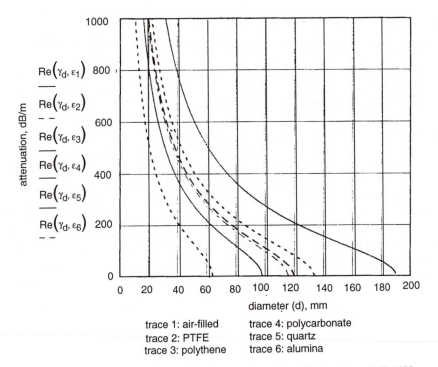

trace 1: air-filled   trace 4: polycarbonate
trace 2: PTFE   trace 5: quartz
trace 3: polythene   trace 6: alumina

Figure 4.11   Attenuation of circular cutoff waveguide, TE_{11} mode at 915 MHz

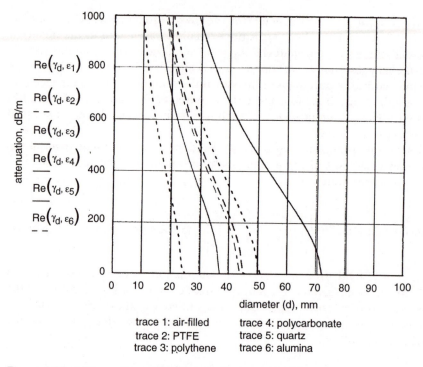

$Re\left(\gamma_d, \varepsilon_1\right)$
——

$Re\left(\gamma_d, \varepsilon_2\right)$
– –

$Re\left(\gamma_d, \varepsilon_3\right)$
⋯⋯

$Re\left(\gamma_d, \varepsilon_4\right)$
—

$Re\left(\gamma_d, \varepsilon_5\right)$
—

$Re\left(\gamma_d, \varepsilon_6\right)$
– –

trace 1: air-filled	trace 4: polycarbonate
trace 2: PTFE	trace 5: quartz
trace 3: polythene	trace 6: alumina

Figure 4.12   Attenuation of circular cutoff waveguide, TE_{11} mode at 2450 MHz

Because of distorted fringing fields at the junction between the cutoff waveguide and the 'hot' microwave zone, there is uncertainty of the attenuation rate in this region and it is advisable to allow a length of at least one diameter of cutoff waveguide before the full attenuation is assumed to develop. Similarly, at the exit end the attenuation may be reduced.

Note that the introduction of a conducting fibre, of any material, however small, immediately changes the structure to a coaxial TEM-mode transmission line with a low attenuation rate. Water in a plastic tube appears to behave in this way, and the introduction of water pipes to a 'hot' microwave zone requires extreme care to avoid high levels of leakage. However, there is also the possibility that the water pipe may form a dielectric waveguide, within a tube, with a surface-wave mode (Ramo *et al.*, 1965). It is also worthy of note that the field intensity developed in a plastic water pipe at the entry to a 'cutoff' tube can be very high, sufficient for local melting of the pipe to occur.

4.7 Waveguide practice

In radar and communication systems waveguide transmission lines have usually not only to convey power with low loss, but also to meet exacting specifications for impedance matching to avoid signal distortion, and for peak power handling. In industrial heating systems, the primary requirement is to convey power with minimum loss, and the matching requirements are much less severe because they are of concern only insofar that they affect power transmission. Consequently, the very high precision of waveguide manufacture familiar to those in the communication and defence industries is inappropriate and unnecessary, and is a very important cost consideration.

Nonetheless, waveguides for heating systems require some care in design and manufacture, and it is important that those involved know the particular, and unusual, requirements.

4.7.1 Waveguide size

Universally, rectangular waveguides are designed to operate in the fundamental, TE_{10} mode. The optimum size of a waveguide for a specific frequency band is a compromise of minimising power loss (and consequent heating) which requires a large cross-section size as indicated by eq. 4.48, and avoiding possible excitation of higher-order modes which can cause resonance and severe impedance-matching problems. The latter requirement sets an upper limit to size: in width to avoid the TE_{20} mode, and in height to avoid the TE_{01}, TE_{11} and TM_{11} modes. In practice it is desirable to have a substantial margin between the operating and cutoff wavelengths, because waveguide discontinuities (bends, junctions etc.) are prone to excitation of the higher-order modes. It is also desirable to operate the TE_{10} mode at a wavelength at least 25% shorter than cutoff, because the waveguide wavelength changes very rapidly close to cutoff, making matching very frequency sensitive; moreover the I^2R losses increase rapidly.

Manufacturing tolerances of waveguide tubing for industrial heating are typically ±0.3% of the principal cross-section dimensions. Bow and twist tolerances, squareness of flanges etc. are dictated by the need of easy mechanical assembly without strain. Wall thickness should not be less than 1.5 mm at 2450 MHz and 2.5 mm at 896 MHz to give reasonable resistance to damage from impact.

These factors affect all waveguide uses, and an international standard of waveguide sizes exists, specifying the optimum waveguide size for operation at specified frequencies throughout the microwave spectrum. All the waveguides have a 2:1 aspect ratio, which conveniently satisfies the above criteria. The waveguide sizes most commonly used in industrial microwave systems are listed in Table 4.9.

Table 4.9   Commonly used waveguide sizes

Microwave frequency	Waveguide designation	Internal dimensions	λ_c (TE$_{10}$ mode) (f_c MHz)
(MHz)		(m)	(m)
896 ±5%	WG4; RG204U; WR975	0.248 × 0.124	0.496 (604 MHz)
915 ±5%	WG4; RG204U; WR975	0.248 × 0.124	0.496 (604 MHz)
2450 ±5%	WG9A; RG112U; WR340	0.0864 × .0432	0.1728 (1735 MHz)

Further details of these waveguides are given in Tables 4.4, 4.5, 4.6 and 4.7.

4.7.2 Materials and construction methods

Materials most commonly used in industrial microwave equipment are shown in Table 4.7. The choice is essentially a compromise between high conductivity to minimise losses, corrosion and, sometimes, high-temperature survival, and ease of manufacture (welding, brazing, machining etc.) (Rollason, 1973). In food and pharmaceutical processing, surface hardness and scratch resistance are important to avoid bacterial traps.

Aluminium is relatively low cost, and has a high electrical, and thermal, conductivity. It is widely used for waveguides transmitting high power, with an upper limit, without forced cooling, of about 100 kW at 900 MHz and 30 kW at 2450 MHz. Most of the alloys have high conductivity, and the choice is determined by mechanical issues: *BS 1470* H15, H30, E91E (high conductivity) are typical alloys. Waveguides at 2450 MHz are usually of drawn tubing, while folding and seam welding are most often used for waveguides at 900 MHz because this method is compatible with the tolerances needed for the larger waveguide and is cheaper than extrusion in the quantities usually required.

From experience, cast flanges are best made from compatible aluminium alloy, welded or brazed to the parent waveguide using an alignment jig. Distortion often occurs, and a final machining allowance is advisable to ensure a flat face square with the waveguide axis.

All waveguides should be assembled without strain, and it is common practice to make one section to fit 'on site' where two large assemblies spaced apart are joined.

On no account should a misaligned flange be 'pulled down' by its bolts. Equally, a waveguide should never be used as a support for heavy equipment unless specially stiffened for the purpose. Long lengths of waveguide must be supported at regular intervals. Where there is relative movement due to vibration or differential expansion, it is good practice to arrange a loop in a waveguide path to allow some flexure.

Stainless steel should not be used for power transmission unless there are overriding reasons for so doing. It has a low electrical and thermal conductivity and so is liable to hot spots and wasted power, as shown by Table 4.7. Its power-handling capability for the same surface temperature is only about 25% that of aluminium, though it can be allowed to run hotter. Usually, stainless steel is used sparingly in a short section leading to a high temperature oven where aluminium would distort.

Plain mild steel is extremely 'lossy', not only because of its low electrical conductivity, but also because it has magnetic-hysteresis loss. However, it can be used successfully with a surface coating. Cadmium plate has been used, with performance as shown in Table 4.7; although cadmium is a good protection against corrosion, it is toxic and must not be used in food applications. Sprayed aluminum has also been used as a coating for mild steel waveguides, but it is difficult to apply except in short sections.

Copper and brass are not often used in industrial microwave plant except in relatively low-power laboratory equipment. The cost is very high and the performance of copper relative to aluminum is marginal; brass has poor conductivity in comparison.

4.7.3 Seams and joints in microwave equipment

As will be appreciated from Section 4.5, the microwave currents flow in a sheet over the inside surface of the waveguide or oven structure. Any discontinuity or imperfection in a joint or seam will interrupt the current flow and cause leakage of energy. Where there is a high-intensity microwave field, the seam must be continuously welded and faultless; it is not acceptable to spot weld such a structure, unless the seam can be caulked with a conducting filler such as a resin heavily loaded with silver particles, and this should only be a temporary solution. Riveting (i.e with pop rivets) is unsatisfactory and should never be used except for temporary laboratory experiments at low power.

Quantifying the leakage from an imperfection is very difficult, and the only safe approach is to seek 100% integrity of all joints where there is significant microwave power, greater than 10 W. Where there is a requirement to meet a RFI specification, all joints must have 100% integrity, whatever the power level.

4.7.4 Waveguide flanges

Most waveguide flanges in industry are plain faced, having bolted flanges with a metallic gasket. The same criteria apply as for the joints discussed in Section 4.7.3. It is particularly important to know that the main current sheet passing between the flanges is at the broad face of the waveguide, which must appear as continuous as possible as a conduction path.

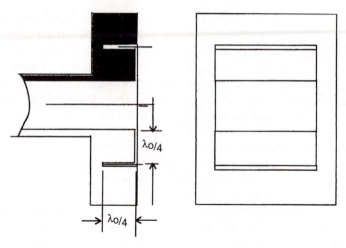

Figure 4.13   Typical choked waveguide flange

The flange faces are machined flat, and up to 16 bolts may be used, particularly if an RFI specification is required. Sometimes a raised narrow 'land' is machined around the aperture to define a more positive seating. A much-used gasket is a copper sandwich with a mineral filler, like an engine cylinder-head gasket: every bolt-hole in the gasket, together with the internal and external perimeters, must be closed with copper because microwave fields can propagate in the mineral filler. At 2450 MHz a plain copper shim is often used. Hardened beryllium–copper or phosphor-bronze sheet gaskets with contact 'fingers' have also been used, but they have a tendency to arc. When a flange is undone the gasket should always be renewed.

Flanges are best joined to the waveguide by allowing a rebate from the rear of the flange of limited depth. The waveguide is inserted from the rear and welded internally and externally. It is not good practice to allow the waveguide to pass right through the flange, welding it at the flange face, because there is a risk of cutting away the weld in the final machining.

It is desirable sometimes to arrange a quick-release flange without bolts. It is then impossible to have a positive metal/metal contact. A choke flange is used, comprising a plain flat flange in one half, mating with the choke flange in the other. The choke flange has a pair of blind slots $\lambda_g/4$ deep in its face, parallel to the broad faces, and located $\lambda_0/4$ from them, as shown in Figure 4.13. The arrangement forms a transmission line in which the transformed impedance is very high at the open end of the slots, which transforms further to a very low impedance at the gap between the broad faces where the main current is required to cross. In effect, the short circuits at the bottom of the slots are transformed through $\lambda/2$ overall to give an apparent short circuit where good contact is desired. Because the impedance is very high at the open ends of the slots the current there is low, and so imperfections in contact are relatively unimportant. Although the leakage from a well designed choke flange is well within safety limits, it will not meet RFI limits such as at 915 MHz in countries where there is no ISM band allocated.

4.7.5 Waveguide bends and twists

Invariably, a waveguide system incorporates bends. Although large-radius bends create negligible reflection, they are expensive and inconvenient, and 'tight' bends are needed. Although small-radius bends are technically excellent, they too are expensive and mitre bends are preferable. They are easily made by fabrication, and can be designed to give low resultant reflection.

The design technique is to use an intermediate section of standard waveguide, effectively $\lambda_g/4$ long measured along the centre line of the broad face, joined between the two straight sections so that its axis is at the half angle of the bend. There are then two impedance-mismatch zones where the reflection coefficients are equal, and spaced apart $\lambda_g/4$ so that they cancel.

Well designed waveguide bends have good VSWR figures with negligible reflected power, and very low insertion loss with negligible extra heat dissipation compared with the standard waveguide.

Bends are designated as E-plane or H-plane:

E-plane is a bend with the side walls of the waveguide remaining in the same plane.

H-plane is a bend with the broad walls of the waveguide remaining in the same plane.

E-plane bends follow closely the above design technique, but H-plane bends require some empirical adjustment because there is more uncertainty about the value of λ_g. Some typical bend dimensions are given in Table 4.10, with reference to Figure 4.14.

Waveguide twists are occasionally required. Again the easiest technique is to make a 'stepped' twist with a set of intermediate waveguides $\lambda_g/4$ long joined end-to-end, with each rotated relative to its neighbour by θ/n where θ is the angle of twist overall and $(n-1)$ is the number of sections. It is desirable

Table 4.10   Typical bend dimensions

Type	Frequency	Waveguide size	Angle	Dimension a	Dimension b
	(MHz)		(deg)	(mm)	(mm)
E-plane	896/915	WG4	90	97.0	—
E-plane	2450	WG9A	90	43.3	—
H-plane	896/915	WG4	90	—	138
H-plane	2450	WG9A	90	—	43.3

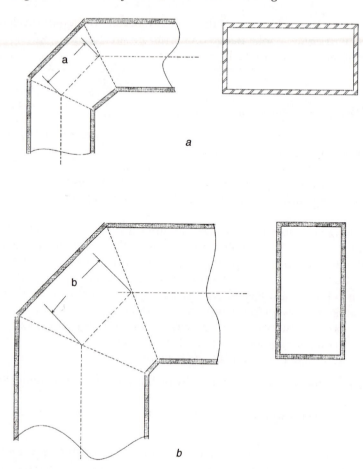

Figure 4.14   Dimensions of waveguide 90° mitre bends

    a E-plane waveguide bend
    b H-plane waveguide bend

that there be an odd number of sections, because this gives an even number of junctions whose reflections cancel in pairs. For example, a 90° twist could be formed of three sections such that at each of the resulting four junctions there is a rotation of 22.5°. Obviously, the construction leaves open gaps at each junction which must be filled.

It is sometimes necessary to preserve a phase relationship in a waveguide system, and care must be taken in choosing twists. A 180° twist clearly gives a phase reversal; however a 90° twist may or may not give a phase reversal, depending on whether it is a right- or left-handed twist. Careful study of the system is required.

The microwave field pattern at a bend is a complicated summation of the TE_{10} and higher-order modes. The latter are evanescent, or cutoff, in the plain waveguide and die away rapidly with distance from the bend. However, if bends are placed too close to each other their respective high-order modes may interact giving an overall degradation in performance. It is good practice to avoid placing bends closer than λ_g apart to avoid this effect, but where this is not possible the VSWR of the combination should be measured and corrected if necessary with a matching element.

4.8 Waveguide components

In this section are brief descriptions of commonly used waveguide components together with outlines of their purpose and performance.

4.8.1 Terminating loads, high-power

A terminating load for a waveguide is often required for absorbing residual power at the end of a heating oven, or as a 'dummy' load for testing a microwave generator. The requirement may be to absorb a large amount of power, e.g 75 kW at 900 Mhz, or 30 kW at 2450 MHz. At the same time, it may be necessary to measure the power. Water cooling is essential, and water is used directly as the absorber, as it has a very high loss factor. Power measurement is then accomplished easily by calorimetry, measuring flow rate and temperature rise accurately. Unfortunately, the dielectric properties of water vary considerably with its temperature, and this results in the VSWR being a function of temperature. Occasionally, the available water is a glycol mixture, which considerably affects the load VSWR.

Various designs of water load exist, divided between compact loads which suffer from VSWR variation with temperature, and long, distributed loads where this effect is less pronounced.

In the first type, the water passes through an array of dielectric pipes mounted at 90° between the broad faces of the waveguide. Various arrangements are possible, but usually there are just two pipes placed one behind the other, the water flowing through them in sequence. The design is empirical, and a matching device is provided which has to be adjusted on low-power test. This water load is very compact but its VSWR performance on load is rather unpredictable. Nonetheless, it is the most widely used because of its size. Preventing water leaks is of paramount importance, and this design has the advantage of relatively short water-seal paths. It is also possible to have a high flow rate which gives it an excellent power-handling capacity.

In the second type, water is introduced into the waveguide over an extended distance, (e.g. 1 m), making the waveguide appear to have a distributed lossy dielectric. This is usually accomplished by introducing the water in a dielectric tube inclined at a shallow angle to the waveguide axis. In this format the overall VSWR is less sensitive to variations in dielectric properties of the water.

Care must be taken with water loads to ensure that air pockets cannot form and that there is sufficient flow that local boiling does not occur. Both these can affect the VSWR. Local boiling can occur even when the outflow temperature is well below boiling point, and can usually be detected by sound, or observing the associated pipework 'kicking'.

4.8.2 Terminating loads, low-power

For low-power testing and in certain monitoring equipment there is a need for a reflectionless broadband compact load, with a power-handling capacity up to about 300 mW.

The least expensive satisfies the above except for size, and comprises a wooden block, machined to be a good sliding fit in the waveguide, having a gentle taper of about 15° cut across the broad dimension at the microwave input end. Beech is a suitable wood, being dimensionally stable. The load has excellent characteristics in that it is easy to determine whether there is a significant amount of power leaking past the load, as movement of the hand near the open end of the waveguide is revealed as a change in impedance match, instantly seen on a network analyser. Moreover, the match of the load itself can readily be separated from other mismatches by sliding the wooden taper along the waveguide, the reflection coefficient of the load being displayed as a small circle on the Smith chart of the network analyser. A metal insert can be fitted empirically to the taper to cancel the residual reflection. This load is excellent for use at 2450 MHz, but is cumbersome at 900 MHz. A good broadband match with VSWR better than 0.95 is readily achievable with this load.

A more compact load comprises a matched transition from waveguide to standard coaxial line (usually 41 Ω line), fitted with a proprietary coaxial matched termination. Over ±2.5% frequency band, it gives VSWR better than 0.9, being dependent on the quality of the transition.

4.8.3 Short-circuit terminations

Adjustable short-circuit terminations are required for some types of heating chamber, and also for impedance-matching devices. The essential requirement is that the sliding piston makes an effectively perfect electrical contact to the waveguide 'cylinder' at the plane of the face of the piston, and nowhere else. It is very important that the piston can be adjusted on-load.

For low-power (< 1 kW) experimental purposes, a simple plate with spring finger contacts is effective. However, if a finger does not make good contact microarcing will occur, burning the surface and annealing the finger. Lubrication is a problem, desirable mechanically to prevent scoring the surfaces, but incompatible with the electrical-contact requirement. For this reason, this design is generally unsuitable in industrial plant.

High-power short circuit pistons are invariably, and paradoxically, called 'noncontact' pistons. The piston usually comprises three sections: two with a small clearance (about 0.5% of the waveguide dimensions or 1 mm max.) to the waveguide wall, and the third section a central rod joining the two to form a rigid assembly. Each section is $\lambda_g/4$ long. Electrically, the assembly consists of three sections of quarter-wave line, the two (A and C) with small clearance having very low characteristic impedance (Z_0) and the centre section (B) having relatively high Z_0 due to its large clearance to the waveguide wall. At the far end from the input, the section C is open-circuit, so the impedance at the junction B/C is very low, having transformed $\lambda_g/4$ in a line of low Z_0. In section B this low impedance transforms a further $\lambda_g/4$, but now in a high Z_0 line, to form a very high impedance at the junction A/B. Again there is a further transformation in section A through $\lambda_g/4$ in low-Z_0 line to give an extremely low impedance at the gap, a virtual short-circuit, in the plane of the face of the piston.

The piston is truly noncontacting, being supported all round by a set of insulating studs standing 'proud' of the surface of the piston which form sliding bearings inside the waveguide. The studs are best mounted at the low-impedance areas where they are exposed to relatively low E-field intensity. PTFE is a suitable material for the studs. For mechanical reasons the waveguide must be of precision bore, the folded-and-seam-welded fabrication method being unsuitable.

This design of piston can give VSWR >200 over ±2.5% frequency band. It is capable of operation in an impedance-matching role at 75 kW at 900 MHz, and >25 kW at 2450 MHz.

Care must be taken in the design of the mechanical drive to ensure minimum 'backlash', and that generous provision is made for the extent of the stroke. The absolute minimum stroke is $\lambda_g/2$ at the highest value of λ_g. However, this can be very inconvenient if the desired operating point is close to the end of the stroke, and it is good practice to provide at least 15% extra travel at each end. It is very helpful to have a readout of the piston position.

4.8.4 Waveguide 3-port T-junctions

Where it is desired to divide microwave power into two equal amounts, a simple 3-port junction can be used, but care must be taken to ensure that the impedances of the two loads *at reference planes equidistant from the junction* are exactly equal; otherwise power division will be in error. This condition is seldom easy to achieve, and so 3-port junctions are only used where equality of power splitting is unimportant, or the balance condition can be guaranteed.

Waveguide T-junctions can be either E-plane or H-plane, as illustrated in Figure 4.15 , which also shows the equivalent circuit of each. In the E-plane junction the same current flows through both loads in series, and obviously

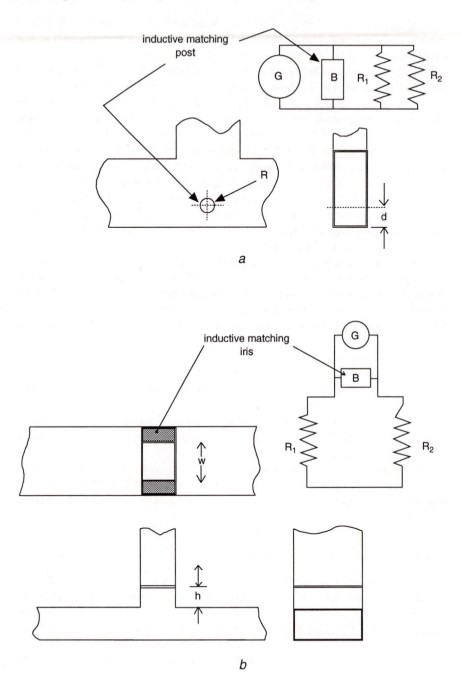

Figure 4.15   Waveguide T-junctions, 3-port, third-angle projection

   a H-plane T-junction showing metal matching post
   b E-plane T-junction showing matching iris

the powers will only be the same if the load impedances are equal. Similarly, for the H-plane junction a common voltage is applied to the loads in parallel, with the same need for equality.

In principle the junctions are designed so that, looking into the 'stem' of the T-junction, the Z_0 values of the branch arms are each $Z_0/2$ for the E-plane junction and $2Z_0$ for the H-plane junction. A simple junction with all the waveguides the same cross-section would therefore have an inherent mismatch; indeed the junction itself creates a further reflection. In practice, matching is best achieved by using a lossless matching iris (E-plane junction) or post (H-plane junction), which substantially cancels all the reflections (Marcuvitz 1986). Although the matching elements can be calculated, it is in practice more reliable to design them empirically with a network analyser.

4.8.5 Directional couplers, 4-port

The 4-port directional coupler (Figure 4.16) is a very important class of waveguide junction which divides input power (to port 1) into any desired output fractions (between ports 2 and 4) so that the sum of the output powers at ports 2 and 4 equals the input power, assuming that all the ports are terminated in matched loads. The power output from port 3 is then, in

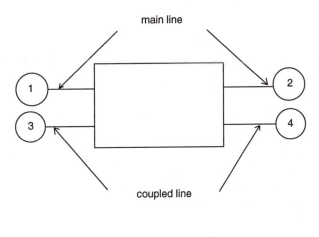

main line

coupled line

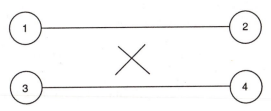

Figure 4.16 Diagrammatic representations of 4-port directional couplers: identification of ports

principle, zero. The junction is symmetrical and lossless, and by inspection of Figure 4.16 it is clear that if the input power is fed to port 2, the output will be between ports 1 and 3, with zero at port 4. This means that the coupler is directional, and if a generator is connected to port 1 and the load to port 2, then the power emerging from port 4 is proportional to the forward power to the load, whereas the power emerging from port 3 is proportional to the reflected power.

The structure therefore has a very important role in forward- and reflected-power measurement, in which case the coupler is weakly coupled so that the coupled forward power is in the order of milliwatts, for the main power of tens of kilowatts.

A subclass of great importance is the coupler with equal power division between ports 2 and 4, with power input to port 1. Unlike the 3-port junction, the power division is not fundamentally affected by the load impedances connected to ports 2 and 4, from which the output powers are constant. Again, the structure is symmetrical and reciprocal. These junctions have special properties which are utilised in onload tuners (Section 4.8.7.1) and variable power dividers (Section 4.8.7.2). There are various forms of this subclass of coupler with equal power split, generically called 'hybrid couplers' or '3 dB couplers', and structurally known as magic-T or magic-Y, multislot, or short-slot couplers.

4.8.5.1 Directional couplers: definitions

The following parameters are used to define the performance of directional couplers, with reference to Figure 4.16. The characteristics of directional couplers are frequency sensitive and the bandwidth and centre frequency should always be stated. All the parameters are given with assumed matched loads connected to all ports except the feed port.

(a) *Coupling Factor:* Coupling factor defines the amount of power coupled from the main to the subsidiary waveguide, and is usually expressed in decibels. Strictly, it is a negative number in decibels, because the coupled power is obviously the lesser; conventionally the negative sign is omitted and understood.

$$\text{coupling factor} = 10 \log P_4 / P_1 = 10 \log P_3 / P_2 \quad \text{decibels} \quad (4.72)$$

(b) *Input VSWR:* This is the VSWR 'seen' looking into any port, with all the others terminated in matched loads.

(c) *Directivity:* Directivity is a measure of the amount of power back-coupled to the supposed zero-output port. It is usually defined as the power ratio in the ports of the subsidiary:

$$\text{directivity} = 10 \log P_3 / P_4 \quad \text{decibels} \quad (\text{power input to port 1}) \quad (4.73)$$

Occasionally it is defined as

$$\text{directivity} = 10 \log P_3 / P_1 \text{ decibels} \qquad (4.74)$$

This latter definition includes the coupling factor and can give a false impression of performance, unless the definition is stated.

(d) *S-parameters:* The S-parameters are scattering coefficients which define the directional coupler precisely in terms of voltage-coupling coefficients in both amplitude and phase. The terminology is simple: the coefficients are equivalent to transmission or reflection coefficients (Section 4.3.1). They are vectors having amplitude and phase. The suffices indicate the ports to which the coefficient applies, e.g. S_{12} means the voltage (and phase angle) of the voltage wave emerging from port 2, with unit input to port 1. The phase angles are specified at defined reference planes at each port. The performance is succinctly defined as a 'scattering matrix' *S*:

$$S = \begin{bmatrix} S_{11} & S_{12} & S_{13} & S_{14} \\ S_{21} & S_{22} & S_{23} & S_{24} \\ S_{31} & S_{32} & S_{33} & S_{34} \\ S_{41} & S_{42} & S_{43} & S_{44} \end{bmatrix} \qquad (4.75)$$

Defining the output voltages as V_{01}, V_{02}, V_{03}, V_{04} from the four ports, and the input voltages as V_{i1}, V_{i2}, V_{i3}, V_{i4}, the following matrix equation determines the overall input/output relation:

$$\begin{bmatrix} V_{01} \\ V_{02} \\ V_{03} \\ V_{04} \end{bmatrix} = \begin{bmatrix} S_{11} & S_{12} & S_{13} & S_{14} \\ S_{21} & S_{22} & S_{23} & S_{24} \\ S_{31} & S_{32} & S_{33} & S_{34} \\ S_{41} & S_{42} & S_{43} & S_{44} \end{bmatrix} \begin{bmatrix} V_{i1} \\ V_{i2} \\ V_{i3} \\ V_{i4} \end{bmatrix} \qquad (4.76)$$

If the directional coupler is perfectly matched, the diagonal elements $S_{11}...S_{44}$ are all zero. Further, if the coupler is lossless it must be symmetrical so that $S_{12} = S_{21}$ etc., and if the directivity is infinite, $S_{13} = S_{31} = 0$ and $S_{24} = S_{42} = 0$. The perfect directional coupler is therefore represented by the matrix S_p where:

$$S_p = \begin{bmatrix} 0 & S_{12} & 0 & S_{14} \\ S_{21} & 0 & S_{23} & 0 \\ 0 & S_{32} & 0 & S_{34} \\ S_{41} & 0 & S_{43} & 0 \end{bmatrix} \qquad (4.77)$$

Since the coupler is lossless, all the input power at one port must emerge at the corresponding ports of the main and coupled lines, and remembering that the scattering coefficients are voltage-coupling coefficients (so that power $\propto S^2$), the following relations also apply:

$$S_{12}^{2} + S_{14}^{2} = 1 \qquad S_{21}^{2} + S_{23}^{2} = 1$$
$$S_{32}^{2} + S_{34}^{2} = 1 \qquad S_{41}^{2} + S_{43}^{2} = 1 \qquad (4.78)$$

There is obvious resemblance of these expressions to the trigonometric relation

$$\cos^2\theta + \sin^2\theta = 1 \qquad (4.79)$$

The angle θ can be used to define the coupling of the coupler so that $\sin^2\theta$ represents the coupled power and $\cos^2\theta$ the power remaining in the main line (Lomer and Crompton, 1957).

Note that the elements S_{14}, S_{41} and S_{23}, S_{32} are the voltage-coupling coefficients of the coupler. Thus $|S_{14}|^2$ is the power-coupling coefficient, and the coupling factor given by eqn. 4.72 can be written

$$\text{coupling factor} = 20 \log |S_{14}| = 10 \log |S_{14}|^2 \ \text{decibels} \qquad (4.80)$$

4.8.5.2 Directional couplers: types

In industrial microwave-heating equipment, directional couplers are used for dividing power between two or more loads, and for reflected-power measurement. The former are usually in the range 3–20 dB (i.e. 50–1% power division), while the latter are very loosely coupled to enable sensors operating in the 1–10 mW range to be used to measure power typically of 1–100 kW (i.e. with coupling factor in the range 50–70 dB).

Directivity is usually achieved in one of four ways:

(a) by an array of slots spaced apart by $\lambda_g/4$ along the waveguide. Consider two slots only, each exciting equal E-field amplitudes in the coupled guide. Individually, the slots are nondirectional, but clearly both 'forward' components in the coupled waveguide will have travelled equal distances and will add beyond the final slot, in the direction of propagation. However, the 'back'-coupled field component from the far slot will have travelled $\lambda_g/2$ further than the 'back' component from the first slot so they are in antiphase and will cancel. Since the phase relationship is dependent on λ_g, the directivity deteriorates as the frequency is changed from the design optimum. The directivity also depends on the coupled components being equal, which is clearly dependent on equality of slot size;

(b) by utilising the fact that the magnetic field is circularly polarised at the transverse plane at $x = x_p$ in the waveguide, where $H_x = H_z$. Circular polarisation is discussed in Section 1.5, and the location of this point is obtained by equating eqns. 4.26 and 4.28. The direction of rotation of polarisation (clockwise or anticlockwise) changes between forward and reflected waves. The coupled waveguide mounted above, and usually at right angles, to the main waveguide is arranged so that the (single) common coupling hole between them is at the same position x_p in both waveguides. A forward wave in the main waveguide then couples a circularly polarised field into the coupled waveguide, which, by reciprocity, excites a unidirectional propagation in the coupled waveguide. The coupling is through a small cruciform or circular hole, which couples weakly, making this form of coupler suitable for forward/reverse-power monitoring.

(c)   by sampling the electric and magnetic fields E_y and H_z in the main waveguide and utilising the fact that H_z changes sign between a forward and reflected wave. The sensor comprises a loop linking with the magnetic field, with a probe (mounted in the plane of the loop) responding to the electric field. The outputs from the probe and loop are added and fed to a diode detector. Two sensors are used, one for the forward wave, the other for the reflected wave. The arrangement is very compact but has the disadvantage that the device is essentially coupled to high field strengths in the main waveguide, giving voltages too high for a normal microwave semiconductor diode, so that a thermionic diode must be used; and

(d)   by using a selected higher-order mode, e.g. TE_{20}, in combination with the fundamental TE_{10}; the differential waveguide wavelengths of the two modes are used to create the necessary phase relationships for directional coupling. This method is particularly appropriate for 3 dB couplers with main and coupled waveguides in a side-by-side configuration.

Figure 4.17 illustrates the various couplers described above.

The tightly-coupled (3–20 dB) 'branch-arm' directional couplers (Figure 4.17a) comprise an array of E-plane T-slots coupling the main to the coupled waveguide. Their dimensions are readily calculated (Crompton *et al.*, 1957), and have 2–5 slots, depending on the desired coupling factor and bandwidth required. For tight coupling, the slot size for a 2-slot design approaches the waveguide height and the simple theory falters, for 10 dB and tighter 3, 4 or 5 slots are preferred to minimise this effect. Tables 4.11 and 4.12 give dimensions of slots in WG 4 and WG9A waveguides: the slot size is substantially independent of frequency, and is the same for 896 and 915 MHz in WG4 waveguide. However, the slot spacing and length are frequency sensitive, and should be adjusted in the proportional change in waveguide wavelength for another frequency. Note that there is a phase change of $\pi/2$ radians between the main and subsidiary waveguides at corresponding reference planes in the output waveguides.

The slot length is $\lambda_g/4$ to achieve a good input match to the couplers (Crompton, 1957); for large slots, a small correction is necessary for optimum performance.

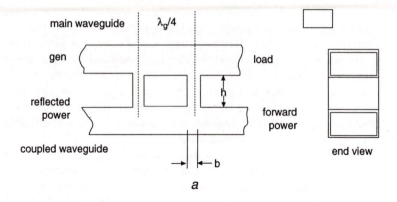

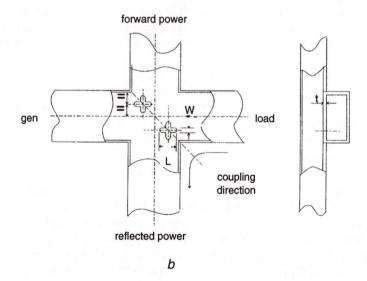

Figure 4.17   Common types of four-port directional coupler

        a Branched-waveguide coupler with two slots
        b 'Moreno' crossed-waveguide coupler

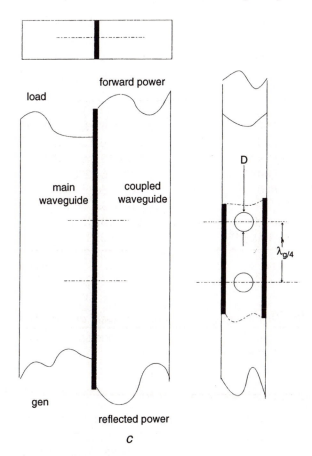

Figure 4.17 continued

    c Sidewall coupler

Note that with three or more slots the slot sizes are in the symmetrical graded ratio of a binomial distribution: 1,2,1 for three slots, 1,3,3,1 for four slots and 1,4,6,4,1 for five slots. This technique broadens the frequency response of the device (Slater, 1942; Crompton, 1957).

Table 4.11   Dimensions of branch-arm directional couplers WG4 (915 MHz)

Coupling factor	Coupling factor: power ratio	Number of slots	Slot sizes	Slot length	Slot spacing
(dB)			(mm)	(mm)	(mm)
3.0	(1/2):1	5	13.3; 53.0; 80.3; 53.0; 13.3	87	117
4.7	(1/3):1	4	19.0; 56.6; 56.6; 19.0	93	115
6.0	(1/4):1	4	16.2; 48.3; 48.3: 16.2	98	114
6.0	(1/4):1	3	32.1; 64.2; 32.1	100	112
7.0	(1/5):1	3	28.5; 56.9; 28.5	103	111
10	(1/10):1	3	19.9; 39.7; 19.9	105	110
20	(1/100):1	2	12.4; 12.4	109	109
30	(1/1000):1	2	3.9; 3.9	109	109

Table 4.12   Dimensions of branch-arm directional couplers, WG9A (2450 MHz)

Coupling factor	Coupling factor: power ratio	Number of slots	Slot sizes	Slot length	Slot spacing
(dB)			(mm)	(mm)	(mm)
3.0	(1/2):1	5	4.63; 18.47; 28.00; 18.47; 4.63	34.5	46.4
4.7	(1/3):1	4	6.62; 19.71; 19.71; 6.62	36.9	45.6
6.0	(1/4):1	4	5.64; 16.82; 16.82; 5.64	38.8	45.2
6.0	(1/4):1	3	11.18; 22.36; 11.18	39.6	44.4
7.0	(1/5):1	3	9.91; 19.82; 9.91	40.8	44.0
10	(1/10):1	3	6.92: 13.84; 6.92	41.6	43.6
20	(1/100):1	2	4.32; 4.32	43.2	43.2
30	(1/1000):1	2	1.37; 1.37	43.2	43.2

For monitoring forward and reflected power, the coupling factor required is usually very loose, in the range 50–70 dB. The slot sizes for branch-arm couplers (Figure 4.17a) become impractically small and couplers are based on small holes between the main and coupled waveguides; the hole size is considerably less than $\lambda/2$ and is therefore cutoff to all modes; it is, however, entirely practical in size for precision machining. Two configurations are shown in Figures 4.17b and 4.17c.

Figure 4.17b shows the Moreno (Moreno, 1948) crossed-waveguide coupler. The dimensions of the cruciform slots are typically as given in Table 4.13, with reference to the Figure 4.17b.

Table 4.13 *Dimensions of Moreno crossed-waveguide directional couplers*

Waveguide designation	Waveguide size (internal)	Frequency	Coupling factor	Common wall thickness, t	W	L
	(mm)	(MHz)	(dB)	(mm)	(mm)	(mm)
WG4	248 × 124	896/915	50	10.0	17.8	57.5
WG4	248 × 124	896/915	55	10.0	17.8	50.3
WG4	248 × 124	896/915	60	10.0	17.8	43.9
WG4	248 × 124	896/915	65	10.0	17.8	38.2
WG4	248 × 124	896/915	70	10.0	17.8	33.2
WG9A	86 × 43	2450	50	3.44	6.19	19.9
WG9A	86 × 43	2450	55	3.44	6.19	17.5
WG9A	86 × 43	2450	60	3.44	6.19	15.2
WG9A	86 × 43	2450	65	3.44	6.19	13.2
WG9A	86 × 43	2450	70	3.44	6.19	10.7

Figure 4.17c shows a directional coupler with waveguides side by side, with circular coupling holes on the centre line of their common narrow face. The holes are spaced apart by $\lambda_g/4$. Table 4.14 gives typical dimensions.

Table 4.14 *Dimensions of side-wall directional couplers*

Waveguide designation	Waveguide size (internal)	Frequency	Coupling factor	Common wall thickness	Centre-line spacing	Hole diameter D
	(mm)	(MHz)	(dB)	(mm)	(mm)	(mm)
WG 4	248 × 124	896	50	9.9	113.3	40.2
WG 4	248 × 124	896	60	9.9	113.3	29.7
WG 4	248 × 124	896	70	9.9	113.3	18.6
WG 4	248 × 124	915	50	9.9	109	40.2
WG 4	248 × 124	915	60	9.9	109	29.7
WG 4	248 × 124	915	70	9.9	109	18.6
WG9A	86 × 43	2450	50	3.44	43.5	13.9
WG9A	86 × 43	2450	60	3.44	43.5	10.3
WG9A	86 × 43	2450	70	3.44	43.5	6.4

There are many sources of error in absolute-power measurement using directional couplers, mainly from the difficulty of verifying, precisely, weak coupling factors over 50 dB and the accuracy of the sensor and its associated circuits. It is relatively easy to calibrate the whole system accurately at high power using a water–load calorimeter, and this procedure is strongly recommended.

4.8.6 4-port 'hybrid' junctions

A very important class of 4-port junction is the so-called hybrid junction, which in reality is a 3 dB directional coupler, giving a precise power split of 2:1, i.e. exactly equal powers emerge from the two output ports, with zero at the fourth port. The branch-arm 3 dB directional coupler described in Section 4.8.5.2 and Tables 4.11 and 4.12 is one particular form. Other structures having the same properties are:

(a)   The 'Magic-T': this is the superposition of an E-plane and an H-plane, as shown in Figure 4.18. Looking into the junction from either the H-plane or the E-plane branch-arms (3 and 4), the junction appears symmetrical about the vertical centre plane. Such symmetry obviously implies an equal power split between the output arms (1 and 2) of the junction. However, there is an important difference between the performance of the two junctions: if the E-field distribution at the junction is sketched it is at once apparent that the H-plane junction launches waves into the output waveguides which are mutually in-phase at reference planes equidistant from the junction, whereas waves excited by the E-plane junction are in mutual phase opposition, viewed at the same reference planes.

Both the junctions are inherently mismatched, and an inductive iris is inserted to match the E-plane junction, a capacitive post achieving matching for the E-plane junction.
Referring to Figure 4.18, the dimensions of the matching post and iris are as in Table 4.15.

Table 4.15   Dimensions of magic-T matching elements (see Figure 4.18)

Frequency	Waveguide designation	A	B	C	D	Thickness of iris plate	h
(MHz)		(mm)	(mm)	(mm)	(mm)	(mm)	(mm)
896/915	WG4	90	96	162	34	3.0	179
2450	WG9A	34.6	28.8	34.2	12.7	1.6	66

The magic-T is the most commonly used hybrid junction in industrial-heating equipment because the configuration of the waveguides is convenient, frequently 'fitting in' to a waveguide layout neatly. Its power-handling capability is well within the capacity of contemporary generators, and its frequency bandwidth is adequate.

(b)   The short-slot broadwall coupler: This (Figure 4.19) comprises two waveguides side by side, having the common sidewall cut away for a specified distance. In the cutaway section, the waveguide is double

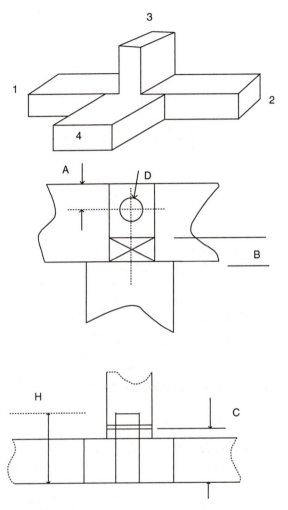

Figure 4.18   Magic-T junction

width, permitting the TE_{20} mode to propagate as well as the TE_{10}. In principle, the length of the common slot is set so that there is a relative propagation phase shift of $\pi/2$ between the TE_{10} and T_{20} modes in that region. At the junction A of the input waveguide with the broad section, the TE_{10} mode of the standard waveguide excites both the TE_{10} and TE_{20} modes in the broad section so that their respective E fields are in phase at the plane of the input junction. Since the E_{20} mode is asymmetrical (i.e. it has a phase inversion on one side of the waveguide relative to the other), the resultant E field cancels at entrance 3 to the coupled waveguide at the input end. However, at the output end (B) the

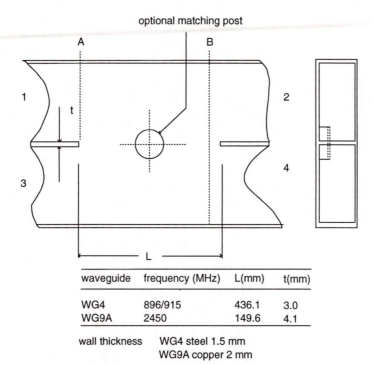

optional matching post

waveguide	frequency (MHz)	L(mm)	t(mm)
WG4	896/915	436.1	3.0
WG9A	2450	149.6	4.1

wall thickness	WG4 steel 1.5 mm
	WG9A copper 2 mm

Figure 4.19   Long-slot narrow-face 3 dB directional coupler

differential $\pi/2$ phase shift, together with the phase inversion in the TE_{20} mode, gives resultant E-field vectors at the two output waveguides 2 and 4 equal to $(1/\sqrt{2}) \exp(+j\pi/4)$ and $(1/\sqrt{2}) \exp(-j\pi/4)$, respectively. Clearly these two vectors have the same resultant amplitude, resulting in equal powers in the two output waveguides. Note, however, that the resultants have a relative phase shift of $\pi/2$ at equidistant reference planes from the junction.

4.8.7 General properties of the hybrid junctions

All hybrid 4-port junctions are 3 dB directional couplers and share the same ideal scattering matrix as eqn. 4.77, where the coupling coefficients S_{12}, S_{14} etc. have amplitude $1/\sqrt{2}$ with a phase angle 0 or π rad. The factor $1/\sqrt{2}$ arises because of the half-power coupling, bearing in mind that power $\propto$ (voltage)^2 where the impedance level is constant. The matrix equations (eqns. 4.76 and 4.77) therefore become

$$\begin{bmatrix} V_{O1} \\ V_{O2} \\ V_{O3} \\ V_{O4} \end{bmatrix} = \begin{bmatrix} 0 & \frac{1}{\sqrt{2}}e^{j0} & 0 & \frac{1}{\sqrt{2}}e^{j\pi} \\ \frac{1}{\sqrt{2}}e^{j0} & 0 & \frac{1}{\sqrt{2}}e^{j\pi} & 0 \\ 0 & \frac{1}{\sqrt{2}}e^{j\pi} & 0 & \frac{1}{\sqrt{2}}e^{j0} \\ \frac{1}{\sqrt{2}}e^{j\pi} & 0 & \frac{1}{\sqrt{2}}e^{j0} & 0 \end{bmatrix} \begin{bmatrix} V_{i1} \\ V_{i2} \\ V_{i3} \\ V_{i4} \end{bmatrix} \qquad (4.81)$$

4.8.7.1 The on-load tuner

On-load impedance matching is an essential requirement of high-power systems: usually the load characteristics are dependent on temperature, free moisture or composition, and these are dynamic parameters, functions of power; and correct matching cannot be determined by low-power tests. Using a magic-T, the on-load tuner comprises a pair of adjustable short-circuit pistons (Section 4.8.3) mounted on ports 3 and 4, with the main waveguide from generator to load being between ports 1 (generator) and 2 (load), as shown in Figure 4.20.

Because of the 100% reflection from the pistons, the input voltages to ports 3 and 4 are equal to their respective output voltages, but phase shifted by $4\pi l_1/\lambda_g$ and $4\pi l_2/\lambda_g$ respectively, where l_1 and l_2 are the (adjustable) distances from the reference planes of the hybrid to the short-circuit pistons. The reference planes are used for mathematical correctness; in practice there is no need to locate them precisely. Using eqn. 4.81 and equating the

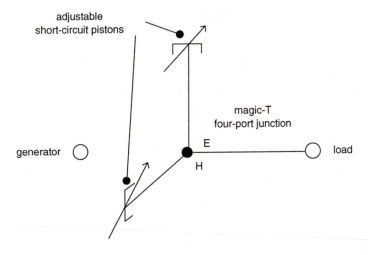

Figure 4.20 Schematic diagram of on-load tuner

input and output voltages to ports 3 and 4 as above, it is readily shown that a reflected wave can be generated by the structure at port 1 equivalent to a reflection coefficient of any amplitude and phase angle, by adjustment of the positions of the short-circuit pistons. In principle, differential movement of the pistons controls the amplitude of the reflection coefficient, while tandem movement controls the phase angle.

It is therefore possible to use the structure for tuning a mismatched load to optimum match. The tuning range is very wide and it is possible to match near-short-circuit loads; in this extreme case, care must be taken because the resultant combination is likely to be highly resonant, having a narrow-band frequency response and very high associated field strengths.

The short-circuit pistons are preferably of the pattern described in Section 4.8.3. Their mechanical movement is achieved by either a leadscrew or rack-and-pinion, driven by a stepper motor with remote control. A 'readout' of piston position is desirable, together with electrical and mechanical endstops to prevent overtravel.

Because the forward-transmission path (input-to-output) includes the distance to the pistons, the device can be used as a phase shifter, when the differential position of the pistons has been set to give substantially zero reflection, and the pistons are mechanically coupled to move as a single assembly.

Obviously other hybrid junctions can be used provided that their scattering matrices have the same symmetrical form as the magic-T. Normally they would only be chosen for a particular case of a more convenient configuration. However, with the increasing power levels of generators, the branch-arm 3 dB coupler or the short-slot coupler may become more widely used as their power handling capacity is greater.

4.8.7.2 On-load high power dividers and attenuators

It is sometimes necessary to reduce the power output of a generator to a level less than can be achieved reliably by electronic control; or the frequency change resulting from electronic control may be unacceptable, for example when operating with a high-Q-factor resonant load. In these cases, it is desirable to be able to adjust the power delivered to the load independently of the generator, which remains operating under constant steady-state conditions. The on-load power divider provides this facility, (Figure 4.21).

Consider a single input at port 3 of the magic-T hybrid (Figure 4.18). The output is two equal power components at ports 1 and 2, with voltages in a defined phase relationship. Reciprocally, if two equal inputs are applied to ports 1 and 2 with the same phase relationship, their vector sum will emerge at the original input port 3, and their vector difference (zero) at port 4. Moreover, by consideration of the scattering matrix (eqn. 4.81) it is readily shown that, if the phase relationship between the inputs 1 and 2 are made adjustable, the output powers can be controlled between ports 3 and 4: a 180°

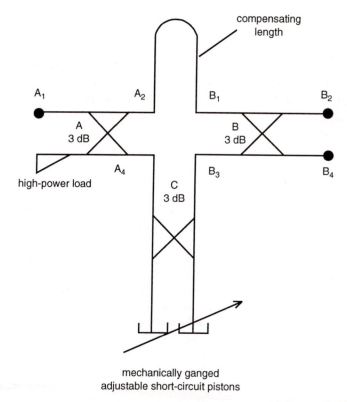

Figure 4.21 *On-load adjustable high-power divider*

phase change in the input to port 3 would switch the whole output from port 1 to port 2.

These characteristics form the basis of a variable power divider comprising two hybrid couplers (A, B) interconnected by a phase shifter as shown in Figure 4.21. The input to port A1 can be divided in any desired ratio between ports B2 and B4 by control of the relative phase lengths of the waveguides joining A2 to B1 and A4 to B3.

The phase lengths may be adjusted by any conventional phase shifter appropriate to the power rating, but for high power use the on-load tuner (Section 4.8.7.1) in its phase-shift role is the most reliable and stable. It is desirable to fit a compensating loop of line in the path A2→B1 to equate with the nominal electrical length of A4→B3 accommodating the phase shifter, to minimise the sensitivity of the power division to frequency changes.

The input to port A1 is normally from a generator having a circulator, and is therefore from a matched source; a matched load also terminates port A3. Under theses conditions, power reflected from the loads connected to ports B2 and B4 is dissipated in the circulator water load and the load at port A3, and does not affect the power division of the device. The actual power dissipated in the load is affected only to the normal extent of its reflected power. Obviously, if a high-power water load is placed on one of the output ports the device forms a high-power attenuator.

The power division law between ports 2 and 4 is readily shown to be in the ratio $(\cos^2\theta) : (\sin^2\theta)$, where θ is the change in phase angle from the setting to give 100% power in port 2, related directly to the movement of the pistons in the phase shifter, which provides a calibration of attenuation in terms of piston movement.

4.9 Multiscrew tuners

A very crude tuner, widely used at low power, comprises a set of adjustable screws set in a thick block forming the broad face of the waveguide, set apart along the longitudinal centre line. Each screw provides a variable, capacitive susceptance whose amplitude is a function of its insertion into the waveguide, and the diameter of the screw. If four such screws are considered, spaced apart by $\lambda_g/8$, then, for small reflections, the screws can be seen to give, at a reference plane located at the first screw, an admittance adjustment of $\pm jb$ for screws 1 and 3, and conductance adjustment $\pm\delta r$ for screws 2 and 4. *Within the adjustment range* it is therefore possible to generate a reflection of any amplitude and phase simply by setting the screws for minimum net reflected power from the load/tuner combination.

In general only two screws should be inserted at one time, because the alternate screws have opposite effects and cancel each other. Although for small mismatches the screws operate in an independent, orthogonal manner, it will be realised from the curvature of the contours of the Smith chart that for large mismatches there is an interdependence between the screws, so that sequential adjustment is usually necessary. Moreover, the screws are large in diameter, and at $\lambda_g/8$ spacing their fringing fields interact, so there is interdependence for this reason also.

As the screw is inserted, its capacitive susceptance increases, but so also does its inductance due to currents flowing in the screw. When the insertion exceeds roughly half the waveguide height, the screw will become resonant so that its electrical property changes completely; the available travel must be limited to prevent this effect.

Some designs use three screws, others have used more than four, but in both cases there is an inherent interaction which makes adjustment more difficult.

The main merit of multiscrew tuners is cheapness. They have the serious disadvantage that the currents induced in the screws; are substantial, and essentially flow into the broad face of the waveguide at the root. In practice it has proved impossible to provide a perfect contact at this point at all times, and so the current flows into the screwed tapping block. Arcing occurs in the thread, causing corrosion, and the screw frequently seizes in its thread.

Attempts at lubrication are usually counterproductive. In some cases the currents may be significant on the outside surface of the tapping block, causing sparking and alarm to the operator making adjustment. Where there is high power and the setting approaches the limit of the adjustment range, the I^2R heating of the screws may cause very high temperatures, accelerating corrosion.

Attempts to provide a sound current path have not been entirely successful. Methods tried have included collet clamps, contact gaskets and electrical choking. All have failed due to arcing or voltage breakdown.

For these reasons, multiscrew tuners should never be used in industrial production equipment, certainly not where there is significant power (e.g. >2 kW at 2450 MHz, or >10 kW at 915 MHz).

4.10 Circulators

The circulator is a three-port (sometimes four-port) waveguide component which has nonreciprocal properties in its treatment of forward and reflected waves. It has important applications in protecting generators from excess reflected power from a load, and is used almost universally with magnetrons with an output of over 5 kW. In diverting reflected power away from the magnetron, the benefits are twofold: first, the reduction in internal heating of the magnetron due to high reflected power; secondly, the improvement in magnetron stability where the load is frequency sensitive.

The circulator is activated by slabs of ferrite material with gyromagnetic properties (Ramo *et al.*, 1965). The ferrite is a sophisticated mixture including rare-earth elements, e.g. yttrium and iron oxides. In the presence of a strong field provided by a permanent magnet, the ferrite has different values of μ' when interacting with microwave magnetic fields of right-hand and left-hand polarisations. There is a region in a waveguide where H_x and H_z are in the correct amplitude and phase relationship to give circular polarisation, the direction of rotation depending on the direction of propagation. It is therefore possible to make a waveguide phase shifter in which the electrical phase length is different in the two directions of propagation. If the phase difference is π radians, the phase shifter can be used in the power divider configuration shown in Figure 4.21. It replaces the variable phase shifter between points A4 and B3: if the generator is connected to port A1, and the workload oven to B2, it is evident that the reverse power incident on B2 from the load is coupled to the absorber load at A3, and no power goes to the generator. An absorber load is fitted to port B4 where there should be little power under normal conditions. This is the four-port circulator: it is of growing importance because its power handling capacity is much greater than the simpler three-port device as the power density in the ferrite is lower and it is easier to cool. For power above 75 kW at 915 MHz and 30 kW at 2450 MHz it is probably essential.

The three-port circulator usually comprises an *H*-plane junction of three waveguides at 120° with a ring of ferrite at the centre. A strong magnet provides a field parallel to the *E*-field in the junction as above. In effect the phase length is different between the opposite directions of propagation around the ring, determined to give ideally lossless coupling from arm A→B, from B→C and from C→A. If the generator is connected to port A and the workload oven to port B, reflected power from port B will couple to port C which has a reflectionless load connected, absorbing all the reflected power so that the magnetron effectively 'sees' a perfectly matched load.

The scattering matrix for the three-port circulator is

$$
\begin{bmatrix} V_{oA} \\ V_{oB} \\ V_{oC} \end{bmatrix} = \begin{bmatrix} S_{AA} & S_{AB} & S_{AC} \\ S_{BA} & S_{BB} & S_{BC} \\ S_{CA} & S_{CB} & S_{CC} \end{bmatrix} \times \begin{bmatrix} V_{iA} \\ V_{iB} \\ V_{iC} \end{bmatrix}
\tag{4.80}
$$

The elements S_{AA}, S_{BB} and S_{CC} represent the input and are ideally zero. S_{AC}, S_{BA} and S_{CB} represent the desired coupling and are ideally unit amplitude. S_{AB}, S_{BC} and S_{CA} represent undesired reverse coupling and are ideally zero. In practice three-port circulators have a forward attenuation (S_{AC}, S_{BA} and S_{CB}) in the region of 0.1 dB, with reverse coupling, or isolation, (S_{AB}, S_{BC} and S_{CA}) greater than 18 dB. The input match is typically better than VSWR = 1.10. Although three-port circulators of 100 kW rating have been reported, 75 kW is a practical limit when derating for an industrial environment is considered. The power rating is often, correctly, quoted with the workload arm terminated in a short-circuit.

Although the attenuation of the circulator is low, the heat dissipation in the ferrite is high in high power devices. The Curie temperature is usually about 150°C above that at which the magnetic properties cease. Liquid cooling is essential and the ferrite, usually in the form of small tiles, is mounted on a water-cooled heatsink. The ferrite has poor thermal conductivity and so the tiles are thin. The waveguide height in the ferrite junction is low and extra care is necessary to keep the waveguide clean. The strong permanent magnet attracts swarf etc. which is very likely to cause an arc. Water ingress must be avoided as this too can cause an arc. Unfortunately the circulator is easily damaged by arcing, and because of its low height it forms a trap where a travelling arc from elsewhere will stabilise. An optical arc detector is an essential safeguard.

Chapter 5

Microwave-heating applicators 1: multimode ovens

5.1 Introduction

Multimode ovens are overwhelmingly the most commonly used form of microwave heating applicator. Mechanically they are very simple, essentially comprising a closed metal box with accessories, and so they are easy to make. The popularity of the multimode oven arises not only from its mechanical simplicity, but from its extraordinary diversity in its ability to process a very surprisingly wide range of workloads, in both size and electrical properties.

However, the simplicity stops here, for the electrical analysis is of the utmost complexity, being the focus of intense contemporary research (Metaxas, 1996; Jia and Jolly, 1992; Dibben and Metaxas, 1995). In principle, the multimode oven supports a large number of resonant high-order waveguide-type modes simultaneously, which add vectorially in space and time to give a resultant field pattern. The rectangular multimode oven may be considered as a waveguide (Meredith, 1993), large in cross-section size compared with the excitation wavelength, and short-circuited at both ends by metal walls, so forming a closed box. The waveguide can support many modes, TE_{mn} and TM_{mn} (Ramo *et al.* 1965; Meredith, 1993); each, in general, having a different waveguide wavelength from the others. Each resonates when the length of the oven is $p\lambda_g/2$ for the particular mode, where p is an integer. The integers m, n, p are, respectively, the numbers of half wavelengths of sinusoidal electric-field variation along the principal axes of the oven.

There is a very loose acoustic analogy with a large building where the reverberation is indicative of the decay of resonances excited by a sound source. There are concert halls with a 'good' acoustic, while others are poor, and it is only in recent times that some understanding has been developed of the important parameters. Similarly, there are multimode ovens with efficient, acceptably uniform heating, while others perform badly; but the dispersive characteristic of electromagnetic waves is a complication absent with acoustic waves and so this analogy can be taken no further than noting a spectrum of resonant frequencies.

There are currently three approaches to the design and analysis of a multimode oven:

(i)   the engineering designer's methods based on laboratory experiment, together with a sound understanding of microwave principles and past experience, to give an empirical design. Quantitative prediction of performance is of doubtful accuracy, and 'scaling up' from a small model presents difficulties;

(ii) An electrical 'equivalent-circuit' approach where the mode structure is modelled, the overall performance being the sum of the individual modes. This approach gives some insight to the important parameters, but it has not been developed to the status of a design procedure except where the load is such that the oven is aperiodic or forms a lamina parallel to one face of the oven and fills the cross-section (Paolini, 1989); and

(iii) Numerical methods, based for example on finite-element methods in the time domain. The main thrust of contemporary research is in these techniques and good progress is being made, though they are not yet (1996) sufficiently advanced to provide a 'user-friendly', reliable design procedure. Already, however, they are able to provide, with limitations on size and complexity of configuration, a guide to heating uniformity and efficiency of a proposed design.

Whatever approach is taken, it is essential to appreciate that the performance of a multimode oven has a first-order dependency not only on its shape, dimensions and configuration of its microwave feeds and accessories, but also on the load, its electrical properties ε' and ε'' (which usually are functions of temperature, moisture etc.), its dimensions and location in the oven. Quite minor changes, particularly in small ovens, can have a radical effect on performance as has been demonstrated in the wide variations in performance of domestic microwave ovens of nominally the same design, operating on the same load (Burfoot *et al.*, 1990). No less important are the characteristics of the microwave generators, their frequency stability both in the short and long term, the effect of incidental and deliberate modulation at power and other frequencies, and output changes due to variations in load impedance.

There are two extreme conditions of workload for a multimode oven. The first is where the load has a microwave penetration depth comparable to its physical size, and minimum reflection from its surface. A high proportion of the microwave energy directed at the workload penetrates it and is absorbed. Under these conditions there is very little stored energy which means that the Q-values of the modes are very low, resonance is virtually suppressed and the oven operates aperiodically. This is the simplest operational regime, for which the design procedure is usually straightforward. The second is where the workload is low-loss, so that the penetration depth is large compared with its size. A complicated situation then arises of the microwave energy being transmitted through the workload with only a small proportion being absorbed at the first 'pass', the balance being reflected from the oven walls for second and further passes through the load. The microwave field 'builds up', storing energy in the electric and magnetic fields. Under these last conditions the oven is operating under resonant conditions, probably with several modes resonant close to the generator frequency.

5.2 Mode structure

In its simplest form the oven is a rectangular box (dimensions a, b, d) in which one or more dimensions are several half wavelengths long at the excitation wavelength λ_0.

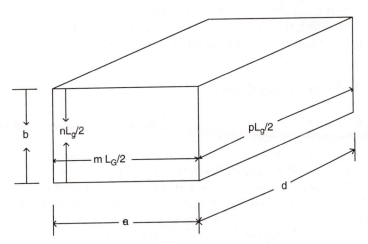

Figure 5.1   Basic rectangular multimode oven showing modes

One method of analysis is to consider the box a waveguide with short-circuit terminations at each end (Ramo *et al.* 1965; Meredith, 1993): actually it is three such waveguides with their axes aligned with the principal axes of the box. These waveguides can each support both TE_{mn} and TM_{mn} modes (Turner *et al.*, 1984) as described in Section 4.5. Considering the waveguide of cross-section $a \times b$ there is, for each mode, a defined cutoff wavelength $\lambda_{c,mn}$ given by eqn. 4.32, and a corresponding waveguide wavelength $\lambda_{g,\,mn}$ by eqn. 4.34. By equating the length of the box d to p half wavelengths of $\lambda_{g,mn}$, where p is a positive integer (including $p = 0$ for TM modes only), the well known equation for the resonant free-space wavelengths $\lambda_{0,Res}$ is obtained for the m, n, p modes within an empty rectangular cavity (Ramo *et al.*, 1965):

$$\left(\frac{2}{\lambda_{o,Res}}\right)^2 = \left(\frac{m}{a}\right)^2 + \left(\frac{n}{b}\right)^2 + \left(\frac{p}{d}\right)^2 \tag{5.1}$$

Note that the maximum possible values for m, n, p are the highest integers satisfying the condition $m < 2 \, a/\lambda_0$, $n < 2b/\lambda_0$ and $p < 2d/\lambda_0$, respectively.

The integers m,n,p are sometimes called eigenvalues, and collectively an eigenvector. Note that the TE and TM modes of the same eigenvalues have the same resonant frequency, given by eqn. 5.1. In practice, the presence of the workload and other perturbations in the rectangular cavity causes the TE and TM modes of the same eigenvalues to separate in frequency. For example, consider a rectangular waveguide short-circuited at each end to form a closed cavity; it resonates in the TM_{111} and TE_{111} modes. If a thin metal rod is introduced a short distance axially at the centre of one end plate it will couple capacitively to the E_z component of the TM_{111} mode, modifying its resonant frequency, because, for the TE_{111} mode, $E_z = 0$, there will be little capacitive effect, recalling that E_y and E_x are weak in the vicinity of the end plates. The TE_{111} mode will therefore change little from the perturbation.

5.3 Mode spectrum density

By definition, a resonant multimode oven must have many modes resonant in the vicinity of the operating frequency of the generator. A single mode would create a sinusoidal heating pattern correponding to its mode number. Many modes are necessary to approach uniform heating. Each mode has the same resonant characteristic as a simple tuned circuit and has a frequency spread which depends on the circuit Q-factor, in turn a function, in this case, of the lossiness of the workload. A lossy load gives a broad frequency response to the mode which may be significantly excited even if its resonant frequency is 10% different from the generator frequency. But with a very low loss load, even a 0.1% frequency difference may mean that the mode is little excited.

As discussed in Section 5.1, the heating pattern in the load is determined by the resultant E-field vector sum of the modes at each point in the workload. It is shown that the more modes excited the more uniform the heating capability, and so the spectral density of the modes, i.e. the number of modes having a resonant frequency within a specified bandwidth, is the important parameter of a multimode oven, indeed it can be considered a figure of merit (FoM) of the oven.

Eqn. 5.1. is readily solved by computer to find all the resonant wavelengths for a given size of rectangular cavity within a chosen bandwidth. A simple program in QBASIC is shown in Appendix 5.

From Eqn. 5.1 it is clearly desirable to avoid cavities where the dimensions have a simple proportional relationship because modes become 'degenerate', i.e. two or more modes have the same resonant frequency and mode pattern: cubes are an obvious example.

Eqn. 5.1 is readily modified to provide a 'universal' FoM, normalised by the wavelength factor $g = a/\lambda_0$, also including the aspect ratio factors u, v to represent the proportions of the rectangular oven, such that $b = ua$, $d = va$. Making these substitutions eqn. 5.1. becomes

$$g = \frac{1}{2}\sqrt{\left\{ m^2 + \left(\frac{n}{u}\right)^2 + \left(\frac{p}{v}\right)^2 \right\}} \tag{5.2}$$

Starting with a near-cube oven (sides 1.0 : 0.95 : 1.10), eqn. 5.2 can be evaluated to find the total number of resonant frequencies within a bandwidth Δg as a function of oven size g, for various proportions for the rectangular box. Moreover, if (uv) is maintained constant a comparison of FoM is made on a constant-volume basis for the various ovens. Figure 5.2 shows a FoM mode count for sets of ovens of constant volume with *nominal* proportions 1 : 1/2 : 2, 1 :1/3 :3, 1 : 1/4 : 4, 1 : 1/5 : 5, 1 : 1/6 : 6, 1 : 1/7 :7 and 1 : 1/10 :10. The exact sizes of the ovens are the above ratios applied to the near-cube oven, an exact cube being avoided to eliminate degenerate modes.

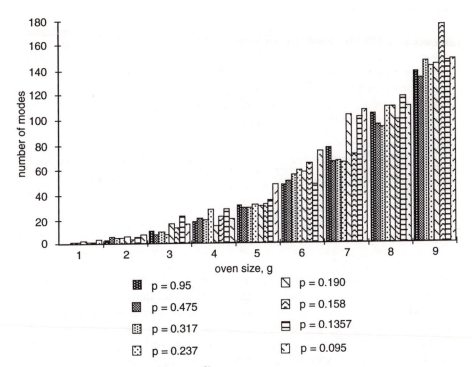

Figure 5.2   Universal mode-count diagram

In Figure 5.2 the FoM is the number of modes resonating within a bandwidth of $\Delta g = \pm\ 0.5\%$ and is shown as a histogram for the above proportions of the oven at various oven sizes $2 < g < 10$. Note the following:

(i)   The FoM is roughly proportional to the volume of the oven (g^3).
(ii)   The FoM is not greatly affected by the oven proportions, rising slightly from the cube to the extreme case cited of $1 : 1/10 : 10$.
(iii) The size of an oven for a specified FoM can be estimated from Figure 5.2. For example, a FoM of 30 at 2450 MHz ($\lambda_0 = 12.2$ cm) would require an oven with dimensions $6 \times \lambda_0 = 73$ cm for the principal, with the other dimensions $0.95 \times 73 = 69$ cm and $1.05 \times 73 = 77$ cm. The volume would be 0.39 m^3, roughly constant for this FoM at 2450 MHz, allowing other oven proportions to be evaluated.
(iv)   The resonant frequencies of the modes are irregularly spaced, and 'holes' can appear, particularly with ovens of low FoM.
(v)   In general the workload will increase the FoM because its permittivity $\varepsilon' > 1$ shortens the effective wavelength, in a complicated manner difficult to quantify. Dielectric hardware in the oven has a similar effect.
(vi)   Metallic hardware within the oven generally reduces the FoM because it reduces the effective volume.

5.4 Mode Q-factor

In Section 5.3, the FoM for an oven is defined as the number of modes resonant within a specified bandwidth. In practice, the number of modes effectively excited is proportional not only to the FoM but also the Q-factor of the modes. In electrical engineering, 'Q' is the quality factor of a simple resonant circuit indicating its damping and, more relevantly here, the frequency bandwidth of its response. Two basic definitions of Q are:

$$Q = 2\pi \frac{total\ energy\ stored}{energy\ dissipated\ /\ cycle} = 2\pi \frac{total\ energy\ stored \times resonant\ frequency}{power\ dissipated}$$

(5.3)

$$Q = \frac{f_0}{\delta f}$$

(5.4)

where δf is the *total* frequency spread between the half-power points on the frequency-response curve, and f_0 is the resonant frequency

From these definitions, a resonant circuit in which the stored energy equals the energy lost per cycle will have $Q = 2\pi$, and the bandwidth will be $f_0/2\pi = 0.16\ f_0$. This is a very low-Q system and is almost aperiodic, representing a cavity with a high-loss load.

The Q-factor often is subdivided into elements representing the losses in various sections of the system, all associated with the same resonant mode. Thus Q-values are ascribed to to the workload alone (Q_{L1}), wall and other losses in the oven (Q_{L2}), and from the loading of external circuits, including open-aperture chokes, if any (Q_{L3}). The overall Q-factor for a given mode is given by

$$\frac{1}{Q}=\frac{1}{Q_{L1}}+\frac{1}{Q_{L2}}+\frac{1}{Q_{L3}}+.....etc. \tag{5.5}$$

If Q is moderately high, representing a low-loss load, e.g. $Q = 500$, the mode bandwidth δf is, from eqn. 5.4, $0.002f_0$. In practice, modes outside this bandwidth are also excited sufficiently to contribute, and a bandwidth $\delta f'$ of at least $(5 \times f_0/Q)$ (Meredith, 1993) should be taken, representing for this example $\delta f' = 0.01$, or 1.0% bandwidth. For a large cavity with 1%-bandwidth FoM of 100, this gives the total number of contributing modes as 100 in a 1% bandwidth. In practice, the microwave generator has a spectral width of about 0.1 %, so the actual number of contributing modes increase further to 110. Such an oven could be expected to give a good uniformity of heating, with the techniques of Section 5.7 applied.

At the other extreme of a small oven with FoM of 4, there would, on this basis, be 4.4 contributing modes. This figure is, of course, a probability; the most likely being between 3 and 6 modes, but could be more or even zero, depending on the precise mode resonant frequencies relative to that of the generator. It means that the oven would perform badly with this load in uniformity, and its efficiency would be low. Such an oven would only be satisfactory with higher-loss/larger loads.

The estimation of Q-factor for workloads partly filling an oven is difficult, but a very approximate relation has been derived by Metaxas and Meredith (1993) using eqns. 1.17 and 5.3. for spherical load:

$$Q=\frac{\varepsilon'}{\varepsilon''}\left\{1+\frac{(\varepsilon'+2)^2}{9\varepsilon'}\left(\frac{1-v}{v}\right)\right\} \tag{5.6}$$

where v is the volume filling factor = (volume of the load)/(volume of the cavity).

Note that this expression shows that the Q-factor increases with ε' at constant ε'', but decreases as v increases. Eqn. 5.6 should be used only as a rough guide; assumptions are made in its derivation that the field in the cavity is uniform, and that the microwave-penetration depth is large compared with the radius of the sphere.

In small ovens, with loads of dimensions small compared with $\lambda_0/4$, the Q-factor depends critically on the location of the load. Generally, workloads have low impedance (because $\varepsilon' > 1.00$) and so match to zones of similar low impedance. This means that they will couple more favourably to zones where there is a standing-wave voltage minimum, and, conversely, least favourably in the vicinity of a voltage maximum. Thus the effective heating varies significantly with position. For loads of sheet form, Paolini (1989) calculates Q-factor variations >30:1 for the extreme, and more generally >6:1 for a set of 16 modes, illustrating the wide variation of coupling between the load and individual modes.

Figure 5.3 is a set of curves from eqn. 5.6 showing the Q-factor for selected conditions.

A method commonly used to limit the effective Q-factor is to make the oven with a metal, e.g. stainless steel, having low electrical conductivity. This results in a significant proportion of the input power being dissipated as I^2R losses in the walls of the oven, but it means that the Q-factor is at a low value, typically 500, even when the oven is empty. The advantages are an improvement in uniformity through more modes contributing, and protection of the generator from damage due to the severe mismatch otherwise occurring when operated with the oven empty. The loss of

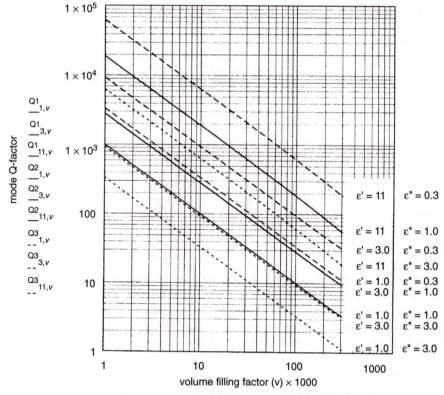

Figure 5.3   Mode Q-factor for a spherical load against volume filling factor and dielectric properties based on eqn. 5.6

efficiency is usually unimportant, because the Q-factor with the workload present is usually much lower, a situation typical of domestic and catering oven operation.

5.5 Equivalent circuit of a multimode cavity

An equivalent circuit of a complicated electrical system is often helpful in understanding and identifying important features affecting its performance, and this is especially true of the multimode oven. Each mode resonating close enough to the generator frequency to affect the performance can be represented as a simple conventional series-tuned circuit. Its input admittance $Y(\omega)_{lmp}$ is given by

$$Y(\omega)_{lmp} = \cfrac{1}{R_{lmp} + j(\omega L_{lmp} - \cfrac{1}{\omega C_{lmp}})} \tag{5.7}$$

where R_{lmp}, L_{lmp}, and C_{lmp} have their conventional meanings, with R representing all the losses coupled to the l,m,p mode, and L and C being the reactive elements whose values are determined by

$$Z_{0\,lmp} = \sqrt{\frac{L_{lmp}}{C_{lmp}}} \qquad and \qquad f_{R\,lmp} = \frac{1}{2\pi\sqrt{L_{lmp}C_{lmp}}} \tag{5.8}$$

where f_{Rlmp} is the resonant frequency of the lmp mode and Z_{0lmp} its characteristic impedance. The characteristic impedances of the TE and TM modes differ for the same values lmp and are given (Ramo *et al.*, 1965; Meredith 1993) by

$$Z_{0TM} = \sqrt{\left[\frac{\mu_0}{\varepsilon_0}\left\{1-\left(\frac{\lambda_0}{\lambda_{c\,lmp}}\right)^2\right\}\right]} \qquad and \qquad Z_{0TE} = \sqrt{\left[\frac{\mu_0}{\varepsilon_0}\cfrac{1}{\left\{1-\left(\frac{\lambda_0}{\lambda_{c\,lmp}}\right)^2\right\}}\right]} \tag{5.9}$$

where $\lambda_{c\,lmp}$ is the cutoff wavelength of the lmp mode.

These values of Z_{0TE} and Z_{0TM} are the mode characteristic impedances Z_{0lmp} given in eqn. 5.10. For each mode, the factor L/C has a particular value depending on its proximity to cutoff.

Note that if $\lambda_0 \ll \lambda_c$ the characteristic impedances converge towards $(\mu_0/\varepsilon_0)^{0.5}$, the characteristic impedance of free space.

However, if λ_0 approaches cutoff (λ_c), $Z_{0TM} \to 0$ and $Z_{0TE} \to \infty$. These important results mean that TM modes excited near their cutoff frequency have very low impedances and the complementary TE modes have very high impedances. Such modes have a low value for one or more integers l, m, p, indicating a long waveguide wavelength. The low-impedance TM modes will tend to couple more tightly to loads of high ε' than the TE modes. Because of their low impedance they will have high H-field components with correspondingly high wall currents, and will therefore reduce efficiency by $I^2 R$ heating of the cavity walls. However, the TE modes will tend to have high E fields which may be a source of arcing in the oven. In a large oven it is certain that there will be some modes excited near their cutoff frequencies; however, the extent of their excitation depends on the feed system and their effect may be reduced if the design favours the excitation of modes (in the majority) far from cutoff.

In a small domestic oven with very few relevant modes, the presence of a mode excited close to its cutoff frequency can be a cause of wide variations in performance under apparently identical conditions, because the cutoff frequency is determined by the dimensions of the oven and the load, the location of the load in the oven, and the values of ε' and ε'' of the load. A small change in any of these parameters will affect λ_c. Inspection of eqns. 5.9 shows that, if the mode is close to cutoff, a small change in λ_c will cause a very large change in impedance and consequently in oven performance.

A simple equivalent circuit of an oven based on the above points is shown in Figure 5.4. Note that this equivalent circuit is an electrical engineer's image of the system using conventional electric-circuit components; none of these components exists in real hardware.

The circuit comprises two sets of resonant circuits representing the TE and TM modes, respectively. Each is fed via an ideal transformer with turns-ratio determined by the coupling to the source, i.e. the configuration of the feed. All the circuits are connected in parallel at the aperture through which the generator feeds the oven. As a parallel circuit, the overall input admittance Y_{in} is simply the vector sum of the admittances of the individual resonant circuits. For a series-tuned circuit, the admittance at frequencies far removed from resonance tends to zero and so the circuit makes no contribution.

For a large oven with a high FoM there will be many modes excited under a given set of conditions of load. The feed will have been designed to give a good match to the generator, implying that the normalised vector sum of the admittances of the modes is close to $(1+j0)$. Because there will be roughly as many contributing modes resonating above the generator frequency (negative susceptance) as below (positive susceptance), the overall load susceptance will tend to zero. Also all the modes are undercoupled because the arithmetic sum of their conductances is nearly unity because of the feed design. As the load conditions change, or if the generator frequency is

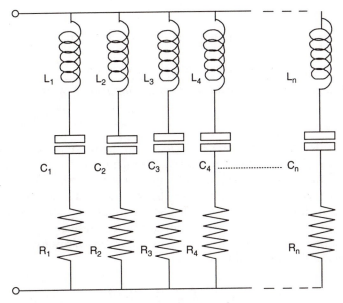

Figure 5.4 Simplified equivalent circuit of multimode oven

changed, roughly equal numbers of modes will advance towards the generator frequency and recede from it. This means that the match 'seen' by the generator is substantially constant, or broadband. In practice, the match of a large multimode oven is usually good, with VSWR < 1.5, over a very wide band; within the 1.5 VSWR circle there are rapid fluctuations in both phase and amplitude. Even when empty of workload, large ovens usually display a good impedance match, due to TM modes near cutoff coupling efficiently to the surface resistance of the walls.

An observed effect with large ovens having multiple feed points is the high isolation, or loose cross coupling, between them, typically in the range –20 to –30 dB. The reason is the same as above, that the overall transmission coefficient between two ports is the vector sum of the coefficients of the individual modes, which are random in both phase and amplitude, giving a resultant tending to zero. This is a very important property, permitting multiple-magnetron operation with minimal cross-coupling effects between the generators.

5.5.1 Microwave efficiency

The microwave efficiency η_m per cent of an oven is defined as

$$\eta_m = \frac{\textit{Power dissipated in the workload}}{\textit{Microwave power incident upon the input port of the oven}} \times 100\%$$

(5.10)

This definition includes power reflected from the oven, power dissipated in the walls and other hardware in heating through the mechanisms of I^2R, $\varepsilon''E^2$ and $\mu''H^2$, and losses in resistive choke absorbers and terminating loads. The definition excludes transmission losses between generator and load, and the efficiency of the generator itself. It also excludes any nonmicrowave heat input to the oven for the control of surface heat transfer.

Calculation of the resistive losses in the oven hardware requires the evaluation of the surface integral

$$P_{I^2R} = \frac{1}{2\delta_s\sigma} \int\limits_{Surface} H_t^2 dS$$

(5.11)

and for the stray dielectric and magnetic field losses the volume integrals, of similar form, over the volumes of the respective parts. In practice, the latter losses are usually negligible and can be ignored; however they may be significant if ferromagnetic materials, ferrites or 'microwave receptors' are used. Microwave receptors are sometimes used in the food industry to provide additional heating in 'cold spots'.

Where the oven is operated aperiodically (i.e. $Q < 10$) the evaluation of eqn. 5.11 is straightforward and with conventional materials usually results in I^2R losses less than 5% of the power input. With high values of Q ($Q > 100$) the value of H_t can be difficult to assess because of the wide variations of coupling between the load and specific modes; where the coupling is weak, an individual mode may have a very high associated magnetic field if excited close to resonance. The resistive losses would then be high, and ovens operated in this way are prone to wide variations in efficiency depending on the proximity to resonance of modes of high Q-value.

Again, with ovens operated substantially aperiodically, the reflected power is small from an optimally designed feed and should not exceed 5%. In high-Q operation, the reflected power is a rapidly varying function of generator frequency, but rarely exceeds 50%, except in a small oven with very low mode-spectrum density when there can be significant gaps in the mode coverage in the frequency domain.

It will be appreciated from the above that estimation of microwave efficiency for a multimode oven is difficult. Numerical modelling provides the best route, but, especially in high-Q operation, great care must be taken to evaluate over a range of operating conditions, particularly of frequency spread.

5.6 Heating uniformity

Intuitively, the greater the number of modes available for excitation the better is the possible uniformity of heating. A helpful approach to this view of uniformity is to consider the modes as a mathematical series, analogous to a Fourier series, where the harmonics (the modes) add to give a desired overall spatial distribution, e.g. a rectangular waveshape. The analogy with a Fourier series is helpful for two reasons. First, it is well known that a harmonic series with a limited number of harmonics can only give an approximation to a desired waveshape; for example the first five harmonics of the Fourier series of a square wave gives a square waveshape with rounded corners, sloping flanks and a ripple on the 'flats'. In a multimode oven, the mode series is limited as shown in Sections 5.3, and so the uniformity of heating cannot be perfect. Secondly, the Fourier series for a symmetrical rectangular waveshape contains odd harmonics only, even harmonics imparting an asymmetry. By analogy, it is preferable to excite, if possible, only those modes with odd-value mode numbers; this is especially relevant for ovens where the workload is a continuous bed on a conveyor band, where the preferential excitation of odd-number modes transversely to the band should be the objective.

Many factors affect the uniformity of heating of a load in a multimode oven, but they can be divided broadly into two groups: those primarily a function of the workload itself, and those primarily associated with the oven, with a broad range of overlap where both are involved.

5.6.1 Workload limitations

Obviously, a workload which is thick compared with the microwave penetration depth D_p (Section 2.4) will have an inherent exponential heating nonuniformity. Such conditions result in aperiodic operation unless the volume of the workload is a very small proportion of that of the oven (e.g. < 10%). Depending on the size of the oven relative to the workload, and the location of the microwave feed points, the exponential may extend from more than one face of the workload. For discrete workloads of simple rectangular shape, the sum of the exponentials from each face results in addition at the edges and especially at the corners, so that a high temperature rise occurs along the edges, peaking at the corners. This effect is often enhanced by workloads of high permittivity ε' due to field-distortion effects at edges and corners (refraction), which cause vector addition of the E-field components. If the workload cannot be changed suitably, reconsideration of the choice of applicator is necessary.

Where the workload has a high penetration depth relative to its size, resulting in relatively high Q-values for the modes, the effect of its permittivity is again to cause refraction at the corners and edges giving enhanced heating in these regions. Moreover, there will be internal reflections from the surface of the workload which will cause standing waves within the volume of the workload. These standing waves, positioned with the workload surface as the datum, create a sinusoidal heating pattern; the peaks at the E-field maxima are separated by $\lambda_0/2\sqrt{\varepsilon'}$, where ε' is the permittivity of the workload material.

Workloads comprising a flat bed of material on a travelling conveyor are commonplace. The linear motion of the band inherently eliminates the effect of longitudinal nonuniform heating. Usually, the band is itself a dielectric material because it offers no constraint to the orientation of the E-field vector. If a metal conveyor must be used because, for example, of temperature limitations of a dielectric conveyor, then the boundary condition that E_{tan} is zero at the surface of the band must be satisfied, permitting only the E_{norm} component to exist. This constraint does not necessarily exclude use of a metal band, and is discussed in Chapter 6. Where the workload is on a dielectric conveyor, illumination can be provided from both above and below the conveyor band. This is especially desirable if the penetration depth is comparable to, or greater than, the thickness. However, in practice feeds below the band should be avoided if possible because of foreign particles falling. If there is freedom to adjust the thickness of the workload on the conveyor to less than the penetration depth, this is preferable to providing a below-the-band feed. The distribution of heat across the band is a function of feed design. The optimum design of feeds for use with conveyor ovens is described in Section 5.6.3

Particulate workloads may suffer nonuniform heating through concentration of displacement current around the points of contact (or near contact) between individual particles, as illustrated in Figure 5.5. The relatively large surface area of the particles 'captures' the E field, and the resulting displacement current is concentrated to flow through the contact point (or narrow gap) between the particles. The concentration may be very intense, depending on size, shape and ε' for the particle; the resulting E field causing intense local heating which can degenerate into a localised arc, as considered in Section 2.6.2 The magnitude of the effect is also dependent on the thermal properties of the workload, which influence how heat is conducted away from the hot spot into the main thermal mass of the particle. When this heat flow equals the rate of heat dissipation at the hot spot, then there is equilibrium. Local geometry and the thermal diffusivity of the particle are critical parameters. This effect places a limit on the maximum safe power-dissipation density in the workload; it is a very difficult effect to quantify and requires experiment to establish. Indeed it is an effect of probability as the particles are likely to be of random shape, and certainly of random disposition on the conveyor. The user must decide what would be an acceptable incidence of the effect in his product, which at worst appears as

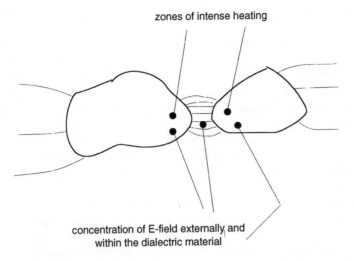

zones of intense heating

concentration of E-field externally and
within the dialectric material

Figure 5.5   Field concentration between particulates in near contact

tiny burn-marks. Usually the desired power-dissipation density is much less
than that initiating this effect. Materials particularly prone to localised
overheating are vegetables, especially brassicas, whose low thermal diffusivity
and irregular shape and structure, with high values of ε^*, combine all the
conditions above to create the effect. Root vegetables, carrots, potatoes etc.
are also liable to localised overheating. For all these, the limit on power-
dissipation bulk density has been found to be typically in the range 30–100
kW/m^2 at 915 MHz.

If the proposed process is drying, the nonuniformity may be unimportant
if the workload dry matter has a low loss, because the heating pattern will
change as drying proceeds, the heat being dissipated mainly at the highest
residual moisture exposed to the microwave field. However, if the drymatter
is not low-loss, heat will continue to be dissipated after dryness, causing
temperatures exceeding 100^0C.

5.6.2 Oven effects

A lone individual mode in a multimode oven will clearly impart on the load
a sinusoidal heat distribution corresponding to its mode number. In practice,
all the modes resonating close to the excitation frequency will be excited with
various amplitudes and phase relationships depending on the configuration
of the oven and workload, and the details of the feed system. At a chosen
point in the oven, the vector sum of all the E-field contributions from each of
the modes is the resultant value E_R, dissipating heat in the workload at a
power density given by eqn. 2.7 where E_{Ri} is substituted as the electric field
within the workload.

From the above, it is clear that the uniformity of heating is inherently dependent on the FoM for the oven; the oven should be as large as possible to secure a high spectral density of modes. However, the oven volume should be commensurate with the volume of the workload because the efficiency may fall (Metaxas and Meredith, 1993). Following this guideline, rectangular ovens for conveyorised workloads are of low-height, of width 20–30% greater than the conveyor, and long in the direction of motion. For a spherical or cubic load, the oven would be of similar volume but a quasicube to avoid degenerate modes. Obviously, the actual dimensions must be great enough to secure the necessary spectral density, and for conveyorised loads to have a width factor $2b/\lambda_0 > 5$, sufficient to allow enough transverse halfcycles to be able to achieve the requisite uniformity.

For the configuration of a laminar workload filling the cross-section of the oven, Paoloni (1989) derives the heating distribution by summation of the modes with resonant frequency close enough to the generator frequency to contribute, calculating the Q-factor of each mode from the loss factor of the load and the losses in the cavity walls. The excitation of each mode is based on matching the aperture distribution of the feed to a summation of the modes at the aperture plane, equivalent to a Fourier series synthesising a waveshape of the feed. Paolini considers a plain waveguide feed and shows that the coupling to each mode is different, and is a function of the position and orientation of the feed in the oven wall. Such variation is to be expected because coupling will obviously be tight to a particular mode where the feed coincides with a position of maximum H-field for that mode. However, another mode having a zero H-field aligned with the centre of the feed would have zero coupling. Both are realisable conditions, and by repositioning the feed the coupling levels could be reversed. The choice of feed and its position have first-order effects on the heat distribution in the workload. Paolini quantifies the coupling coefficient to each mode and shows that it is also dependent on the loss factor of the load.

A variation on Paolini's procedure, made practical by the advent of mathematical computer programs such as Mathcad 6 Plus®, is to consider his configuration in three zones as shown in Figure 5.6: first the empty cavity between the feed face and the laminar load, secondly the load, and thirdly the space between the load and the remote face of the cavity. Each zone is a waveguide transmission line for each mode, lossy for the centre zone comprising the load, but otherwise lossless in the outer zones, unless wall losses are to be included. Thus for each mode there are three cascaded transmission lines, the third terminated in a short-circuit, for which the propagation characteristics are readily calculable. Using the lossy-transmission-line equations the input admittance of each mode is easily calculated, and for each the coupling to the feed using Paolini's procedure.

It is evident from the above waveguide-mode model that there will usually be some resonant modes close to their cutoff frequency. The contribution of these modes may be significant, especially if the oven is lightly loaded; a very

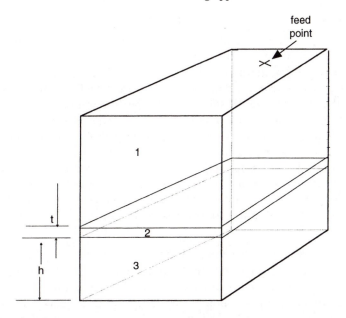

Figure 5.6   Multimode oven with laminar load

minor perturbation in the internal oven profile may change the cutoff frequency, altering the amount of contribution from that mode. Such a perturbation can be introduced by an access door in the oven wall and has been observed to cause a major asymmetry in the heating pattern of a conveyor oven. Moreover, the characteristic impedance of the modes changes very rapidly near cutoff (Meredith, 1993), and so their coupling to both feed and load may change critically.

Considering the transverse heating uniformity, the primary aim is to secure symmetry relative to the axial centreline of the conveyor, and then to minimise the residual nonuniformity. Axial symmetry is achieved by ensuring that the axis of the conveyor and its workload coincide with the longitudinal axis of the oven. Structural asymmetry introduced from brackets, supports or other hardware must be avoided. Access doors must be designed to appear as a continuous sheet coplanar with the oven wall, with no electrical discontinuity around the periphery. Manufacturing tolerances require careful scrutiny to ensure, inter alia, flatness of surfaces; stability is an essential feature and the walls must be made of thick enough material, adequately supported, to withstand industrial duty without distortion.

5.6.3 Feed systems

In practice most multimode-oven feed systems are simple plain waveguides feeding through the wall of the oven, possibly with a dielectric window to prevent steam or corrosive gases entering the waveguide. The location within the oven wall is often determined by mechanical constraints without consideration of electrical optimisation, resulting in an oven with poor heating uniformity and low efficiency.

The feed requirements of a batch oven are different from those of a continuous, conveyor oven, and the latter are considered first. The rectilinear motion of the load inherently gives, overall, perfect uniformity of heating along the axis of motion. The only reservation to this assertion is where a time-dependent process is involved, e.g. a chemical process; the final temperature rise may be very uniform across the width of the conveyor, but the progress of the reaction may vary because the time profiles of the temperature rise of the workload as it passes through the oven may be different along different parallel paths. For these ovens, achieving good heating uniformity oxial to the conveyor is important.

5.6.3.1 Feed systems for continuous-flow ovens

Symmetry of structure is an essential requirement for good uniformity of heating in a conveyor oven, as discussed in Section 5.6.2. Symmetry of the feed arrangements is equally important. If the oven is heavily loaded to operate aperiodically, a single feed mounted on the centreline above the conveyor will give a heating pattern roughly corresponding to the radiation pattern of a waveguide radiating into space: maximum heating on the centreline, falling towards the edges of the conveyor. Where the oven is fed from several generators each of low power, placing their feeds symmetrically and equispaced about the centreline provides an initial arrangement. It is good design practice to provide a mounting flange across the width of the oven to which a plate incorporating the feed system can be bolted, enabling feeds of differing design to be fitted easily as experimental development proceeds.

For heating a thin laminar load, the coupling is strongest to those modes with E-field vectors in the plane of the load; the option remains of arranging the feeds to provide the E-field parallel to the axis or orthogonal to it, the latter being marginally preferred because the inherent sinusoidal distribution of the E-field at the feed aperture is then aligned with the motion of the conveyor and so is self-cancelling. Experimental trials are essential for optimising the feed system; numerical methods may also prove helpful.

Particularly if high-power generators are used, the feed system can be formed of slotted waveguides, similar to the slotted waveguides used in radar antenna practice. With a slotted waveguide it is possible to create almost any desired distribution of power radiated by adjustment of the positions of the slots in the waveguide wall. This is especially applicable to aperiodic conveyorised ovens where the axis of the slotted waveguide would be mounted transverse to the direction of travel. An array of slots is cut in the waveguide (Silver, 1984), spaced apart by $\lambda_g/2$ where λ_g is the waveguide wavelength in the feed waveguide, the slots themselves being nominally $\lambda_0/2$ long, to be resonant. The essential principle is that the slots, being resonant, have a purely conductive admittance and, being spaced apart by $\lambda_g/2$ in the feed waveguide, their conductance values g_s add arithmetically to give the input admittance match 'seen' at the input, i.e:

$$y_i = 1 + j0 = \sum_1^S g_s \qquad (5.12)$$

The radiation pattern from the feed is determined by the number of slots, and the efficiency by their conductance, because this controls the proportion of the total power radiated from the slot. Silver gives relations for the conductance of resonant slots as functions of their position in the waveguide. Typical slots and their conductance relations are shown in Figure 5.7 and eqns. 5.13 and 5.14.

Two types of resonant slot are used as radiators from a waveguide: the longitudinal broadwall slot, and the inclined sidewall slot as illustrated in Figure 5.7. Both slots radiate by intercepting the transverse wall current in

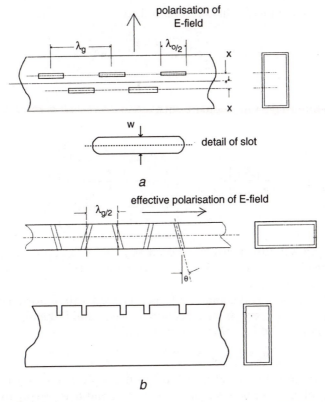

Figure 5.7   Slotted radiating feeds

a Broad-wall radiating slots
b Narrow-face radiating slots

the waveguide, or the grazing magnetic-field components H_x and H_y. The longitudinal broad-wall slot has zero coupling when located on the centreline of the broad face of the waveguide because $H_x = 0$ at this position. Moving the slot sideways towards the narrow wall increases the radiated power because H_x increases in value sinusoidally with distance. Moreover, H_x has a 180° phase shift from one side of the centreline to the other; by placing the slots alternately about the centreline the 180° phase shift within the feed waveguide between locations $\lambda_g/2$ apart is cancelled so that all the slots radiate in the same phase. The longitudinal broad-wall slot radiates with E-field polarised normal to the axis of the feed waveguide.

The inclined side-wall slot has zero coupling if the slot is normal to the broadface and intercepts the grazing magnetic field H_y progressively as the angle of inclination to the vertical is increased from zero. There is a phase reversal between positive and negative inclination angles, and this effect is used to achieve in-phase radiation from all slots, similarly to the broad-wall slot.

The slot conductances g_s are given by (Silver, 1984):

(i)   Broad-wall longitudinal resonant slots:

$$g_s = 2.09 \frac{\lambda_g}{\lambda_0} \frac{a}{b} \cos^2\left(\frac{\pi \lambda_0}{2 \lambda_g}\right) \sin^2\left(\frac{\pi x_1}{a}\right) \tag{5.13}$$

(ii)   Inclined side-wall resonant slots:

$$g_s = \frac{30}{73 \pi} \left(\frac{\lambda_g}{\lambda_0}\right) \frac{\lambda_0^4}{a^3 b} \left\{ \frac{\sin\theta \, \cos\left(\frac{\pi \lambda_0}{2 \lambda_g}\sin\theta\right)}{1 - \left(\frac{\lambda_0}{\lambda_g}\right)^2 \sin^2\theta} \right\}^2 \tag{5.14}$$

The two types of slot shown are chosen as giving E-field polarisation parallel to the feed axis (Figure 5.7a) or orthogonal (Fig 5.7b), and the slot conductances in Figure 5.8.

For the simplest and commonest case of a feed with uniform illumination, all the slots would have the same conductance g_s, and if the width of the oven requires N slots the feed would be matched if

$$N g_s = 1 \text{ or } g_s = 1/N \tag{5.15}$$

The slot width is uncritical in its effect on coupling, but affects the power handling of the slot as discussed below. The width is usually in the range 10–20% of the length. The resonant wavelength is more precisely twice the perimeter of the slot, so that wide slots are a little shorter than narrow slots

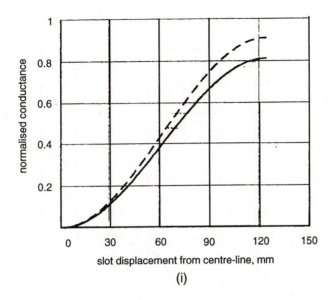

(i)

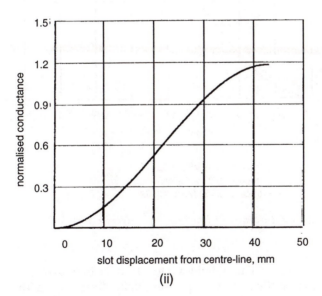

(ii)

Figure 5.8a   Slot conductance: radiating slots in broad face of rectangular waveguide

(i) WG4: ——— 896 MHz;   ---- 915 MHz
(ii) WG9A, 2450 HHz

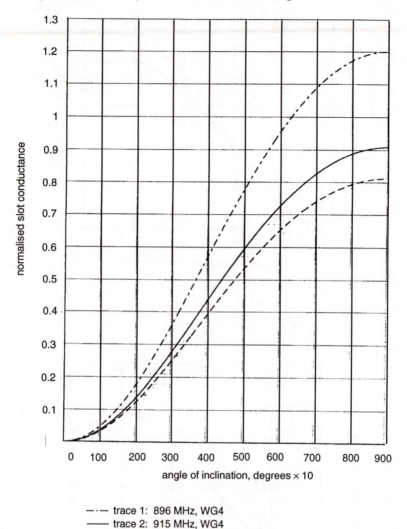

normalised slot conductance

angle of inclination, degrees × 10

—·— trace 1: 896 MHz, WG4
——— trace 2: 915 MHz, WG4
---- trace 3: 2450 MHz, WG9A

Figure 5.8b   Slot conductance: inclined resonant slots in narrow face

which tend to $\lambda_0/2$. The E-field vector is polarised orthogonal to the longitudinal axis of the slot, varying sinusoidally along the length of the slot to be at a maximum at the centre. The power P_s (watts) radiated from a resonant slot is derived by Silver (1984):

$$P_S = 73 \left(E\,w\right)^2 \left(\frac{\varepsilon_0}{\mu_0}\right) \tag{5.16}$$

where E is the peak instantaneous voltage stress (volts per metre) at the centre of the slot and w is the slot width (metres).

(NB: the constant 73 has the dimension ohms, making the equation consistent dimensionally.)

Table 5.1 shows some resonant slot sizes at the principal frequencies, and the power radiated at a peak E-field value of 10^5 V/m (1000 V/cm).

Table 5.1 Radiating slot sizes

f	w	l	P (at 1000V/cm)
(MHz)	(mm)	(mm)	(W)
2450	2.9	58.1	43
2450	5.5	55.4	155
2450	7.9	53.0	320
2450	10.2	50.8	534
2450	12.2	48.8	764
915	7.8	156.0	312
915	14.9	148.9	1140
915	21.4	142.4	2352
915	27.3	136.5	3827
915	32.7	131.0	5492
896	8.0	159.3	329
896	15.2	152.1	1187
896	21.8	145.5	2441
896	27.9	139.4	3998
896	33.4	133.8	5729

Slot ends may be semicircular for ease of cutting, retaining the total slot periphery at λ_0 as above. The edges of the slots should be rounded to minimise field concentration at sharp edges

An alternative to the slotted array is a waveguide-horn feed, which has been widely used in conveyor ovens, giving good uniformity of heating, especially where the oven is loaded aperiodically, or with $Q < 20$. The horn extends across the oven from one side-wall to the other, and is designed to give a uniform illumination at the entry wall of the oven, with the E-field polarised normal to the side-walls, i.e. transverse to the conveyor. In principle, the horn tapers the feed waveguide as shown in Figure 5.9 so that the broad dimension remains constant and the E-field is 'stretched'. Recalling that the E-field E_y in the TE_{10} mode in the waveguide is of constant amplitude between the broad faces, the taper will continue to give a substantially constant value E_y-field within the horn, provided that higher-order modes are suppressed. As in the waveguide, the value of E_y varies sinusoidally in the x direction, but this is aligned with the axis of travel and its variation is therefore self-cancelling. As the cross-section area of the horn

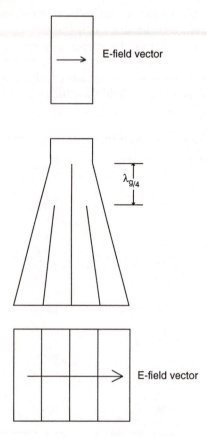

Figure 5.9   Transverse E-field horn feed with three vanes

increases, the power density falls, and so the amplitude of the E-field falls, but this does not affect its uniformity.

As the cross-section size of the horn increases, progressively more unwanted modes can propagate. These are easily eliminated by vanes mounted inside the horn, extending from the throat to the aperture, mounted parallel to the broad faces of the feed waveguide. In effect, they divide the horn into a stacked set of subhorns whose apertures have dimensions corresponding to a rectangular waveguide supporting the TE_{10} mode only. It is important to maintain mechanical symmetry and tolerances in the horn, because the power radiated from each subhorn is proportional to its fractional height at the throat; it is also a function of the inequality of the individual impedances reflected from the loaded oven at the throat by each of the subhorns.

In practice, the vanes are staggered so that they do not all converge on the feed waveguide at one plane. If they were to do so, a substantial fraction of the waveguide height would be blocked by the combined thickness of the vanes, resulting in a large amount of reflection, and reducing the power-handling capacity at this point. Typically, there is a centre vane bisecting the structure, then each half so formed is bisected again by a second vane, starting from a point before the dimensions are such as to support the first possible, unwanted mode, ideally a distance $(2n + 1)\lambda_g/4$ from the start of the centre vane so as to cancel the respective reflections.

The path length from throat to aperture obviously increases from the centre of the aperture of the horn to its outer extremities, giving rise to a taper in the phase angle across the aperture. In practice, this has not been significant, but could be corrected by adjusting the waveguide wavelengths in the subhorns to equalise their electrical lengths; this can easily be acheived by adjusting the broad dimension of the subhorn, or by fitting dielectric phase-shifting slabs.

The advantages of the horn feed are its simplicity and its high power-handling capacity, not much less than the feed waveguide itself. It is, however, a large component frequently incompatible with equipment layout, although a folded version in which the subhorns terminate in a set of H-plane bends has proved successful in mitigating this problem.

5.6.3.2 Feed systems for batch ovens

There is no single design of feed system for a batch oven which will optimise performance irrespective of the load, and a 'general-purpose' microwave oven is essentially a compromise. In general, the best compromise is to provide a feed system which gives a good performance in heating uniformity when the oven is loaded aperiodically with a load having a volume-filling factor of about 20%, and then accepting the resulting performance on lighter loads with the oven responding as a multimode with modest Q-factors for the modes. Again, to improve heating uniformity, multimode ovens are often fitted with mechanical means to rotate or oscillate the load, and also 'mode stirrers', described in Section 5.7; these techniques also affect the design of the feed system.

Most industrial batch ovens have a rotating table, with a slow rotation speed but sufficient to give at least 30 turns during the microwave exposure. For a circularly cylindrical load mounted concentrically with the rotation axis, this results in excellent circumferential heating uniformity. Heating uniformity along the axis is then determined principally by the feed system. Many ovens have overhead feeds, which, in a heavily loaded oven, tend to heat the top surface of the load preferentially. If this is the normal load, additional feeds in the side walls improve uniformity. Some control of heat distribution is possible by adjusting the distance of the feed waveguide or horn from the surface of the workload; it is not a fundamental requirement that the feed aperture be in the plane of the oven wall. Figure 5.10 shows this arrangement in which the clearance between the feed-aperture plane and the

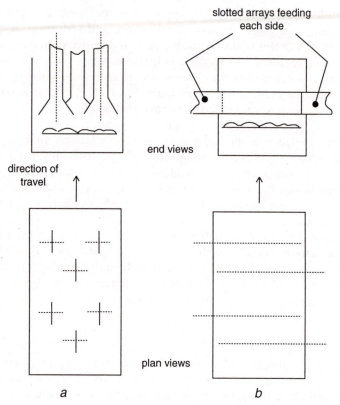

Figure 5.10   Extended feeds into multimode oven

     a Horns
     b Slotted arrays

workload is adjustable by adding or removing short sections of waveguide. Several feeds, or a slotted array waveguide (Section 5.6.3.1), may further enhance performance; design is essentially an empirical procedure, with intelligent interpretation of test results leading to further modification of the feed system until an acceptable performance is achieved.

5.7 Techniques for improving heating uniformity

Various secondary methods are used to enhance heating uniformity, but it is essential that any oven be designed at the outset to give good uniformity in their absence. They can then be used for final refinement.

5.7.1 'Mode stirrers'   徒文式

So-called 'mode stirrers' are rotating/reciprocating mechanical devices which take a variety of forms at the whim of the designer. In a lightly loaded oven they have the effect of modulating the resonant frequencies of the modes as they move, and in so doing bring into effect modes marginally outside the operating spectrum. In a quasiaperiodic oven they are used as scanning deflector plates, mounted near the feeds.

A mode stirrer may be considered as a secondary antenna within the oven, coupling to the existing fields and reradiating a secondary pattern which constantly changes with rotation. Alternatively, it can be considered as a structure with an admitttance $Y(\theta) = 0 + jB(\theta)$ where θ is the angle of rotation of the stirrer, and the magnitude of the effect of the stirrer is the range of $B(\theta)$ as the stirrer rotates. Obviously the effect of the stirrer is a function of its size, in comparison with both the wavelength and size of the oven. Considering the stirrer as an antenna, an effective diameter of $\lambda_0/2$ would give optimum coupling. However, the stirrer is usually larger than this as its size in proportion to the oven is also important; the diameter may be up to 70% of the shortest side of the oven. More than one mode stirrer may be used, preferably rotating at nonintegrally related speeds. There is no formal design procedure for mode stirrers, and experience and experiment are necessary.

Care must be taken in positioning mode stirrers, especially where a conveyor oven has been designed with symmetry as the principal feature. The mounting position must retain the symmetry. Mode stirrers should not be mounted so close to a feed as to obscure it at any angular position, because this may cause large fluctuations of reflected power as the stirrer rotates, affecting the efficiency, and possibly the generator, adversely.

Usually mode stirrers are rotating swash plates, or multibladed fans, the latter also providing some air movement. The drive shaft passes through the wall of the oven to a motor–gearbox assembly, via a microwave choke to prevent leakage. The choke must reduce leakage to a safe level for personnel exposure, and in some applications to radio-interference level; it must also provide a path for currents induced in the shaft to flow to the wall of the oven without arcing. There is no precise calculation for the required attenuation because the local field near the shaft inside the oven is very dependent on the load in all its respects. The worst possible case would be for all the generator power, reduced by the cross coupling between the generator port and the shaft (considered as another port), to couple to the shaft. From practical observation this cross coupling would be about –20 dB, and a factor of safety of 10 dB should be taken, giving the power coupled to the shaft as –10 dB relative to the generator, i.e. 10% of the generator power. For a 10 kW oven this gives 1 kW coupled to the shaft. Taking the easiest case of personal exposure of 10 mW/cm^2 at 50 mm distance, this power intensity is given by a point source of 1.6 W radiating uniformly into the half space, which is –28 dB relative to 1 kW. On this basis the choke should have at least 28 dB attenuation. In practice, it is easy to achieve this attenuation, and the usual criterion of choke design is to provide as high an attenuation as possible (typically 50 dB) without a significant effect on cost. Where radio-interference limitation is the objective, a very much higher attenuation is required, and the best practice is to enclose the whole drive assembly in a fully screened box with electrical connections individually filtered at the entry to the box.

In the simplest choke the shaft is made of low-loss dielectric material passing through an extended metal sleeve (which may also be a journal bearing), forming a cutoff waveguide as discussed in Section 4.6. In Figures 4.11 and 4.12 are curves showing the attenuation rate in decibels per metre of circular waveguides operating in the fundamental TM_{11} mode, for various dielectric fillings. The length of the choke is determined from the required attenuation, to which should be added a further length at least equal to the diameter of the choke, because the field distortion in the vicinity of the open end reduces the attenuation locally. Note that higher-order modes will have higher attenuation than the fundamental and so can be ignored. This is a very effective system because it is easy to design, and there are no surface currents returning from the shaft to the oven wall. Its disadvantage is the small diameter of shaft which may not, as a plastics material (e.g. polypropylene or PTFE), be sufficiently robust to withstand the duty. Ceramics and glass can be used but they are prone to damage from shock.

An alternative electrical choke is to consider the metal shaft as the inner of a coaxial line propagating the TEM mode, and to form a low/high/low-impedance choke as described in Section 4.8.3. The low-impedance $\lambda_0/4$ sections have a small radial clearance between the shaft and outer sleeve, which may incorporate a dielectric ring as a bearing and concentricity locator. The high-impedance section has a large-diameter outer sleeve, and the inner may also be reduced in diameter if mechanical design permits.

Simple mechanical-contact 'chokes' can also be used where the current return from shaft to oven wall is within the capability of a sliding contact. Lubrication is a problem because an oil film impairs the contact, which must have 100% integrity all round the shaft. Various systems have been used including woven stainless-steel contact rings, carbon (graphite) rings, and spring fingers of beryllium–copper, and ball-bearing races. Wear and risk of mechanical seizure are obviously prominent design considerations. Current density at the contact, as a 'rule of thumb', should not exceed 0.1 A/cm. For the above 10 kW oven, the current density would be in the order of 1 A/cm for a 20 mm shaft.

5.7.2 Modulation of the feed system

It is observed in Section 5.6 that the coupling from the feed to the individual modes is dependent, inter alia, on the position and orientation of the feed.

If the feed comprises two ports which have different coupling factors to the modes, then an electrical means to divide the input power cyclically between the ports has the effect of changing the relative excitations of the modes. One method is to provide a power divider (Figure 4.20) arranged so that the output ports 2 and 4 are connected to the two feeds in the oven; the phase shifter is driven continuously to oscillate the power between the feeds through at least 50 mechanical cycles during the microwave exposure of the load.

Another technique is to rotate the plane of polarisation of the wave emitted from a circular feed operating in the fundamental TM_{11} mode. In this case, the single feed port is effectively two ports with mutually orthogonal planes of polaristaion. The feed is driven from a mechanical 'Faraday rotator'

The above two systems have the advantage of no moving parts within the oven. Another method is to arrange a mechanically rotating feed in which the electrical centre of the feed is displaced from the mechanical axis. It requires a waveguide rotating joint (Harvey, 1963), which is an expensive component.

Modulation of the feed is a useful and effective technique where the workload is necessarily static inside the oven; however, if this is not the case the complication is probably not worthwhile.

5.7.3 Movement of the workload

Moving the workload inside the oven is a very effective way of combating nonuniform heating; obviously, if the standing waves present are mainly located relative to the walls of the oven, the effect of movement will be to smudge the pattern within the load. Linear movement of a load on a continuous conveyor band will eliminate variations in field intensity along the axis of movement. However, field variations along the two orthogonal axes are not compensated. Likewise, field concentrations along edges and especially at corners and protuberances of the workload are not compensated as their reference is the workload itself.

The above illustrates the principle that movement in one axis will only compensate for nonuniformity along one axis. To improve uniformity further, movement in the orthogonal axes is necessary, but this is only effective if the movements are nonintegrally related and several excursions along each axis occur during microwave exposure. One example of apparent two-axis movement is in certain meat-tempering equipment, where the workload is mounted on a platform driven by a pair of cranks rotating synchronously. There is simultaneous movement along both vertical and horizontal axes with the same amplitude, equivalent to the locus of movement of the load being a circle with horizontal axis. The above condition of nonintegrally related motion is not satisfied by the arrangement: it is equivalent to a rotating table.

The horizontal rotating table clearly provides good circumferential uniformity, but does not inherently compensate radial and vertical field variations. However, some compensation along these axes is possible by introducing asymmetry to the oven walls relative to the rotation axis. For example, the rotation axis would be off centre in the oven, and the top wall may have a step so that the oven height is different in two halves.

Unfortunately, there has been no published systematic, quantitative study of the efficacy of these techniques. There is no analytic procedure. Numerical methods would give an approximate solution by first determining the heat distribution in a cylindrical body, and then rotating the body about its axis. However, this approach is limited to circular cylinders rotating on-axis on a table, a configuration which does not give an inherent modulation of the field as the workload rotates. If this last condition is not satisfied, many plots have to be made, which becomes prohibitive in time. As with most multimode problems, systematic experiment is the only satisfactory procedure.

5.8 Doors, chokes and open entry ports

Industrial microwave ovens are provided with access doors for cleaning and maintenance, as well as loading doors for batch ovens, or open aperture ports for continuous-flow ovens. All must provide effective containment walls to sustain the correct multimode operation, which implies that they should not introduce any microwave perturbation to the surfaces of the oven. Clearly this is not possible for open-aperture ovens, and for these the criteria are of maintaining symmetry and consistency of microwave admittance looking outward from the oven. Moreover, they must limit microwave leakage under worst-case conditions to a safe level for personnel, and in many cases to the extremely low levels demanded for RFI.

5.8.1 Doors

Doors for industrial microwave ovens must be robust mechanically and are required to operate with high surface currents, much greater than those in a domestic oven. There is a much greater risk of arcing and overheating of lossy material incorporated in choking systems. For these reasons, an industrial oven door differs from its domestic equivalent.

The ideal microwave requirement is for the door to appear as an unimpeded interior surface, as though it does not exist. Inside it should be coplanar with the rest of the wall, and the contact with the door jamb should appear a short-circuit in that plane at all points around its periphery. These and the requirement for minimal leakage are the prime design objectives of the door; there are often others such as avoidance of crevices and bacterial traps for food-processing ovens, provision of a viewing window, or constraints in mechanical design for automatic loading.

The cheapest door is a plain metal plate screwed to the wall of the oven, with a conducting gasket forming the microwave seal. It is essential that the gasket be compressed fully to specification limits at all points around the periphery. If this condition is not obtained, arcing will probably occur, destroying the gasket and damaging the contact surfaces. There must be a sufficient number of fixings or clamps to ensure compression, which is inconvenient for all except doors for occasional access for maintenance.

Time spent loading/unloading the oven represents lost production, and must be minimised, so doors for normal access must have as few clamps as possible. They must still provide compression of a gasket as above, or closure of the mating surfaces in accordance with the requirements of a noncontact choke system, or both. To achieve this both the door and the door jamb must have sufficient stiffness not to distort on clamping, and the choice and location of clamps, locks and hinges require careful consideration. Hinges are necessarily robust precision components with tolerances and stiffness appropriate to the above demands. Many production ovens have large heavy doors pneumatically opened with powered clamps.

5.8.2 Door chokes

In ovens of high power density (>20 kW/m), the wall-current density may be too high for a simple contact gasket to handle safely. In assessing the performance of the gasket, wear and compression set must be considered, as well as the effect of foreign matter trapped between the gasket and the mating surface, thus impairing the contact. The latter redistributes the current flow locally, causing excess current adjacent to the fault, possibly leading to arcing and damage. The duty on the gasket can be much reduced by using a choke system in which the impedance-transformation properties of quarter-wave lines is used to form a virtual short-circuit at the joint face inside the oven, with the gasket positioned $\lambda_0/4$ (A) behind it, and a further $\lambda_0/4$ short-circuited (B) line behind the gasket, as shown in Figure 5.11. The short-circuited $\lambda_0/4$ line (B) forms a very high impedance at the gasket location, corresponding to a low current, relatively high voltage. The line (A) transforms the high impedance at the gasket to a very low impedance at the working gap, effectively allowing the wall currents to cross it. In principle, an imperfect contact between the gasket and its mating faces is unimportant because the current is ideally zero; in practice there will be a residual current which may still cause arcing, so it remains desirable to ensure good mechanical compression of the seal.

In some chokes the gasket is replaced by a strip of lossy material (e.g. carbon-loaded rubber or plastics) which does not have to form a direct contact to the mating surfaces. The equivalent circuit is shown in Figure 5.11. This is a good system provided that the heat generated in the lossy strip can be conducted away to avoid a significant temperature rise, representing a clear limitation in power handling. A disadvantage of this door-choke design

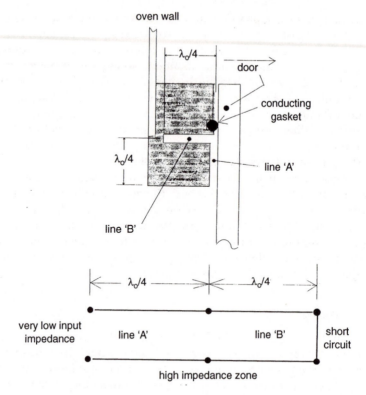

Figure 5.11   Simple door choke and equivalent circuit

is that, as the door opens, the faces immediately separate, impairing the choke action. Mechanical distortion, or any other cause of the door failing to close fully, results in a potentially high leakage.

Figure 5.12 shows a folded choke with the internal surface flush with the oven wall. It allows the door to open slightly before the choke action becomes impaired seriously, but the additional complexity, the mechanical tolerances and stiffness of components adds substantially to the cost.

5.8.3 Open entry ports

Multimode ovens are often used in continuous production with a conveyor band passing horizontally through open ports each end of the oven. The aperture size, especially the height, is critical because the leakage of microwave power rises rapidly with increasing height, particularly as the height approaches $\lambda_0/4$ if reactive (reflective) chokes are to be used.

There are four methods of choking large open apertures:

(i)   to allow the workload itself to absorb the power leaking into the choke tunnel;

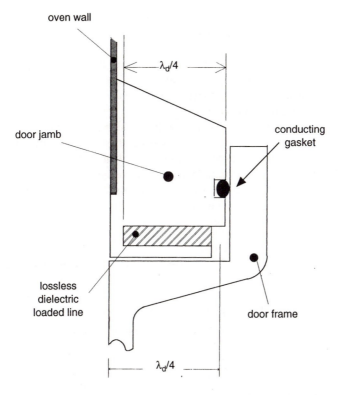

oven wall

door jamb

conducting gasket

$\lambda_d/4$

lossless dielectric loaded line

door frame

$\lambda_d/4$

Figure 5.12   Folded door choke

(ii)   to apply some reactive choking in conjunction with (i) above;
(iii)   to line the walls of the choke tunnel with absorbent material; and
(iv)   for powders and small-size particulates, to provide an array of vertical cutoff waveguide choke tubes through which the workload falls at entry and exit.

If none of these is effective in a particular case, it becomes necessary to use a semicontinuous conveyor with automatic doors to the oven.

Method (i) is often used, and clearly is critically dependent on the presence of the workload, which must fill the aperture to a chosen fraction. The length of the tunnel is then calculated from the attenuation rate of the partially filled choke tunnel with the workload giving least attenuation. This procedure is considered in Section 8.5.

Secure, failsafe interlocks of the highest integrity must be provided to prevent high-power microwave being switched-on in the absence of workload. Techniques used for such interlocks include optical (or infrared) beams interrupted by the workload, microwave sensors near the aperture of the choke tunnel set to trip an emergency stop if the level of leakage exceeds a preset limit, and a microwave fuse near the aperture though which a pilot current passes. If the fuse blows, the pilot current is interrupted, tripping the microwave generator. Lightbeams are impaired by dirt, and a test lamp is often provided to check the system automatically at regular intervals (e.g. at every start-up). The integrity of the trip system is paramount, and it must be scrutinised in extreme depth to audit its fail-safe criteria under all conditions, provide automatic self-testing, and desirably to comprise separate, independent, duplicated systems.

Method (ii) involves creating a high reflection coefficient to energy escaping through the tunnel using passive, lossless reflectors. These may be reflector plates or an array of resonant slots or posts forming a bandstop filter (Van Koughnett and Dunn,1973), as illustrated in Figures 5.13 and 5.14.

The reflector-plate choke is very simple and comprises a section of radial waveguide terminated in a circular reflector having an aperture equal to the

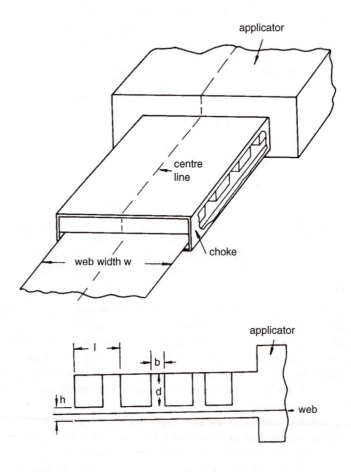

Figure 5.13   Choke-pin structure (Van Koughnett and Dunn, 1973)

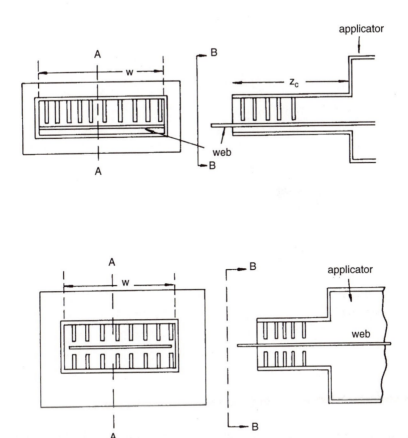

Figure 5.13   continued

desired headroom , as shown in Figure 5.14. In principle, energy leaking into the radial waveguide reduces in intensity as it propagates towards the circular reflector, such that the power density at the aperture is reduced in the ratio of the area of the aperture to the area of the reflector. Power incident on the reflector is reflected back to the oven. The attenuation is simply the above area ratio, and can be up to 15 dB per section. However, the structure is large and overmoded and the frequency response displays peaks and troughs of attenuation about the predicted mean. Although the hardware is cheap, the development cost in modelling may be high.

Van Koughnett's choke is based on the principle that any mode present in a large aperture can be synthesised from a set of coherent plane waves propagating along inclined axes within the tunnel, causing currents to flow

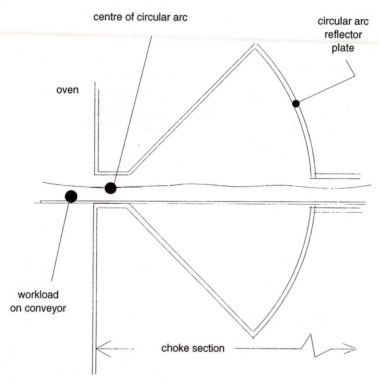

Figure 5.14   Choke reflector plates

in the walls. By providing two arrays of slots in mutually orthogonal patterns, each forming a bandstop filter, the wall currents are intercepted irrespective of their direction because they can be resolved into components at right angles aligned with the axes of the slot patterns. The slot patterns can be formed from a set of square posts mounted on the wall of the tunnel, spaced apart to form the slots. In practice, circular posts form a much simplified construction; though apparently departing far from the ideal model, they nonethless provide excellent choking. Note that the attenuation rate falls sharply as the workload aperture height approaches or exceeds $\lambda_0/4$. A more detailed analysis of Van Koughnett's 'choke-pin' structure is given in Section 8.4.

The third method of choking is to line the walls of the tunnel with microwave absorbent material, as outlined in Section 8.3. Because of fire risk, care must be taken in applying this method in regions of the choke tunnel close to the oven where the power density may be high, particularly in abnormal conditions. Optimum lossy dielectric materials are often plastic foams having very poor thermal diffusivity; excessively high temperatures can be attained with low power-dissipation density, particularly if conventional heating is applied to the oven.

Water or glycol loads have been used where there is high power density, but the attenuation rate is low due to the high values of ε' and consequent difficulty of matching. The water flows through an array of plastic pipes of diameter comparable to the penetration depth. Reliability problems of water leaks, and of condensation forming with humid atmospheres in the tunnel, are commonplace. Extreme care must be taken in design and manufacture to avoid leaks which usually arise from embrittlement or 'plastic flow' of the pipework with time. A further problem is that a dielectric water pipe is a waveguide of modest attenuation, and will transmit power in the TEM mode when housed in even a tight-fitting metal tube. Very high field concentration can occur at the point of entry of such a water pipe through a metal wall, which can result in local heating and melting or burning of the plastic pipe. Water absorbers should not be used unless they are totally unavoidable.

The best practice is to use reflective chokes adjacent to the oven to reduce the power density, and then to alternate between absorptive and reflective techniques until the desired attenuation is achieved.

5.9 Numerical methods of analysis

The subject of computational electromagnetics is devoted to the numerical solution of Maxwell's equations to fit a given set of boundary conditions corresponding to the configuration of a microwave structure, ranging from a simple waveguide to a loaded multimode oven with entry chokes. While numerical methods are well known and widely used for the solution of problems in heat flow, mechanical stress and electrostatic fields in electrical engineering, the time dependency of the Maxwell equations at high microwave frequencies creates a considerable further difficulty in numerical modelling.

Because of the complexity of the subject and its current high rate of advancement, it is beyond the scope of this book to do more than outline some of the advantages and disadvantages of these procedures for the design engineer.

Much contemporary research is devoted to these techniques and good progress is being made; most of the work is directed at verifying the integrity of the analyses, comparing computed results with those experimentally measured, or those where a conventional analytic solution exists (Paolinin, 1989). Early success was achieved by Lau and Sheppard (1986) using a 'time-domain finite-difference' (FDTD) procedure. This involves forming a rectangular three-dimensional mesh conforming to the boundaries of the structure and then computing field intensities at the crossing points of the mesh to satisfy Maxwell's equations. Initially ($t = 0$) there is no field present. At $t = 0$ a field is injected into the input port and thereafter the field is computed repeatedly throughout at time intervals of about $0.1/f$, where f is the microwave frequency. The field 'builds up' progressively, taking into account the reflections from the boundaries, and computation continues until a predetermined and permitted residual error is reached. Dielectric materials of complex permittivity can be included. The heat distribution within the workload can then be evaluated from the computed values of E-field using eqn. 2.7, a procedure readily incorporated into the program. Some researchers have taken the further refinement of including the temperature variation of complex permittivity, and yet further to include the effects of heat and mass transfer within the workload.

Obviously, the accuracy is related to the fineness of the mesh. Too fine a mesh wastes computer time, too coarse gives inaccurate and misleading results. Unfortunately, inaccurate results are frequently very plausible, a problem particularly for multimode ovens where the engineer has only a rough qualitative view of performance. Where a rectangular mesh is used, boundary surfaces which are not parallel to the principal axes of the mesh are represented by a 'staircase', which is again an approximation.

Extreme care must be taken in modelling resonant structures where a frequency change small compared with f_0/Q causes a first-order change in field intensities. This translates to a small wavelength change compared with λ_0/Q, equivalent to a small dimensional change to the structure. Clearly, the mesh size must be small compared with such a dimensional increment for the model to be representative. Lightly loaded multimode ovens are a specific example where totally misleading but plausible results are obtained by using an insufficiently fine mesh. Similarly, periodic filters require a fine mesh to ... attenuation.

... odelling is essentially based on a pure unmodulated ... ned frequency and amplitude. In practice, the power ... ncy and amplitude modulated at power frequency and ... with peak deviations around 0.2%, depending on ... In structures such as those described above, the field ... otally over this operating range, and so it would be ... he fields at several frequencies within the spectrum to ... tive heating pattern.

... FDTD approach is that the mesh must be regular, and ... l with a mesh of the same size as that required for the ... n as a workload of high value ε'. This creates a fine mesh in zones where it is not required, so that much unnecessary computation is done, resulting in a slow program.

Dibben and Metaxas (1995) describe a time-domain finite-element procedure in which the mesh size is chosen locally to suit the perceived sensitivity of the problem. Thus a fine mesh would be used within the volume of a workload with high ε', around critical sharp corners, or in the vicinity of choke-filter elements. Elsewhere a coarse mesh is used, so avoiding unnecessary computation. A further feature is that they formulate the problem in the time domain rather than the frequency domain, which enables a Gaussian time pulse to be used for the excitation. It is then relatively straightforward to transform the result into a series of closely spaced frequencies which immediately shows the sensitivity to frequency. As an example, a small plastic block 200 × 200 × 25 mm, (ε = 2.5 - j.01) was modelled, located symmetrically in the base of a rectangular oven 391 × 292 × 200 mm. Figures 5.15 shows patterns of computed power-dissipation density distribution at the four frequencies shown; note the small increments of frequency (0.4%) and the marked differences in computed heating pattern between them. Figure 5.16 shows a measured heating pattern by thermal imaging of an identical block and oven, at a mean frequency of 2455 MHz. The correlation of the computed pattern at 2460 MHz with the experiment is excellent, and highlights the extreme sensitivity of the heating uniformity to operating frequency, which also can be interpreted as dimensional accuracy or stability of the oven.

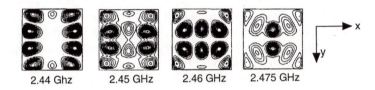

| 2.44 Ghz | 2.45 GHz | 2.46 GHz | 2.475 GHz |

Figure 5.15   Numerical modelling of heating plastics blocks in multimode oven (Dibben and Metaxas, 1995)

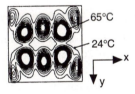

Figure 5.16   Measured results compared with Figure 5.15 (Bows, 1995)

Although these techniques are, at present, in the realms of academia, there is no doubt that in time they will become commonplace in industry. Meanwhile, there is a concentrated effort in developing the programs to be readily useable by design engineers. Metaxas (1996) gives an excellent review of the present state of these techniques.

5.10 Conveyor systems

Continuous-flow equipment usually requires a conveyor band to carry the workload through the oven. The choice of conveyor-band material, and the means provided for its drive, tensioning, centring (tracking) and joining are critical to the successful operation of the installation. Conveyor bands are expensive and are, with one possible exception, short-life items.

5.10.1 Conveyor types and materials

Most conveyor bands for microwave ovens are required to be microwave transparent so that energy can enter the workload from below. Many materials have been used.

(a) Glass cloth and Kevlar®

Cloth woven from glass-fibre thread and impregnated with PTFE has extremely low loss factor and is excellent in this respect. It also provides a smooth, easily cleaned, nonstick surface. However, although strong, if accidentally damaged e.g. cut, its low tear strength results in rapid deterioration. This is a particular risk at the edges, which are easily damaged through poor tracking allowing the belt to chafe against the side of the oven. Reinforced edges can be used, which are an improvement but they tend to stiffen the edges. The material can be used at temperatures up to about 220°C depending on the life required.

Glass cloth has very high tensile stiffness (i.e. it does not stretch significantly up to its maximum permitted tensile force). This means that great care must be taken in mechanical alignment of rollers to ensure that the tensile force is uniformly distributed across the width. It also means that the joint must be made with precision so that the perimeter length is equalised across the width; a badly jointed belt can never be correctly tensioned.

Glasscloth is cheaper than Kevlar but much inferior in tear strength. Kevlar otherwise has all the advantages of glass cloth and is generally the preferred material.

Joints in Kevlar or glass cloth can be either castellated with an inserted pin, or the ends may be fused together with PTFE using a special tool. The former permits easy replacement since the a belt end can be threaded through the machine with little dismantling, mated to the other end and the pin inserted. The fused belt is effectively trapped in the machine and the joint has to be made 'on site'. Experience has shown that the fused belt is usually preferable and has a longer life. Unequal stress on a castellated joint leads to quick failure, and the smoothness of the fused joint is often advantageous.

Occasionally, it is desired that air or steam should pass through the conveyor band, a requirement satisfied by using a stiff net of the above materials.

(b) Polypropylene

Polypropylene and polyethylene have been used as conveyor bands where the temperatures are low, such as in tempering and thawing machines. As a plain band they are not satisfactory, because the joint is too prominent, and they are difficult to drive.

However polypropylene has proved extremely satisfactory as a chain-link conveyor, where the links are injection-moulded items pinned together as a conventional chain. The links are thick, but the chain is flexible, providing a very strong chain capable of hauling heavy loads, such as are encountered in meat-tempering equipment.

The links have projections on their underside at the pitch of the chain, and the drive roller is provided with longitudinal grooves to match. This provides a positive nonslip drive, essential for heavy loads up to 0.3 t. It is also possible to provide links forming a longitudinal projection under the conveyor which engage circumferential grooves in selected rollers, providing a positive tracking action.

(c) Terylene and Nylon

Nylon is not recommended as a belt material because its loss factor is too high, and there is a risk of melting. Terylene and other man-made fibres may be used subject to the provisions noted in (a) and (b) above

5.10.2 Conveyor drives

The importance of accurate control of conveyor speed is emphasised in Section 3.5, from which it is clear that slippage between the belt and its drive is totally unacceptable.

Plain flat belts with PTFE impregnation are difficult to drive because of the low coefficient of friction. Techniques used successfully are a large-diameter drive roller (300 mm typical), preferably rubber-covered, with a floating 'nip roller' which applies pressure to the belt against the drive roller. The rubber covering helps to equalise tension in the belt across its width, and indents under the nip roller to spread the pressure beneath it, minimising the crushing pressure on the belt.

Drives for chain belts are described in Section 5.10.1.

5.10.3 Belt tensioning

Belt tensioning is critical for achieving good life. It is essential to be able to set the tension to a defined limit, and the most satisfactory method is a roller in sliding bearings with compressed-air cylinders providing a thrust to each side. The tension in the belt is then set by a pressure-regulating valve, ensuring precise uniform tension.

This system is excellent in maintenance and cleaning since the air pressure is easily removed and the belt can be relaxed.

5.10.4 Belt tracking

All belts degrade rapidly if they chafe against anything. The first point in design is to ensure that there is a reasonable clearance to allow the belt to wander from its optimum position, because no belt tracker is perfect.

The sideways force which a belt develops in tracking off-centre is very large, and it can never be pushed back into position, but it can be pulled from the other side. One tracking system using this technique uses a pair of pneumatically operated nip rollers fixed each side of the belt, inclined at an angle of about 5° to principal roller axes. Normally the nip is open, but if the belt wanders a sensor is activated which closes one nip which, with the motion of the belt, drags the belt back into position.

Another belt tracker is a steering roller. The roller is positioned so that the belt bends through about 90° over it, and the roller is mounted in a carriage allowing it to be turned around an axis at 45°, located at the nominal centre line of the conveyor. The rotation of the carriage provides a direct steering action without imparting undue nonuniformity of tension in the belt. The steering is controlled by an electric drive controlled by a sensor in a conventional feedback loop.

Two such steering rollers are usually required, ganged together mechanically so that the belt may be returned to the other end of the machine through 180°.

PTFE coated belts can also be tracked using slipping rollers. The principle is that if the belt is *sliding* over a surface it can easily be pushed sideways with a relatively low force. The arrangement comprises a fixed, nonrotating roller at the nondrive end over which the belt slides in normal motion. The roller is flanged, and if the belt wanders against one flange the force exerted is sufficient to prevent further sideways traverse, without damaging the edge of the belt. At the drive end a similar slipping roller is provided, but here the roller is geared to run at a peripheral speed about 15% faster than belt speed. This provides the same tracking action, and assists the drive.

5.10.5 Belt-speed measurement

It is unwise to measure belt speed from the drive roller because of the risk of slippage, and good design practice is to drive a sensor from an idling roller. Some belts are very slow moving, and the data rate of updating the speed readout requires consideration.

5.10.6 Belt and equipment maintenance

Belts are frequently damaged through lack of maintenance and cleaning. Sticky workloads are particularly prone to cause a build-up of solids on conveyor rollers, resulting in uneven tension and severe tracking problems. The nature of the belts used causes electrostatic attraction of particles to aggravate the problem. Machine design should include easy provision for cleaning, and an alarm should a gross tracking error occur.

Chapter 6

Microwave-heating applicators 2: aperiodic structures

6.1 Introduction

This chapter considers the class of microwave-heating applicators which essentially have a broadband, nonresonant frequency response. They are usually terminated in an absorbing load which dissipates the residual power not usefully dissipated in the workload, and in a well designed system the proportion of the input power in the terminating absorber is less than 10%. The principal feature is that the impedance match 'seen' from the generator is always favourable under normal operation (i.e. nonfault conditions), whether or not the workload is present. If the input match has VSWR < 1.3 under all nonfault conditions, these applicators can be fed directly from the magnetron generator (i.e. without a circulator) without instability, especially if a switch-mode power supply is used with a high-speed reverse-power detector for rapid response to faults.

In some cases the terminating load is replaced by a short-circuit plate and a 3-port circulator is used at the input to the applicator, its third-arm load functioning as the terminating load. There will then be a standing wave distribution along the axis, at a high-value VSWR adjacent to the terminating plate, diminishing towards unity at the generator end, at a rate according to the attenuation of the workload. Due to the constant-velocity movement of the workload, this standing wave is self cancelling in its effect on uniformity.

These applicators are used mainly in conveyor ovens, where the workload has cross-section dimensions and loss factor to give an attenuation in the range typically 2–15 dB/m in the direction of travel, to realise an applicator of practical overall length. Typical loads are particulates and powders on a conveyor band, sheet materials and webs, and extrusions. Processes include drying, heating/baking, polymerisation, pasteurisation, tempering and thawing.

In industrial practice, quick and easy access to the inside of the applicator for cleaning and maintenance is of paramount importance, no less than its microwave functioning. Food processing demands inherent freedom from 'bug traps': sharp inside corners to be replaced by smooth curves of radii allowing easy cleaning, joints and gaps to be easily separated. These requirements often exclude otherwise elegant microwave solutions.

6.2 Axial travelling-wave structures

The main feature of these applicators is a transmission line in which microwave energy propagates along the same axis as the workload motion, either in the same direction, or counterflow. Counterflow is usually chosen in dryers because the driest (lowest loss) material is then exposed to the highest field intensity and the drying rate is more evenly distributed. Where the dielectric properties are sensitive to temperature, giving a rising attenuation with temperature, counterflow may cause instability and difficulty in control. The appropriate direction to choose requires careful consideration on individual merit.

6.2.1 Longitudinal TE$_{10}$ waveguide

This is the simplest applicator, comprising a long rectangular waveguide supporting the fundamental TE$_{10}$ mode only. At each end there is a port for passage of workload through a choking system, and a waveguide transition for connection of the power source and the terminating load. The waveguide is oriented with the conveyor band parallel to the E-field vector as illustrated in Figure 6.1. With care to avoid excitation of high-order modes, the cross-section size of the waveguide can be increased beyond that supporting only the fundamental TE$_{10}$ mode.

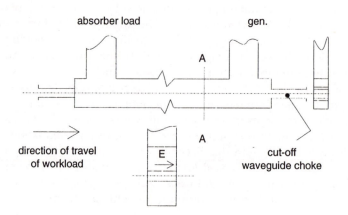

enlarged view in section AA

Figure 6.1   Longitudinal TE$_{10}$ waveguide applicator

An approximate estimate of attenuation rate can be made by the methods of Chapter 8 for calculating the attenuation in waveguides partially filled with lossy dielectrics. For a thin sheet workload mounted at the centre of the waveguide parallel to the E-field vector (i.e. extending between the broad faces, and at the position of maximum E field), Metaxas and Meredith (1993) give the following relations for attenuation, derived from the work of Altman (1964):

at 915 MHz $\alpha = 894\, t\varepsilon''$ decibels per metre   (6.1)

at 2450 MHz $\alpha = 7.35 \times 10^3\, t\varepsilon''$ decibels per metre   (6.2)

where t is the thickness in metres.

The workload distorts the sinusoidal transverse-field distribution of the empty waveguide progressively as the relative permittivity ε' rises, and the effect of this can only be established experimentally or by numerical solution. However, where the workload is a thin sheet aligned in the plane of the E field the uniformity is excellent.

This applicator is relatively inexpensive and simple. As there is no wall current passing across the plane of the centres of the broadfaces of the waveguide, this is a plane for separating two halves of the applicator, convenient for cleaning. However, care is required in the vicinity of the generator and terminating load ports where the perturbation causes transverse currents across the split line, and gaskets or chokes must be used.

6.2.2 Longitudinal slow-wave structures

These are transmission lines with corrugated surfaces, or formed as helixes, giving an extended path length for the wall currents, and causing the effective propagation velocity of the microwave energy to be slowed. More importantly for heating, they have an axial electric field which can give strong coupling to a workload allowing a high power-dissipation density. Another form of slow-wave structure is an array of conductors spaced apart less than $\lambda_0/4$ and driven so that the phase difference between them is about π radians. Such structures are sometimes called 'fringing-field' applicators (Bows, 1995).

6.2.2.1 Corrugated planar-surface structures (periodic arrays)

Microwave energy will propagate along a surface having an array of slots (Lamont, 1942; Ramo *et al.*, 1965; Lines *et al.*, 1949), the fields extending into the space outside the surface where the E field can interact with a workload. A typical arrangement is shown in Figure 6.2, for which Lamont shows that, if the slot depth is in the range $0 < l < \lambda_0/4$, the amplitude decreases exponentially with distance from the surface, but for $\lambda_0/4 < l < \lambda_0/2$ the wave diverges and radiates into space. Clearly, the field intensity in the workload can be controlled by adjusting its distance from the slotted surface, enabling the attenuation rate and power dissipation density to be adjusted. Further, by tilting the applicator relative to the workload, the distribution of heat along the length of the applicator is adjustable.

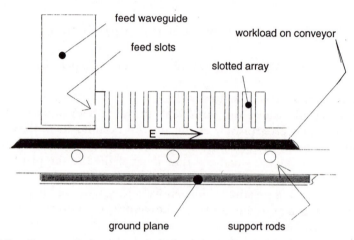

Figure 6.2   Corrugated slow-wave applicator showing slotted feed system

In principle, the propagation is in the TEM mode so that transverse variation in heating intensity is small provided that higher-mode contamination is avoided. Transverse wall currents cause nonuniform transverse heating and can be minimised by longitudinal slots. Uniform illumination from the source can be provided by a slotted-array antenna.

To keep stray leakage within required safety limits the whole assembly is enclosed in a metal screen with choke tunnels each end for passage of workload

Corrugated waveguides and other periodic structures have not had the widespread use they merit, and most of the reported applications have been experimental (e.g. Bowes, 1995).

6.2.2.2 Helical-waveguide and corrugated-tubular-waveguide applicators

Propagation in helixes and corrugated tubular waveguides is well known (Pierce, 1950), these structures being essential components in travelling-wave tubes (TWTs) used widely as microwave amplifiers and power sources. As with the planar arrays, there is a longitudinal component of electric field, which in TWTs interacts with an electron beam, and in heating interacts with the workload. Typical workloads are filamentary materials such as yarn, extruded rubber or plastics, and low-pressure plasmas.

The analysis and design procedure for helix applicators has been developed by Carter and Wang (1995), with an example of an applicator operating at 2450 MHz on wool fibre. The helix has a mean radius of 4.8 mm with a pitch of 6.4 mm, and is mounted in a shield of radius 26 mm. For an air-breakdown voltage of 2.0 kV/m, the maximum voltage between turns of the helix is 3.4 kV, allowing a maximum power input of 27.4 kW. The wool fibre would be heated at a rate of 80°C/s. The attenuation rate of 0.023 dB/m is very low, resulting in an extremely long applicator with low efficiency.

Van Koughnett *et al.* (1974) describe the design procedure for a disc-loaded circular waveguide (Figure 6.3), in effect a corrugated tubular waveguide, for heating filamentary materials. At 2450 MHz the diameter of the workspace is about 25 mm, and the E field is substantially uniform at 3 kV/cm over the cross-section for an input power flow of 5 kW. Design methods for transitions to rectangular waveguide at input and output are given, and it is suggested that the applicator can be resonated by a terminating short-circuit and appropriate matching at the input.

In practice, care must be taken to avoid arcing in these applicators, which can cause damage due to the small interaction space. An inorganic dielectric sleeve affords some protection from arc damage, but may itself be consumed. At the high-voltage stresses involved with these applicators, arcing can be initiated by irregularities in the workload (cross section, or properties),

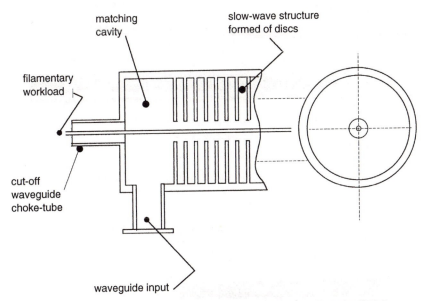

NB the structure may be terminated in a short-circuit plate, or a matched load (high-power) fitted to a matching cavity and waveguide similar to the input arrangement

Figure 6.3   Disc-loaded circular slow-wave structure

breaks in the workload, condensation droplets and accretion of dirt and foreign matter. Special care must be taken to match the throughput to the microwave power, and this is especially necessary where the final temperature is close to a degradation/decomposition temperature of the workload.

6.2.3 Parallel-plate transmission-line applicators

As their name implies, parallel-plate transmission lines comprise a pair of conducting planes, spaced apart and parallel to each other. In mathematical terms, by treating them as of infinite extent, a simple TEM wave propagating between them is a solution to Maxwell's equations. The electric field is polarised normal to the plane surfaces, the magnetic field is orthogonal to it (i.e. parallel to the planes) and the wave propagates in a direction normal to the electric and magnetic fields. The space between the planes is the interaction space for the workload which couples to the electric field. Figure 6.4 shows the general arrangement. Note that the parallel-plate line is not the same as a parallel-plate capacitor widely used as an applicator in RF heaters,

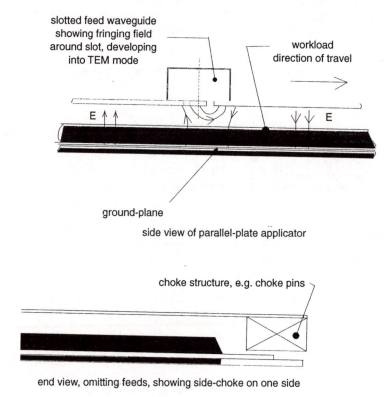

Figure 6.4   Parallel-plate applicator

where the RF voltage between the plates is the same everywhere because the size of the plates is small compared with the wavelength. In the parallel-plate transmission line the size of the plates is large compared with a wavelength, so that the electric-field intensity varies sinusoidally in the direction of propagation, although it is constant transversely if the TEM wave is the only mode present.

In practice, parallel-plate lines are used in continuous flow applicators for sheet materials or thin objects on a conveyor band. Of special interest is a horizontal parallel-plate line in which the lower plate is a metallic conveyor band (Meredith *et al.* 1993), where volumetric heat is to be injected into dough pieces within a continuous baking oven having a metal band. The temperatures within the oven are far in excess of the maxima for dielectric bands of conventional plastic materials, and metal bands are the approved baking-industry standard.

Although wide compared with a wavelength, the plates are of limited width so the boundary conditions for the TEM wave are violated; nonetheless TEM-wave propagation can be made a close approximation. The assembly can be considered a rectangular waveguide supporting the TE_{n0} modes only, whose vector sum at any cross-section add as 'space harmonics' to give a substantially constant E field across the width, in similar manner to the Fourier harmonics of a rectangular pulse. Of course, this can only be an approximation because a limited number of modes is possible, but the resulting heating uniformity ($\pm 5\%$) is adequate for the purpose.

The open sides of the parallel-plate line require careful consideration. The assertion of equivalence to a rectangular waveguide implies total reflection at the edge boundaries. Mounted 'in the open' there would be considerable radiation loss from the edges because the distance between the plates is a significant part of a wavelength, obviously unacceptable for both efficiency and safety reasons. It is therefore necessary to close the edges either mechanically, to form a true rectangular waveguide, or virtually, by choking. The requirements of access are a determining feature in design. Whatever the treatment of the edges, it is very important to secure symmetry about the centre line of the oven; in this context symmetry of reflection coefficients 'seen' looking outwards from the edges. In effect, this means that the impedance profiles looking outwards from the two sides of the applicator must be complementary in both their real and their imaginary parts.

The presence of the workload causes distortion of the TEM-field pattern, because the propagation velocity in the empty headspace is higher than in the dielectric material of the workload, by a factor $\sqrt{\varepsilon'}$. This results in the electric field in the workload 'dragging' behind, so a longitudinal component E_z is created. The boundary conditions require that $E_z = 0$ tangentially to the planes, so E_z increases from zero with distance from the plates, giving a gradient in heat intensity in the workload rising towards its upper surface. This can be advantageous in a baking oven because the lower plate is a hot surface, efficiently injecting heat energy by thermal conduction.

Estimation of the attenuation rate, from which the length of the oven can be determined, can be made by the methods of Section 8.3, specifically using eqns. 8.13 and 8.15. More precise data on heat distribution in the workload and rate of attenuation require numerical modelling by the methods of Lau and Sheppard (1986) or Dibben and Metaxas (1995).

In principle, excitation of the TEM wave requires injecting a uniformly distributed wavefront across the width of the applicator. In practice, this is complicated by the need to provide free passage of the workload, so the excitation means must be at or above the workspace height. The distance between the plates is required to be as small as workload clearances allow to give a practical attenuation rate, and in practice the spacing is a small fraction of a wavelength. Small radiating slots in one plate will have a very high reflection coefficient because of the close proximity of the other. The desired wavefront is injected via an inclined tunnel with an aperture in the upper plate, and its directional properties inject most of the energy unidirectionally. The wave front can be created by a set of sectoral horns fed equally and phased appropriately. Alternatively and more compactly, a slotted waveguide can be used with an array of slots chosen to have equal radiated power, by the methods discussed in Section 5.6.3. This technique has the added advantage that it is relatively easy to change the design of the feed experimentally in development.

6.3 Transverse travelling-wave structures

In this section microwave applicators for continuous-flow operation are considered, in which the microwave energy is directed solely at right angles to the direction of motion. In the simplest form (Section 6.3.1), they are particularly useful for initial experimental sample tests where efficiency is not of concern, although high efficiency can be achieved by multiple-passage operation (Section 6.3.2). Their principal disadvantage is the inherent exponential decay of heat energy across the width of the workload, and more especially the difficulty in control of standing waves due to the numerous reflections inherent in the structure.

6.3.1 Transverse waveguide, TE_{10} mode

This is the simplest of all microwave applicators both mechanically and analytically. It comprises a rectangular waveguide with a pair of slots formed along the centre line of the broad faces for the passage of workload. These are inherently nonradiating slots because there are no wall currents crossing the centre lines of the broad faces. However, the presence of workload does modify the field pattern, and it is desirable to arrange a choke system (e.g. Section 8.4) to prevent high leakage in extreme conditions. The waveguide is usually terminated in a matched absorbing load. The general arrangement is shown in Figure 6.5.

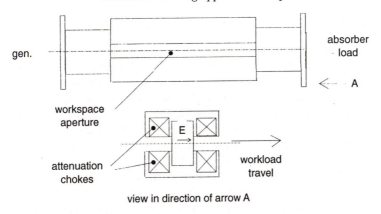

view in direction of arrow A

Figure 6.5   *Transverse-slotted-waveguide applicator*

Again, because there are no wall currents crossing the centre lines of the broad faces, the waveguide can be split along the plane of the centre lines without interrupting current flow, giving minimal leakage and no arcing at the joint face. This conveniently allows the applicator to be formed in two pieces readily separable for cleaning. However, depending on the feed arrangements, there may be transverse currents at the joint face where the generator feeds the applicator, usually necessitating a conducting gasket.

The attenuation rate, as a function of loss factor and thickness of sheet, at the two common frequencies and waveguide sizes (WG4, RG204U, WR975 248 × 124 mm; and WG9A, RG112U, WR340, 86 × 43 mm), is given by (Metaxas and Meredith, 1993):

For 896 MHz $\alpha = 891\, t\, \varepsilon''$ decibels per metre   (6.3)

For 915 MHz $\alpha = 894\, t\, \varepsilon''$ decibels per metre   (6.4)

For 2450 MHz $\alpha = 7.35 x 10^3\, t\, \varepsilon''$ decibels per metre   (6.5)

An important feature of these applicators is that it is possible to estimate the field intensity in the waveguide using eqn. 4.46, because only one mode is present. An allowance must however be made for the relative permittivity of the workload.

Standing waves occur primarily because of reflections from the terminating load, the ends of the slots and the edges of the workload. The first two can be readily matched, but the reflection from the edge of the load adjacent to the terminating load is variable and less easily managed. The standing wave is at the loaded waveguide wavelength λ'_g; if the standing wave is too great for the desired uniformity it must be matched by a susceptance located between the trailing edge and the terminating load; see Section 4.3.

6.3.2 Serpentine or meander-line applicators

These applicators were popular in the 1970s, comprising simply an array of slotted waveguides, as described in Section 6.3.1, assembled side-by-side and connected in series so that the energy flow is in opposite directions in adjacent waveguides, as shown in Figure 6.6. In principle, the attenuation is additive, and a sufficient number of 'passes' is provided to realise a desired efficiency. Ideally, the attenuation should be in the range 0.5–2 dB per pass, which gives acceptable uniformity of heating with a practical number of passes (up to 20) to secure greater than 90% efficiency.

The problem of reflections for the single waveguide of Section 6.3.1 is compounded in the serpentine, resulting in a sinusoidal heat pattern across the workload, which is difficult to control and stabilise. Many attempts have been made to correct the nonuniformity but none has proved entirely successful because the workload itself is a principal source of reflections and is a variable. By choosing the width of the applicator (which means the electrical width, including additional phase length introduced by the relative permittivity of the workload) Metaxas and Meredith (1993) showed that some of the fixed reflections can be made to cancel in pairs.

Various arrangements are possible of groups of serpentine applicators fed from opposite sides as mirror images, either as separate applicators or mutually interleaved, in an attempt to improve uniformity.

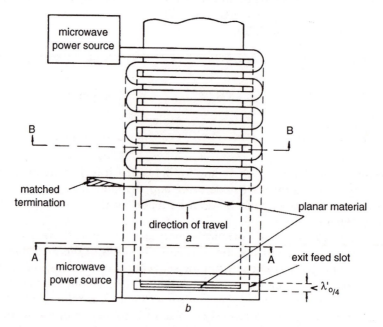

Figure 6.6   Serpentine applicator (Metaxas and Meredith, 1993)

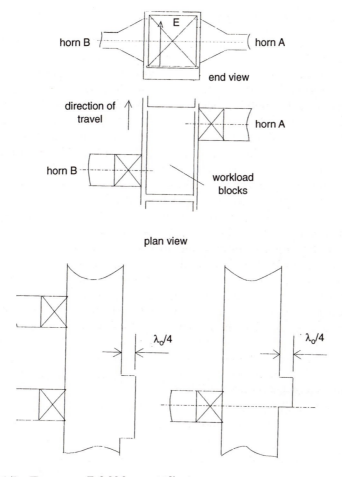

Figure 6.7   Transverse E-field horn applicator

a Plan view
b Two horns on each side
c Self-compensating single horn

The 180° waveguide bends between the passes can be formed as semicircular waveguide E-plane bends, though this is a very expensive method. Meredith (1969/72) proposed a coupling-strap arrangement in which the waveguide 'passes' were formed of folded sheet metal welded at the ends to a pair of side channels. The coupling straps bridged the space between adjacent waveguides, forming coupling loops to transfer energy from one to the other. This system is much less expensive than forming bends, and many applicators were made using it. Serpentine applicators have been used successfully in the food industry (final drying of potato crisps) and rubber industry (vulcanising extruded profiles).

6.3.3 Transverse E-field applicator

In this applicator a constant-value E-field intensity is developed by enlarging the aperture of a rectangular waveguide supporting the TE_{10} mode only, using a horn tapered to increase the height of the waveguide to suit the workload. For the TE_{10} mode, eqn. 4.37 shows that the E-field intensity E_y is independent of y although it varies sinusoidally with x. Provided that this field pattern can be expanded without distortion, this means that a workload traversing the face of the horn aperture in the x direction, at constant velocity, will receive a uniform energy input on that face. Meredith (c. 1976) developed this principle for tempering and thawing foodstuffs and treatment of organic waste, by alternately feeding energy on opposite sides of a workload of rectangular cross-section (Figure 6.7a).

Each injection of energy results in an exponential decay of the forward wave entering the workload, the residual wave on passing right through being reflected by the opposite tunnel wall to re-enter the workload creating a standing wave therein. The dissipated energy is therefore distributed as the sum of two mirror-image exponential distributions, with a superimposed standing wave. The total attenuation through the workload is determined by its thickness and the penetration depth, and is ideally in the range 4–8 dB, which is typical for commercial frozen-meat blocks and frozen-butter blocks at 915 MHz. The dashed curve of Figure 6.8 shows a typical heat distribution computed for a block of 266 mm square section with 6 dB total attenuation (4:1 power ratio) at 915 MHz, and with $\varepsilon' = 3.0$, typical of a butter block. The standing wave is clearly visible.

The electric field intensity E_{ri} within the block is readily calculated from the vector sum of the forward and reflected waves, as in Section 4.3, giving

$$E_{ri} = E_{01}\left(e^{-\alpha x}\, e^{-j\beta x} + e^{-\alpha(2L-x)}\, e^{-j\beta(2L-x)}\, e^{-j\phi} \right) \qquad (6.6)$$

where ϕ is the phase angle of the reflected wave re-entering the block on the far side of the horn, and is proportional to the distance to the tunnel wall. It is set to provide compensation for standing waves as shown below.

In practice $\phi_1 = \phi_2 + \pi$, where ϕ_1 and ϕ_2 are the phase angles of each of a pair of mutually compensating horns.

The standing wave is readily cancelled by using two horns on each side, with the tunnel wall on the far side recessed $\lambda_0/4$ deeper for one than the other (Fig. 6.7b). This displaces the standing waves due to the reflections from the wall such that the minima of one superimpose upon the maxima of the other, as shown by the dotted and dashed curves of Figure 6.8. There is also a lesser standing wave formed at the workload/air interface which in practice does not noticeably affect the uniformity of heating. The solid line of Figure 6.8 shows the resultant distribution from the two horns, where the two standing waves mutually cancel, leaving a simple exponential distribution.

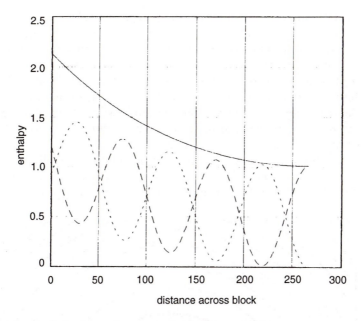

Figure 6.8   Enthalpy input due to a pair of compensated horns

For the second x-axis horn, $ExR2_{x,\alpha,\beta} = \{\exp(-\alpha x)\}\{\exp(-i\beta x)\}+$
$[\exp\{-\alpha(2L - x)\}][\exp\{-i\beta(2L - x)\}] - 1$

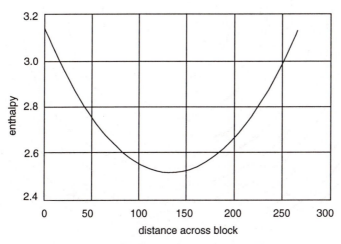

enthalpy ratio centre to face: 1.25:1
6 dB attenuation

Figure 6.9   Resultant enthalpy input from two pairs of compensated horns feeding from opposite sides of the block

Figure 6.7c shows a self-compensating single horn which gives a good approximation to cancelling the standing wave, having the advantage of using one horn only. However, where the power density approaches a limit with one horn, the configuration of Figure 7.7b is advisable.

Figure 6.9 shows the resultant due to adding an identical second pair of horns feeding from the opposite side to the first pair, so that there are two mirror-image exponential curves, like the solid curve of Figure 6.8, adding. The heat input to the exposed faces is only 25% higher than to the centre, although the attenuation through the block is 6 dB.

Figure 6.10 is a contour plot of the heat distribution in a block, as above, in which an identical second set of four horns project a field at right angles to the first. The isothermals are circles and, for the same properties as the

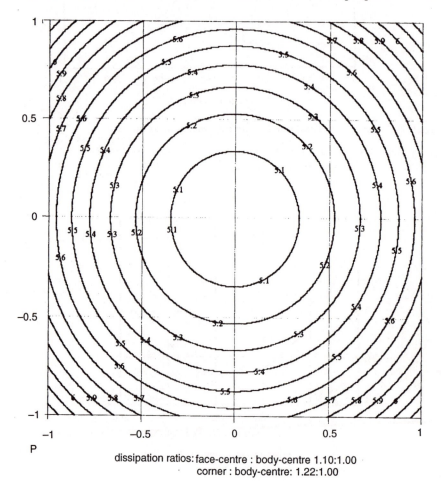

dissipation ratios: face-centre : body-centre 1.10:1.00
corner : body-centre: 1.22:1.00

Figure 6.10   Contour plot of enthalpy input to square block from two pairs of compensated horns on each axis O_x and O_y

above example, the face–centre enthalpy is only 10% more than the body–centre enthalpy, while the edges are 22% greater. These results are in good agreement with results obtained with production equipment used for thawing butter.

The junction of the horn with the process tunnel forms an overmoded T-junction from which some energy propagates in both directions along the tunnel. However, for the relatively high rate of attenuation of the typical workload, the rate of decay of these wavefronts is high provided that the tunnel is substantially filled with the workload, e.g. greater than 80% of the cross-sectional area. This allows the horns to be placed close together, a centre-line spacing of 4 times the block thickness giving a mutual cross-coupling between horns in the order –20 dB.

The high rate of attenuation along the tunnel permits a simple choke design, but great care must be taken to allow for loads only part-filling the tunnel, e.g. less than 80%, when leakage may become high.

It is obviously necessary for the power inputs to the horns to be equal, and this is readily achieved using 4-port hybrid junctions (Section 4.8.6). The magic-T provides a convenient configuration for power splitting.

The efficiency of this horn applicator is high: with 6 dB through attenuation of the workload the return loss is –12 dB, corresponding to 94% efficiency. The self-compensated horn of Figure 6.7c is inherently self-matched, yielding a higher efficiency.

Chapter 7

Microwave-heating applicators
3: resonant structures

7.1 Introduction

Some workloads have a such a low value of loss factor ε'' that the heat-dissipation density using applicators like those described in Chapters 5 and 6 is too low at the electric-field intensities they create. By using resonance, the field intensity can be raised considerably, giving satisfactory heat dissipation. In other cases, it is desired to generate a very high rate of rise of temperature (e.g. ≥ 100 deg C/s) in materials of modest or high loss factor, where again resonance may be used.

The form of the workload is important in choosing the applicator: filamentary materials, small in cross-section size compared with a wavelength, are straightforward to heat in a variety of resonant cavities, particularly the circular cavity operating in the TM_{010} mode (Metaxas and Meredith, 1993). Where the load is a wide sheet material, it is necessary to address the problem of standing waves: the VSWR in the workspace of a resonant applicator may be very high, as discussed in Section 7.5.1.

Except for plasma generation, resonant systems have not been widely used in production industry because of the difficulties of control. As the workload properties and other parameters of the system change with temperature and wear, both the cavity resonant frequency and the generator frequency drift, and power transfer falls. It is necessary to retune to restore the correct operation point, and in most cases this is required so frequently as to demand an automatic system. In the past, automatic frequency control has been difficult to accomplish reliably in a normal production environment, but this situation has now changed with the advent of microprocessor control systems. It is worth observing that plasma generation is a recent and important development of microwave heating in a very high-technology application where production is in the hands of highly qualified operators who readily understand frequency-control systems.

This chapter is a review of the basic principles of resonant-system operation with particular emphasis on practical features. Design procedures are adequately described elsewhere (Metaxas and Meredith, 1993).

7.2 Input impedance and matching

All single-mode resonant cavities have the same general frequency response as a simple elementary tuned circuit, such as the parallel resonant circuit shown in Figure 7.1 which can be considered an equivalent circuit of the cavity, connected to an input transmission line through a matching transformer of turns ratio n^2 :1. Note that this is just an equivalent circuit which enables the response to be evaluated using standard electrical theory; the resistors, inductor, capacitor and transformer do not exist in reality. The turns ratio of the transformer represents the size of the coupling hole of the cavity, and the resistors of conductance G_C and G_L represent the cavity losses and workload, respectively. The inductor and capacitor represent the resonant properties of the cavity.

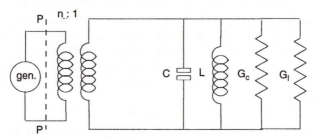

Figure 7.1   Simple equivalent resonant circuit with transformer coupling

The performance of the cavity represented is best visualised by considering the reflection coefficient ρ_i 'seen' looking into the cavity from the transmission-line feeder terminal at P. From the diagram, the input admittance can at once be written

$$Y_i = n^2\left(G_C + G_L + j\omega C + \frac{1}{j\omega L}\right) \tag{7.1}$$

Making the substitutions

$$\omega_0 = \frac{1}{\sqrt{(LC)}} \tag{7.2}$$

and

$$Q_{L1} = \frac{1}{\omega_0 L\left(G_C + G_L\right)} \tag{7.3}$$

$$\rho_i = \frac{Y_i - Y_0}{Y_i + Y_0} \tag{7.4}$$

and the approximations for small frequency excursion about the resonant frequency ω_0:

$$\delta\omega = \omega_0 - \omega \tag{7.5}$$

$$\omega_o = \frac{\omega_0 + \omega}{2} \tag{7.6}$$

the input reflection coefficient is, after some algebraic manipulation,

$$\rho_i = \frac{(g_C + g_L)\left(1 + j 2 Q_{L1} \dfrac{\delta\omega}{\omega_0}\right) - 1}{(g_C + g_L)\left(1 + j 2 Q_{L1} \dfrac{\delta\omega}{\omega_0}\right) + 1} \tag{7.7}$$

where

$$\frac{n^2 (G_C + G_L)}{Y_0} = g_C + g_L \tag{7.8}$$

If the cavity is a perfect, reflectionless match at the resonant frequency $\omega = \omega_0$, then $\rho = 0$, and in eqn. 7.7 $(g_C + g_L) = 1$. For a cavity thus matched at resonance, eqn. 7.7 gives the reflection coefficient ρ' off resonance as

$$\rho' = \frac{j Q_{L1} \dfrac{\delta\omega}{\omega_0}}{1 + j Q_{L1} \dfrac{\delta\omega}{\omega_0}} \tag{7.9}$$

Eqn. 7.9 is the particular case of a cavity, perfectly matched when on tune ($\delta\omega = 0$). Its reflection-coefficient ρ' locus, as the frequency difference $\delta\omega$ changes, corresponds to the $(1 + jb)$ admittance circle on the Smith chart. For small changes of frequency about the resonant point, $j\rho'$ changes linearly, changing sign as the frequency passes through ω_0. A detection circuit measuring $j\rho'$, and responding to this phase reversal, therefore provides a signal for measuring the frequency error, and can be used for automatic frequency control. For greater frequency error, the signal departs from linearity, giving a voltage output as shown in Figure 7.2, which is the classic shape of a frequency-discriminator response.

If the cavity is not matched at resonance, the reflection coefficient given by eqn. 7.7 still follows a circular locus on the Smith chart, but it intercepts the $(g + j0)$ axis at the value of g corresponding to the circuit conductance at

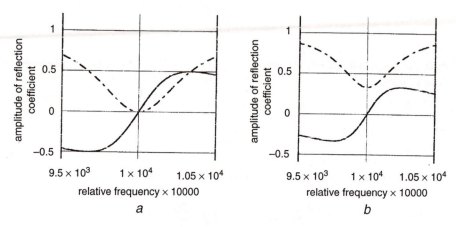

Figure 7.2   Real and imaginary parts of reflection coefficient of resonant circuit against frequency close to resonant frequency

   a Cavity matched at resonance, $g_0 = 1.0$, $Q = 30$
   b Cavity undercoupled, $g_0 = 0.5$, $Q = 30$
   ———— Im(ρ)
   — · — Re (ρ)

resonance. If g is greater than 1.0, the cavity is said to be undercoupled, and overcoupled if g is less than 1.0.

7.3 Open-loop frequency stability

Automatic frequency control (AFC) is essential for satisfactory operation of a resonant applicator if Q_{L1} is greater than 50. At this Q-value, a frequency drift between cavity and generator of 1%, by definition, gives up to a 2:1 change in the power dissipated in the workload; the inherent frequency stabilities of cavity and generator require careful assessment, relating the desired power stability to the predicted frequency errors which may occur from all causes. Typical causes of frequency error are:

Generator:
(i)   warm-up, short-term and longer-term with coolant temperature, especially if closed-circuit cooling system is used;
(ii)   frequency-pushing of magnetron due to anode-current variation;
(iii) frequency-pulling of magnetron due to mismatched load;
(iv)   aging of magnetron (usually very small);
(v)   mechanical vibration and shock;
(vi)   frequency modulation at power frequency and harmonics thereof due to EHT ripple, via (ii) above; and
(vii) frequency modulation at power frequency and harmonics thereof, due to filament-current modulation.

Load:
(i)   warm-up drift of cavity body;
(ii)   warm-up drift of tuner and drive mechanism;
(iii)   variation of cross-section dimensions or density of workload in cavity;
(iv)   variation of dielectric properties of workload ε' and ε'', due to composition variation, temperature (start, finish and gradient), moisture content etc.;
(v)   mechanical vibration and shock;
(vi)   lost movement and 'backlash' in tuner drive mechanism;
(vii)   friction in drive mechanism;
(viii) inertia of drive mechanism;
(ix)   excess clearance in tuner piston guides; and
(x)   accretion of condensation and foreign matter.

Where a closed-loop AFC system is used, there can also be frequency errors introduced by spurious signals in the electronic control system, for example a DC offset in an amplifier.

7.4 Tuning techniques

Reducing the frequency error between generator and resonant applicator can be done in several ways, based on controlling the generator frequency, or the cavity frequency, or both, by using microwave feedback. In general, controlling the generator frequency is the less satisfactory because it is difficult to modulate the frequency without introducing some amplitude modulation at the same time: with a magnetron, modulating the anode current gives direct amplitude and frequency modulation together. It is possible, but difficult, to control magnetron frequency over a limited range by introducing a variable susceptance to its load, modulating the load impedance in a locus parallel to the constant-power lines on the Rieke diagram. Tunable magnetrons would be the basis of a good method, but none are available and they would be very expensive to develop.

7.4.1 Applicator tuning

Tuning the applicator cavity gives the most practical AFC system, with the generator well isolated from the impedance variations of the cavity by circulator(s), and operated from a stable power supply to give as 'clean' a spectral output as possible.

The cavity may be tuned directly by adjusting its overall length with a tuning piston; the position accuracy is critical and requires a precision mechanism to drive the piston rectilinearly without imparting any side thrust or torque, and the piston itself must be guided with minimal mechanical side clearance. The tuning piston should preferably be noncontacting, as described in Section 4.8.3. These are heavy pistons, and the drive motor must have sufficient peak torque to give the estimated maximum acceleration required.

It is readily shown that the piston movement $\Delta l'$ to tune the cavity between the halfpower (-3 dB) points on the resonance curve is

$$\Delta l' = \frac{l_0}{Q_L}\left(\frac{\lambda_g}{\lambda_0}\right)^2 \tag{7.10}$$

For a typical rectangular waveguide cavity operating in the TE_{101} mode at 2450 MHz, with $\lambda_g = 160$ mm, $l_0 = 80$ mm, and $Q_L = 300$, eqn. 7.10 gives $\Delta l' = 0.46$ mm. Typically, for satisfactory power stability, the position accuracy would need to be less than 3% of this, i.e. <0.014 mm ($\pm <0.007$ mm). This is obviously an expensive system because of its precision. However its wide tuning range is advantageous if the workload has widely variable properties, enabling any to be brought to resonance within reasonable limits.

If the workload is specific, with minimal variation in properties, a perturbation tuner is preferable because it is less critical in tuning rate. Such tuners may take a variety of forms, generally comprising a rod which couples to the electric field in the cavity by an amount depending on its insertion. The rod may be metal or dielectric. In practice, a dielectric rod is simpler because it can be made of a diameter small enough to form a cutoff waveguide choke (Section 4.6.1) where it passes through the wall of the cavity. A metal rod requires choking, which can be difficult for the high field intensities present, which may cause arcing. Another form of the dielectric tuner is a vane which is rotated to vary its coupling to the E field. Both these tuners may be visualised as perturbators in a parallel-plate capacitor where, inserted through one of the plates, they clearly increase the value of the capacitance; if the capacitor is in resonance with an inductor the resonant frequency will be reduced as the rod is inserted.

An estimate of the frequency shift due to a small perturbation may be made by the energy method of Slater (1950). As the excitation frequency rises from a low value, the energy in the magnetic field falls, while that in the electric field rises; resonance occurs at the frequency at which the energies stored in the two fields are equal. If a perturbation is introduced which affects the energy balance, the resonant frequency will change to restore the status quo. Such perturbations as those described above affect principally the electric field.

Slater shows that the perturbation causes a small change in the stored energy balance ΔU in the cavity which can be related to the incremental change $\Delta f / f$ in resonant frequency as

$$\frac{\Delta f}{f} = \frac{\displaystyle\int_{\Delta V}\left(\mu H^2 - \varepsilon_0 E^2\right)dV}{\displaystyle\int_{V}\left(\mu H^2 + \varepsilon_0 E^2\right)dV} = \frac{\displaystyle\int_{\Delta V}\left(\mu H^2 - \varepsilon_0 E^2\right)dV}{4U} \tag{7.11}$$

where U is the total stored energy.

For the simple case of a rectangular waveguide cavity operating in the TE_{101} mode, consider a small lossless dielectric rod (ε') inserted at the centre of a broad face, midway along the axis. As the magnetic field is zero and the electric field is at its maximum at this point, there will be maximum disturbance to the energy balance. At resonance, the total stored energy U equals the time maximum in the electric field, because the magnetic field at that instant is zero. Thus U can be obtained by integrating E over the volume of the cavity:

$$U = \frac{\varepsilon_0}{2} \int_0^a \int_0^b \int_0^L E_0^2 \sin^2 \frac{\pi x}{a} \sin^2 \frac{\pi z}{L} dx\, dy\, dz \qquad (7.12)$$

$$= \frac{\varepsilon_0\, a\, b\, L}{8} E_0^2 \qquad (7.13)$$

The change in stored energy in the electric field caused by the rod is

$$\Delta U = -\frac{\varepsilon_0}{4} \frac{\varepsilon' - 1}{\varepsilon'} E_0^2\, \Delta V \qquad (7.14)$$

where ΔV is the volume of the inserted rod.

Combining eqns. 7.11 with eqns. 7.13 and 7.14, the fractional frequency shift due to the rod becomes

$$\frac{\Delta f}{f_0} = -2 \frac{\varepsilon' - 1}{\varepsilon'} \frac{\Delta V}{V} \qquad (7.15)$$

If the rod is metal, eqn. 7.15 applies with $\varepsilon' = \infty$. It is worth noting that an alumina rod ($\varepsilon' = 8.9$) is almost 90% as efficacious as a metal rod of the same volume, but eliminates the problem of choking.

For a circular cylindrical cavity operating in the TM_{010} mode, Ramo *et al.* (1965) derive a similar expression, modified for a dielectric rod ε' as above:

$$\frac{\Delta f}{f_0} = -1.85 \frac{\varepsilon' - 1}{\varepsilon'} \frac{\Delta V}{V} \qquad (7.16)$$

Using eqn. 7.15 the tuning rate of a perturbation tuner 12.5 mm in diameter can be estimated for the cavity example in Section 7.4.1. Three materials are chosen, and the tuning rod is assumed inserted at the point of maximum E-field. Calculated tuning rates are as in Table 7.1

These tuning rates should be compared with that for the short-circuit piston discussed in Section 7.4.1 which has a tuning rate of 17.4 MHz/mm.

Table 7.1   Comparative tuning rates of perturbation tuners: TE_{101} mode rectangular waveguide cavity at 2450 MHz

Material	ε'	Tuning rate
		(MHz/mm)
Metal	∞	2.0
Alumina	8.9	1.8
PTFE	2.03	1.03

7.4.2 Generator tuning

Tuning the generator to the resonant frequency of the load cavity has severe limitations, either in tuning range or in cost, depending on the sophistication of the system.

Contemporary industrial magnetrons are all fixed frequency, and the only scope for frequency control is via the pulling figure, using a controllable mismatch, or the pushing figure where the anode current is modulated to control the frequency. Neither is entirely satisfactory because they both affect the power output, which is undesirable.

Mechanically tuned magnetrons are well known for radar use, but the costs of development and manufacture are high, and would be difficult to justify.

High-power klystrons and travelling-wave tubes (TWTs) used as amplifiers have a broad frequency band of substantially constant power output, and can be driven from a variable-frequency oscillator with rapid slew rate. Although this system forms the technically ideal system, klystrons and TWTs are very expensive and could be justified in but few applications.

7.5 Frequency discriminators

For an AFC system to operate, it is necessary to derive a signal which determines the amount of frequency error between the generator and cavity resonant frequency, and in addition to know whether the latter lies above or below the generator frequency. The method, described in Section 7.5.1, uses the change in reflection coefficient of the cavity as a function of frequency.

7.5.1 Microwave frequency discriminators

The variation of cavity reflection coefficient ρ_c can be used to provide all the data required for an AFC system, the cavity having a response as given by eqn. 7.7, and by eqn. 7.8 for a cavity perfectly matched at its resonant frequency. Figure 7.2 shows how the real and imaginary parts of ρ_c vary as a function of frequency offset from resonance for the cavity represented by eqns. 7.7 and 7.8. Two cases are shown:

(*a*)   for a cavity matched at its resonant frequency, and

(*b*)   for an undercoupled cavity having a VSWR of 0.5 at resonance.

It will be seen, in particular, that the imaginary part is zero at resonance, has a negative value below resonance, and is positive above, both increasing in amplitude with the amount of the offset until their respective peak values are reached, and then slowly declining. These are amplitudes of signals at microwave frequency and must be processed to preserve their data as equivalent DC signals.

This signal processing comprises combining signals proportional to Im (ρ_c) with two microwave reference signals, as shown in the vector diagrams in Figure 7.3, rectifying the resultants in a pair of microwave diodes, and then adding their DC outputs to give an output signal for power amplification in closed-loop control of the cavity tuner. Figure 7.4 shows a schematic microwave circuit providing the necessary signals. The microwave signals are at milliwatts level appropriate for high-level rectifier diodes, and are derived from the main waveguide feed from the generator to the load cavity using directional couplers typically with coupling factors in the range 40–70 dB, with directivity >30 dB. The subsequent microwave circuit can be realised in conventional waveguide, but in volume production would be much less expensive, and certainly more compact, in microwave stripline.

For simplicity, Figure 7.3*a* shows the vector diagram for a cavity perfectly matched at resonance, while Figure 7.3*b* shows the general case of a cavity overcoupled and therefore mismatched at resonance. Referring to Figure 7.3*a*, a pair of oppositely phased reference signals V_{ref1} and V_{ref2} are added vectorially to the signal [Im(ρ_c)], the phase angle adjusted to be in-phase and phase-opposition, respectively, with Im(ρ_c). The resultant signals [V_{ref1} + Im(ρ_c)] and [V_{ref2} - Im(ρ_c)] are rectified in the diodes A and B and compared. If the cavity is on tune, the signal $\rho_c = 0$, and the diodes have equal output, giving zero differential output from the diodes, as desired. If the cavity is off-tune, one diode will have higher output than the other, the polarity of the output from the comparator then depending on the 'sense' of the offset. The amplitude of the reference signal should normally be 6 dB greater than the largest possible value of Im(ρ_c), and the largest resultant should not cause the rectifier diode to depart more than 1 dB from linearity. Two diodes are used in a differential arrangement, as shown, to minimise spurious errors from changes in power output which directly affect the amplitude of the reference signal. If a single diode were used, such a change in generator power would cause an apparent frequency-error signal, resulting in a frequency 'squint'. The 3 dB hybrid junctions inherently provide the required phase reversals of the reference signals to the diodes: for a magic-T via the phase inversion of the outputs from its E-plane T-junction, and for a branch-arm coupler via the $\pi/2$ phase delay in traversing through the slots. In the latter, the signal and reference each cross from one waveguide to the

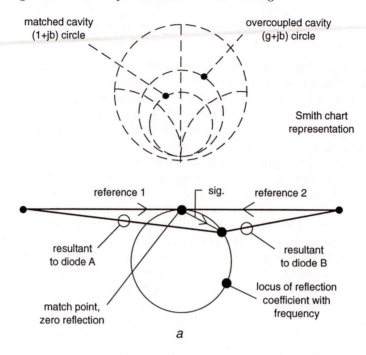

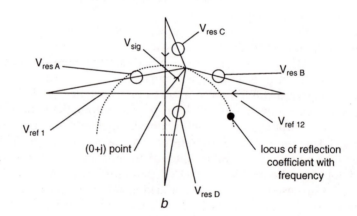

Figure 7.3   Vector diagrams of frequency discriminator

a Cavity matched at resonance
b Derivation of frequency and coupling-error signals

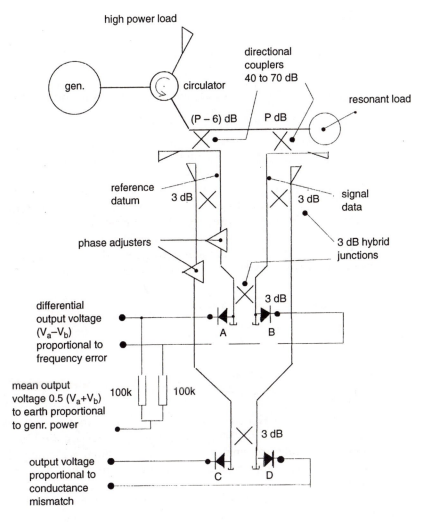

Figure 7.4 Schematic microwave-circuit diagram of frequency discriminator for resonant loads

other, the signal being delayed for diode A, while the reference is delayed for diode B; the effect is $2 \times \pi/2$ phase delay.

In Figure 7.3*b* the vectors are drawn for an overcoupled cavity, although this is not a restriction in the analysis, and the output signals from the diodes can be determined by trigonometry, using the symbols as shown in the diagram, i.e.

Diode A:

$$V_{DA} = (V_{ref1} + \rho_c \cos \phi)^2 + (\rho_c \sin\phi)^2 \qquad (7.17)$$

Diode B:

$$V_{DB} = (V_{ref2} - \rho_c \cos\phi)^2 + (\rho_c \sin\phi)^2 \qquad (7.18)$$

Hence, assuming that $|V_{ref1}| = |V_{ref2}|$, the output from the comparator is

$$V_{comp\,1} = V_{DA} - D_{DB} = 4 \, V_{ref}\rho_c \cos\phi = 4 \, V_{ref}\,\mathrm{Im}(\rho_c) \qquad (7.19)$$

The output from the comparator is a DC signal proportional to the frequency offset, suitable for amplification and filtering before applying to the tuner drive.

Also shown in Figure 7.3*b* is a second detector system but with the reference vectors in quadrature to those previously. With identical processing, they provide a signal proportional to the reflection coefficient of the cavity when tuned to resonance, and can be used as a signal for optimally matching the cavity coupling independently from the frequency-tuning system. The outputs from diodes C and D and comparator Comp 1 are:

Diode C:

$$V_{DC} = (V_{ref3} + \rho_c \sin\phi)^2 + (\rho_c\cos\phi)^2 \qquad (7.20)$$

Diode D:

$$V_{DD} = (V_{ref4} - \rho_c\sin\phi)^2 + (\rho_c\cos\phi)^2 \qquad (7.21)$$

Again assuming that $V_{ref\,3} = V_{ref\,4}$,

$$V_{comp\,2} = V_{DC} - V_{DD} = 4 \, V_{ref}\,\rho_c\sin\phi = 4 \, V_{ref}\,\mathrm{Re}(\rho_c) \qquad (7.22)$$

In practice, the cavity is designed to be matched at resonance when loaded with a typical average workload, by adjusting the size of its coupling hole to its feed waveguide (Metaxas and Meredith, 1993). Final matching then becomes a relatively fine adjustment and can be accomplished by a single stub tuner, for example a series or shunt T-junction with a terminating adjustable short-circuit piston, driven from the error signal of eqn. 7.22 via an amplifier.

In principle, it is possible to use the on-load tuner described in Section 4.8.7.1, but in practice this is not satisfactory unless the cavity Q_L is low (< 30) because of the high fields present, which would extend into the magic-T and waveguides. Voltage breakdown may be a problem in the magic-T, and power is wasted in additional I^2R heating. The effective volume of the cavity is much enlarged, increasing the stored energy, and possibly raising the Q_L value, depending on the ratios of stored to dissipated energy in the main cavity and the extended cavity

These closed-loop control systems are subject to the usual stability criteria of feedback systems, for which an extensive literature exists (e.g. James *et al.*, 1947; Distefano *et al.*, 1990). Having drive motors, they are in principle second-order servomechanisms, or velocity-error systems. The principal delay is in the inertia of the mechanical drive to the tuning device; usually all other systems are, in comparison, almost instantaneous in response.

Care must be taken to avoid saturation of the electronic circuits by large signals due to noise and residual FM of the magnetron due to power-supply ripple; some filtering may be required to combat this problem, which will introduce a further delay in the control loop.

At initial start-up the cavity may be so far off tune that there is insufficient error signal $V_{comp\,1}$ for the system to respond. This is more likely with high-Q_L cavities, or where there is a wide range of possible workloads. A solution is to provide a field-intensity-sensor diode coupled to the cavity to detect the onset of resonance when the cavity is tuned unidirectionally, in open-loop, at a modest rate. When the sensor detects the onset of resonance, the system automatically reverts to closed loop.

7.6 TM$_{010}$ circular waveguide cavity

The TM$_{010}$ circular waveguide cavity (Metaxas and Meredith, 1993) is particularly suited to heating filamentary materials in continuous flow, having its E field polarised parallel to the cavity axis, and being at its maximum value on the axis, as shown in Figure 7.5. The E field diminishes radially from the centre as $J_0(r/R)$ giving a good uniformity of heating near the axis. Metaxas (1974) analyses the heat distribution in detail, making due allowance for the dielectric properties of the workload. In principle, there is no circumferential variation in heating intensity. An interesting, but inconvenient, feature is that the resonant frequency is determined by the effective electrical diameter of the cavity, i.e. the mechanical diameter with an allowance for the dielectric filling. Tuning can only be done by perturbation methods, usually by inserting an array of dielectric rods through one end wall of the cavity.

The microwave feed system is via a slot or hole, the size being determined experimentally to give a good impedance match at resonance with the projected workload, as described by Metaxas and Meredith (1993).

In practice, the presence of the perturbation tuners, and more especially the microwave feed system introduces nonuniform heating circumferentially. For small-diameter filaments the effect is negligible; however, for large diameters ($r/R > 0.05$) it may become significant, especially if $Q_L < 50$. Two or more feed slots may be used to reduce the nonuniformity.

There is a risk of arcing where the workload enters and leaves the cavity through the end walls because the displacement current in the workload 'returns' to the cavity wall through the capacitor formed by the workload (one 'plate') and the cylindrical hole in the end wall. A very high field strength may exist at this point with a risk of voltage breakdown. Good practice is to have a large-radius edge where the hole intersects the cavity end wall, and for the hole itself to be of large diameter, tapering to a smaller

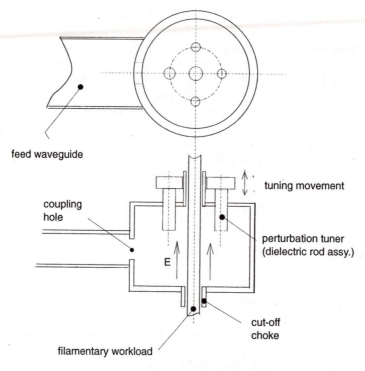

feed waveguide

tuning movement

coupling hole

perturbation tuner (dielectric rod assy.)

E

cut-off choke

filamentary workload

Figure 7.5   TM$_{010}$ circular-waveguide resonant applicator

diameter away from the cavity at about 30° included angle. The effect obviously increases with ε', and workloads with loose fibres require special care.

If the cavity is to be used for processing wet materials, e.g. dye fixation or drying of textile fibres, the power density may be so high as to cause spray from violent boiling which may precipitate arcing.

7.7 Transverse rectangular waveguide cavity, TE$_{10N}$ mode

Essentially, the transverse rectangular waveguide cavity, TE$_{10N}$ mode, is a plain rectangular waveguide with short-circuit plates each end, and a slot or small hole in the centre of each broad face to allow passage of the workload. One short-circuit plate is adjustable axially for tuning and is usually a noncontacting choked piston (Section 4.8.3). The other has a coupling hole for matching at resonance. The length of the cavity is N half wavelengths at resonance.

For $N = 1$ this cavity becomes a rectangular equivalent of the circular TM$_{010}$ cavity (Section 7.6) and has the advantage of tuning directly with a piston.

The heat distribution falls sinusoidally from the peak position, and for small filamentary workloads is substantially uniform. However, the axial position of the standing-wave peak is midway between the end plates, and clearly moves as the tuning piston is adjusted. The position of the workload-entry holes should be set midway between the end plates when the cavity is loaded and tuned to resonance, when the uniformity will be optimised. However, this is sometimes difficult to arrange where there is a wide variety of workloads. A further development is to have a pair of pistons mechanically coupled to be in contrary motion so that the midpoint location of the voltage maximum remains fixed in position, overcoming the above problem.

This cavity may also be used as one of two or more for heating a wide sheet (e.g. $W > \lambda/10$), when $N > 1$. The axial standing wave gives a heating pattern varying as $\sin^2(\pi z/\lambda_g)$ across the width of the sheet. A complementary waveguide alongside, with its standing-wave pattern displaced $\lambda/4$, gives a heating pattern in the sheet varying as $\cos^2(\pi z/\lambda_g)$. As long as the amplitudes in the waveguides are equal, the resultant heating uniformity is, in principle, uniform, because $\sin^2(\pi z/\lambda_g) + \cos^2(\pi z/\lambda_g) = 1$.

This superficially attractive arrangement of two cavities has several difficulties in practice. Equalising the effective amplitudes requires that the cavities have identical input-impedance matches and equal values of Q_L. Where the dielectric properties of the workload are affected by processing they are obviously different in the two waveguides. Moreover, the sinusoidal heat distribution impressed by the first cavity may cause warping of the workload due to the temperature gradients. It may be possible to tune the two cavities from a single drive if $Q_L < 50$, but otherwise separate tuners may be required.

The same precautions must be taken against arcing as described in Section 7.6.

7.8 Travelling-wave applicators

The standing waves inherent in the resonant applicators described above can, at least in principle, be avoided by using travelling-wave resonance where a closed-ring transmission line is fed via a directional coupler so that energy is coupled unidirectionally and builds up through multicirculation. When the ring mean perimeter length $s = n\lambda_g$ (where n is an integer), the stored energy in the ring builds up because the wave travelling around the ring is precisely in phase with that entering through the coupler. Figure 7.6 shows the arrangement, and it is readily shown (Miller, 1960; Suzuki *et al.*, 1975) that there is an optimum value of coupling factor for the directional coupler as a function of the attenuation of the workload to give zero power transferred to the dummy load (i.e optimum efficiency).

The field intensity in the ring is calculated by adding the circulating components of field in the ring, forming an infinite convergent geometric

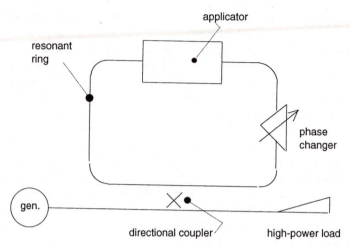

Figure 7.6   Travelling-wave, or resonant-ring, applicator: schematic waveguide system

series, the sum of which is well known (e.g. Pipes, 1946) . This procedure gives the voltage 'magnification' of the circuit M, defined as the E field in the ring divided by the E field incident from the generator, as

$$M = \frac{C}{1 - \tau e^{j\theta} \sqrt{(1 - C^2)}} \qquad (7.23)$$

where C is the voltage-coupling factor of the directional coupler (i.e. 20 log C decibels), τ is the voltage attenuation in the ring (i.e. 20 log τ decibels) and θ is the phase shift around the ring; at resonance $\theta = 2n\pi$.

Figure 7.7 shows the amplification as a function of attenuation in the ring and the coupling factor of the directional coupler. Note that the peak (optimum) coupling is a function of the attenuation of the ring. Also note that, if the coupling is too tight for a given insertion loss in the ring, there is a looser coupling which gives a similar value of amplification. This may present a problem where the effective directivity of the coupler (which may be enhanced by reflections within the ring) is low, resulting in the excitation of a backward wave with a consequent unwanted standing wave in the ring. This latter effect is the main operational difficulty with travelling-wave resonant applicators, and care must be taken to ensure a good match at all times in the ring, and also to use a high-quality directional coupler. A special ferrite isolator giving a small reverse attenuation of about 3 dB and low forward loss (<0.02 dB) would substantially eliminate this problem.

Figure 7.8 shows the resonant response of the ring as a function of the overall electrical phase angle around the ring. It shows a classic resonant response, and the Q_L is readily evaluated.

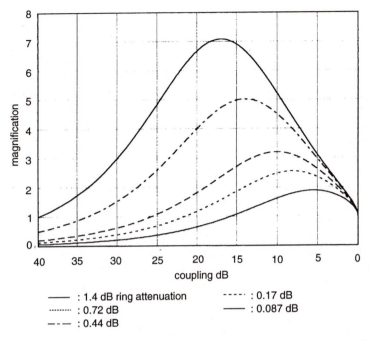

Figure 7.7 Amplification of travelling-wave resonator against ring attenuation and coupling factor of directional coupler

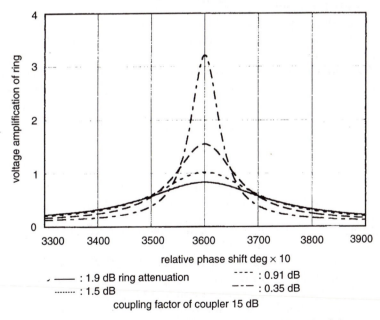

Figure 7.8 Amplification of resonant ring against electrical length of ring

Tuning of the ring can be accomplished by an adjustable phase-shift dielectric slab, taking care to ensure that the high field strength is compatible with the loss factor of the chosen dielectric material to avoid overheating.

Suzuki *et al.* (1975) show that an effective 'power amplification' of up to ×10 is realisable with a travelling-wave resonator, compared with an aperiodic waveguide, and report equipment for the continuous heat tempering of polyester rope, providing heat setting with 25% increased tensile strength and 30% decrease in elongation. The heat setting prevents the rope from untwining when cut.

7.9 Near-cutoff TE_{10} resonant rectangular waveguide

An elegant resonant applicator (Twistleton, 1997) was designed by the late Dr. J. Curran of the former BT-H Research Laboratory, Rugby, UK, circa 1960, for drying very thin continuous-flow web materials. It comprises a rectangular waveguide operating very close to cutoff in the TE_{10} mode so that the waveguide wavelength is considerably longer than the width of the web. The waveguide forms the centre section of a TE_{101} resonant cavity, the end sections, of normal waveguide size, being nominally $\lambda_g/4$ long, as shown in Figure 7.9, and closed at their extremities. The impedance 'seen' looking outwards from the centre section is therefore very high, allowing a high-value E field of cosine distribution axially, which, by symmetry, is centred at the midpoint of the centre section. Thin slots are cut along the centre line of the broad faces of the centre section to allow passage of the web.

The standing wave formed in the centre section causes the heating intensity to vary across the width of the web as $(E_0^2 \cos^2 4\pi x/\lambda_g)$ where x is the distance from the centre of the web. If, for example, the web width $w = 0.5$ m and the uniformity required allows 20% reduction of heating at the edges (adequate in a drying operation) where $x = w / 2$, λ_g is given by:

$$\cos^2\left(\frac{4\pi w/2}{\lambda_g}\right) = \cos^2\left(\frac{2\pi w}{\lambda_g}\right) = 0.80$$

whence $$\lambda_g = \frac{2\pi\ 0.5}{\cos^{-1}\sqrt{0.80}} = 2.2361 m$$

(7.24)

Using eqns. 4.34 and 4.35, and recalling that for the TE_{10} mode $\lambda_c = 2a$, the width of the centre section can be calculated.

For a system operating at 915 MHz and $\lambda_0 = 0.3276$ m, this gives the waveguide width of the centre section $a = 0.1656$ m (165.6 mm). The cutoff width at 915 MHz is 163.8 mm, i.e 1.8 mm less.

In practice, a tuner is used for adjustment to resonance, comprising a dielectric rod along the length of the centre section, adjustable in position to

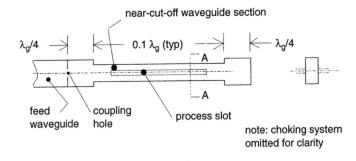

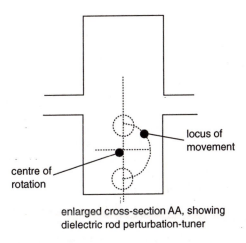

enlarged cross-section AA, showing
dielectric rod perturbation-tuner

Figure 7.9   Near-cutoff rectangular-waveguide resonant applicator

control the coupling to the E field. It operates as a perturbation tuner (Section 7.4.1). The centre section then becomes a dielectric-loaded waveguide, and may be dimensioned to be beyond cutoff in its completely empty condition.

Although the tuner relieves the tight tolerance required on the absolute dimensions, the applicator must be made with precision, particularly in respect of parallelism of the centre waveguide section. Moreover, it must be constructed as a rigid assembly adequately braced to avoid modulation from mechanical vibration.

Care must be taken to ensure that symmetry is preserved about the centre line of the centre section to avoid unbalancing the wall currents near the process slot, as this would result in increased microwave leakage. Residual leakage can be controlled by an array of choke pins (Section 8.4).

Chapter 8

Choking (attenuation) tunnels for continuous-flow applicators

8.1 Specifications

Continuous-flow, conveyorised applicators necessarily have open apertures for the passage of workload through the oven. Choke tunnels are provided at the entrance and exit to reduce the high field strength in the oven to an acceptable value at the open ends. The objectives of the choke tunnel are threefold:

(i)   to reduce the residual microwave leakage to a level less than prescribed as the maximum for health and safety of personnel. For many years, in the developed countries this level has been agreed for domestic microwave ovens at 10 mW/cm^2 (100W/m^2), measured at 50 mm from the surface of the equipment. This level was arbitrarily adopted for industrial equipment too, but it is open to interpretation because of the size of the equipment. Recently the safe level has been more objectively established in the United Kingdom (NRPB, 1993) for industrial equipment, in terms of bodily exposure; see Section 8.1.1 and Chapter 13.

(ii)   to minimise the waste of power lost from leakage, and within the choke tunnel from resistive absorbers;

(iii)   in some cases, to reduce the residual leakage to a level below the maximum   permitted for avoiding radio interference with communication and other services. This requirement applies particularly in countries where a 900 MHz frequency has not been allocated for industrial, medical and scientific use (an ISM band). Principally this is the whole world except north and south America, and Great Britain. It is a problem which has arisen from lack of foresight by planners in the 1950s; it has resulted in a specification by far the most exacting, giving a theoretical total maximum leakage of 0.3 µW.

Details of the relevant specifications and their implications are reviewed in Chapter 13 and are summarised below.

8.1.1 Safety

Safety in this context refers to the level of continuous microwave field intensity emanating from the equipment and presenting a possible hazard to persons in the vicinity. The domestic-oven specification formerly adopted for industrial equipment is not meaningful because the zones from which leakage predominates are not normally accessible in the context of 50 mm, for example, the open ports of a conveyorised oven. The operator is usually at some distance from these zones, certainly more than 50 mm. Moreover, the hazard is now agreed to be thermal, and the exposure of the body is more objectively set in terms of bodily-temperature rise, so that the energy absorbed is the critical parameter. The matter is considered further in Chapter 13 but in summary the whole-body exposure-level limit is set at 100 W/m^2 (10 mW/cm^2) at 2450 MHz and 50 W/m^2 (5 mW/cm^2) at 900 MHz, averaged over an exposure time of 6 min (NRPB, 1993). This limit is an 'investigatory' limit and could in specific cases be allowed to increase on official scrutiny.

8.1.2 Radio-frequency interference

The principal interference problem arises in the 900 MHz bands outside north and south America, where the frequencies 915 MHz and 896 MHz coincide with those allocated to the mobile-telephone network. Elementary calculations of the tolerable level of an interfering signal, based on the receiver noise figure, antenna gain and transmitted power resulted in a specification limit of 100 µV/m measured at a distance of 30 m from the building housing the equipment. It corresponds to the signal from an isotropic source with a power of 0.3 µW, indicating the magnitude of the task of choking typical industrial equipment operating at a power output of many kilowatts. This limit is CISPR specification 16 (CISPR, 1993), and is measured with a receiver having pre- and post-detection characteristics specified in CISPR Document 11 (CISPR, 1990). The derivation of this figure assumes that the ISM equipment is truly monochromatic in its spectral purity, and is absolutely stable in frequency. In practice this is far from true, and the power spectrum speads over a significant frequency band, diluting the interference effect. There is a continuing debate among the interested parties about a more realistic specification. Experience in the UK over some ten years is that the incidence of interference with the mobile-analogue-telephone system from equipment which was designed with safety as the principal objective is minimal.

In the UK, the frequency 896 MHz has been allocated for ISM use since the early 1950s and the permissible leakage on an RFI basis was set at 1 V/m;

this generally exceeds the personal-exposure safety limit, which is the relevant limit, and so in practice is of academic interest only.

8.2 Classification of choking methods

Many techniques are available for choking open tunnels, and usually several are employed together:

(i)   reactive choking which uses lossless-resonant-filter techniques to create a high-value reflection coefficient ($\rho\rightarrow1.0$), effectively reflecting leakage power back into the oven. As the aperture increases in size, these reactive filters become less effective and more difficult to design. Shaped reflectors provide useful reflection where the aperture is large, supporting several propagating modes.

Reactive choking must be used first (next to the oven) because the primary leakage power is usually too great for absorptive (lossy) materials to be used without risk of their overheating;

(ii)   resistive absorption by the workload. This is a very simple method requiring a plain tunnel. It is used extensively, but care must be taken to ensure that there remains sufficient attenuation when the equipment is operating with a workload whose cross-sectional area is significantly less than that of the tunnel. There must be a backup safety system for protection with an empty tunnel;

(iii)   resistive attenuation within the choke tunnel by lining it with lossy material;

(iv)   a dielectric diffuser placed around the aperture, which allows the leakage to escape, but over a greater area than that of the aperture itself so that the leakage-power density is reduced. Note that it does not reduce the total leakage power it merely spreads it over a greater area and prevents people from making close contact with the high field intensity at the aperture. It is a useful device for gaining up to 10 dB extra attenuation in cases where the leakage from the aperture is close to the limit allowed. This technique is ineffective in reducing leakage for interference, where the total power radiated is the prime consideration, rather than the power density;

(v)   an active microwave electronic sensor (a diode detector) which is placed to couple to the field in the choke tunnel, and which automatically clamps the generator to zero power output if a preset threshold level is exceeded. The clamp can operate in about 50 ms, or less with a switch-mode power supply. It can also be arranged to allow the power to rise again after an appropriate delay to resume production if the cause of the leakage was a temporary effect;

(vi)   a microwave fuse which is, in principle, a fail-safe protection. The fuse is fitted near the open aperture of the choke tunnel with an antenna (usually a simple probe or loop) feeding a microwave current through the fuse. Simultaneously, a pilot current flows through the fuse which activates the microwave power generator(s); if this current is interrupted for any reason, the generator(s) shuts down; and

(vii) where leakage must meet RFI specifications for 900 MHz equipment in those countries where this band has no allocation, housing the plant in a fully screened room. This has many disadvantages: it is inconvenient, very expensive requiring manufacture to an exceptional standard, and claustrophobic to operators. The requirements of workload transfer may seriously reduce the throughput of the plant through an extended off-time while transfer is taking place through open doors.

8.3 Attenuation of large, partially filled choke tunnels

Large choke tunnels, several half wavelengths long in at least one and usually both transverse dimensions, are a common requirement in an industrial system. They are often operated with the workload only partly filling the cross-section, and where the workload creates the desired attenuation it is very important to be able to estimate its magnitude. Also, the tunnel may be lined with lossy material, and calculation of the resulting attenuation is similarly important to determine the optimum dielectric (or magnetic) properties of the lining.

Because the tunnel can support a large number of propagating TE and TM modes, and also evanescent modes due to the highly lossy filling, a precise analytic solution is virtually impossible. Numerical solution (e.g. Dibben and Metaxas, 1995) is a valuable approach but requires formulation by a person skilled in the technique. However, an approximate estimate is of considerable value in the initial design stage, and by making certain reasonably valid assumptions and approximations it is possible to derive relations for the attenuation of a tunnel in terms of the fractional filling factor and the material's dielectric properties.

The main approximations and assumptions are

(i)   the E field is of substantially constant amplitude across the aperture, corresponding to a TEM parallel-plate mode for slabs normal to the E vector, or with a sinusoidal taper to zero at the sidewalls for slabs parallel to the vector. This last condition implies a sinusoidal distribution of E field within the slab and a constant amplitude E field in the space between them.

(ii) the calculated attenuation is in a section of tunnel well removed from the ends, where there are anomalies due to field distortion.

(iii) for the slabs parallel to the E vector, ε'' is sufficiently small compared with ε' that the sinusoidal field distribution is approximated as:

$$\sin\left(2\pi x\sqrt{\varepsilon' - j\varepsilon''}\right) \approx \sin 2\pi x\sqrt{\varepsilon'}$$

This last condition is not mandatory to the function of the choke, but it much simplifies the analysis, enabling a quantified, if only approximate, result to be obtained.

This analysis is essentially concerned with dielectric loss, and with propagation modes with characteristic impedance $Z_0 \approx Z_{0TEM} = 377\ \Omega$. There will be present higher-order TE and TM modes which may approach cutoff, and therefore have extreme values of Z_0. While the TE modes have higher impedance and therefore higher E fields with stronger dielectric coupling to the lossy material, as in eqn. 4.42, the converse applies to the TM modes (Ramo *et al.*, 1965). It follows that the inclusion of magnetic lossy material μ'' would further enhance the attenuation of the tunnel.

Marcuvitz (1986) gives relations for the propagation characteristics of waveguides partly filled with a dielectric material. Although he concentrates on loss-free dielectrics, it is valid, as he states, to substitute the complex permittivity $(\varepsilon' - j\varepsilon'')$ in his equations. This results in a complex value for the wavelength $(\lambda' + j\lambda'')$ in the tunnel, where the imaginary part $j\lambda''$ represents attenuation. Recall that the wave propagation along the z axis is of the form

$$E_z = E_0 e^{-j\beta z} \tag{8.1}$$

where $\beta = 2\pi/\lambda_g$. Clearly, if $\lambda_g = \lambda_g' - j\lambda_g''$, then β itself becomes complex, so the exponential index includes a negative real part, signifying attenuation. Specifically,

$$\beta = \frac{2\pi}{\lambda_g} = \frac{2\pi}{\lambda_g' - j\lambda_g''} = 2\pi \frac{\lambda_g' + j\lambda_g''}{|\lambda_g|^2} \tag{8.2}$$

where

$$|\lambda_g|^2 = (\lambda_g')^2 + (\lambda_g'')^2 \tag{8.3}$$

The attenuation is given by the imaginary part of β:

$$\text{attenuation coefficient} = 2\pi \frac{\lambda_g''}{|\lambda_g|^2} \qquad \text{nepers per metre}$$

$$\tag{8.4}$$

$$\text{or, attenuation coefficient} \qquad 8.686 \times 2\pi \frac{\lambda_g''}{|\lambda_g|^2} \qquad \text{decibels per metre}$$

Marcuvitz considers two cases of partially filled rectangular waveguides operating in the TE_{10} mode, first with the dielectric part filling the width so that the air/dielectric interface is parallel to the E field, and secondly part filling the height with the interface now orthogonal to the E field. With a lossy dielectric, both lead to the difficulty of solving equations in complex number of the form

$$f_1(\theta)\tan(\theta'+j\theta'')=f_2(\phi)\tan(\phi'+j\phi'') \tag{8.5}$$

The total attenuation of a tunnel lined all round with lossy material is assumed to be the sum of the attenuation values (in decibels per metre or nepers per metre) of the two cases described in Section 8.3.1 and 8.3.2, shown diagrammatically in Figures 8.1 and 8.2.

8.3.1 Part-filled height

For the dielectric part filling the height, Marcuvitz gives an approximation for λ_g, which for a very wide waveguide ($a \gg \lambda_0$) further simplifies to

$$\lambda_g = \frac{\lambda_0}{\sqrt{1-\dfrac{t}{b}\left\{1-\left(\dfrac{1}{\varepsilon'-j\varepsilon''}\right)\right\}}} \tag{8.6}$$

where t is the height of the dielectric, and b is the height of the aperture.

Note that the TM_{01} mode in a very wide waveguide tends to a parallel-plate transmission line, and this analysis is relevant to the parallel-plate applicator discussed in Section 6.2.3. Eqn. 8.6 clearly gives a complex value to λ_g which enables the attenuation to be calculated using eqn. 8.4. Note that if $t = 0$ is substituted into eqn. 8.6 then $\lambda_g = \lambda_0$ and the attenuation is zero; if $t = b$ (i.e. 100% filling) is substituted the attenuation (nepers per metre) corresponds to the reciprocal of the skin depth D_p (metres per nepers) as given by eqn. 2.9. These limit results give credence to eqn. 8.6.

Figures 8.3 and 8.4 show sets of curves of attenuation (in decibels per metre) as a function of fractional filling height for dielectrics with permittivity $\varepsilon' = 1.5$ and $\varepsilon'' = 1.5$. for a very wide waveguide with the air–dielectric interface normal to the E-field polarisation. Figure 8.3 shows curves for 915 MHz, and Figure 8.4 those for 2450 MHz.

Eqn. 8.6 is valid for predicting the attenuation rate of workloads partially filling the choke tunnel, and also for the attenuation in a parallel-plate-transmission-line applicator.

A suitable lossy material would be a foam plastic incorporating a lossy substance such as carbon (FEF- or HAF-grade powder), or silicon carbide, adjusted in proportions to give the above dielectric properties.

It can be seen that the attenuation is not very high for a lined tunnel, which illustrates the difficulty of designing large open-aperture ovens.

Water or glycol loads have been used where there is high power density, but the attenuation rate is low due to the high values of ε'. The water flows

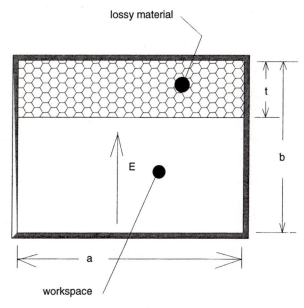

Figure 8.1   Choke tunnel with lossy slab perpendicular to E-field vector

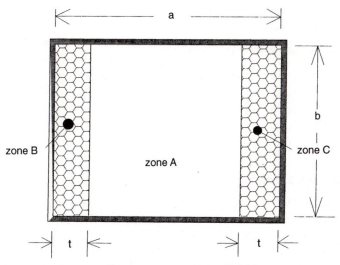

Figure 8.2   Choke tunnel with lossy slabs parallel to E-field vector

through an array of plastic pipes of diameter comparable to the penetration depth. Reliability problems of water leaks, and of condensation forming with humid atmospheres in the tunnel, are commonplace. Extreme care must be taken in design and manufacture to avoid leaks which usually arise from embrittlement or 'plastic flow' of the pipework with time. A further problem

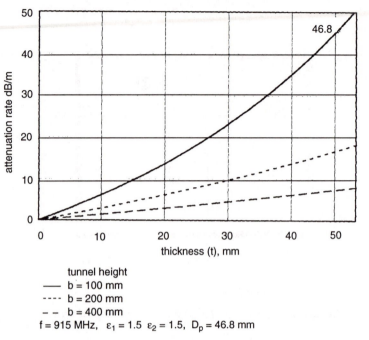

Figure 8.3   Attenuation of choke tunnel loaded with two lossy slabs normal to the E-field vector

is that a dielectric water pipe is a waveguide of modest attenuation, and will transmit power in the TEM mode when housed in even a tight-fitting metal tube. Very high field concentrations can occur at the point of entry of such a water pipe through a metal wall, which can result in local heating and melting or burning of the plastic pipe. Water absorbers should not be used unless they are totally unavoidable.

8.3.2 Part-filled width

For a wide choke tunnel with side walls lined with lossy material, the material–air interface is assumed to be parallel to the E-field vector for TM modes. The tunnel is divided into three zones, the first (A) being the open aperture through which the workload passes, the others (B and C) being the two side zones comprising the dielectric-filled regions. In zone A, the electric field is assumed to be constant across the width; this is a reasonable assumption because there would otherwise be acute nonuniformity of heating in the oven. The electric field in zones B and C is assumed to have a sinusoidal distribution, to be zero at the metallic side walls, and to have a value equal to that in zone A at the interfaces between zones A and B, to satisfy the boundary conditions. Figure 8.2 shows the model, defining the zones and the parameters.

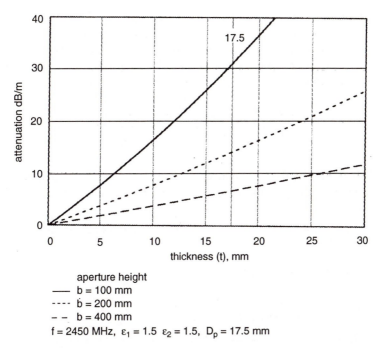

Figure 8.4 *Attenuation of choke tunnel loaded with two lossy slabs normal to the E-field vector*

The calculation, which is nonrigorous but gives sufficient accuracy for the prime purpose, involves three steps:

(i)   evaluation of the total power flow P_F in zones A, B and C in terms of the electric-field intensity, yielding:

$$P_F = \frac{bE^2}{Z_0}\left\{(a-2t)+\frac{2\sqrt{\varepsilon'}}{\sin^2\frac{2\pi\sqrt{\varepsilon'}\,t}{\lambda_0}}\int_0^t \sin^2\left(\frac{2\pi\sqrt{\varepsilon'}}{\lambda_0}\right)x\,dx\right\} \quad (8.7)$$

Evaluating the integral and applying the interface boundary condition for the E field gives

$$P_F = \frac{bE_A^2}{Z_0}\left\{(a-2t)+\frac{\lambda_0}{\pi\sin^2\frac{2\pi\sqrt{\varepsilon'}\,t}{\lambda_0}}\left(\frac{\pi\sqrt{\varepsilon'}\,t}{\lambda_0}-\frac{1}{4}\sin\frac{4\pi\sqrt{\varepsilon'}\,t}{\lambda_0}\right)\right\} \quad (8.8)$$

(ii)   calculation of the power-dissipation gradient dP_L/dz in the lossy material in zones B and C in the direction of propagation z, which is the total power-dissipation gradient giving $dP_L/dz = dP_F/dz$. The power dissipation is given directly by eqn. 2.7, integrating the E field through the thickness t:

$$\frac{dP}{dz} = \frac{2bE_A^2}{q^2}\sqrt{\frac{\varepsilon_0}{\mu_0}}\,\varepsilon''\left(\frac{\pi t}{\lambda_0} - \frac{1}{4\sqrt{\varepsilon'}}\,p\right) \tag{8.9}$$

(iii)   Calculation of the attenuation coefficient α_N (nepers per metre) from eqns. 8.8 and 8.9, and using the relation (Metaxas and Meredith, 1993):

$$\alpha_N = -\frac{1}{2}\frac{dP}{P} \qquad \text{nepers per metre} \tag{8.10}$$

Inserting 8.686 dB = 1 Np, gives:

$$\alpha_{dB} = 8.686\frac{\varepsilon''\left(\pi\dfrac{t}{\lambda_0} - \dfrac{1}{4\sqrt{\varepsilon'}}\,p\right)}{\left\{(a-2t)q^2 + \dfrac{\lambda_0\sqrt{\varepsilon'}}{\pi}\left(\dfrac{\pi t}{\lambda_0} - \dfrac{1}{4\sqrt{\varepsilon'}}\,p\right)\right\}} \qquad \text{decibels per metre} \tag{8.11}$$

where $p = \sin\left(\dfrac{4\pi\sqrt{\varepsilon'}\,t}{\lambda_0}\right)$ and $q = \sin\left(\dfrac{2\pi\sqrt{\varepsilon'}\,t}{\lambda_0}\right)$ \qquad (8.12)

Figures 8.5 and 8.6 show curves of attenuation rate as functions of the thickness of the lossy slabs for three widths of tunnel, and for dielectric properties $\varepsilon' = 1.5$ and $\varepsilon'' = 1.5$.

8.4 Reactive (reflective) choking

Reactive choking involves creating a high reflection coefficient to energy escaping through the tunnel, using passive, lossless reflectors. These may be an array of resonant slots or posts forming a bandstop filter (Van Koughnett and Dunn, 1973), or reflector plates, illustrated in Figures 5.13 and 5.14.

Van Koughnett's choke is based on the principle that any mode present in a large aperture can be synthesised from a set of coherent plane waves propagating along inclined axes within the tunnel, causing currents to flow in the walls. If two arrays of slots are provided in mutually orthogonal patterns, each forming a bandstop filter, the wall currents are intercepted

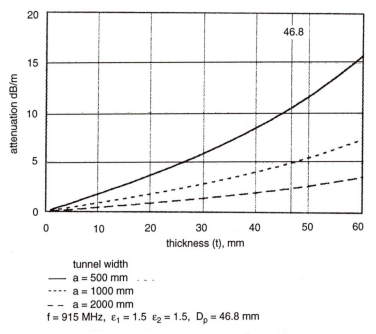

Figure 8.5   Attenuation of choke tunnel loaded with two lossy slabs parallel to the E-field vector

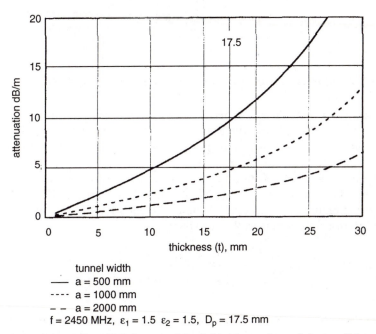

Figure 8.6   Attenuation of choke tunnel loaded with two lossy slabs parallel to the E-field vector

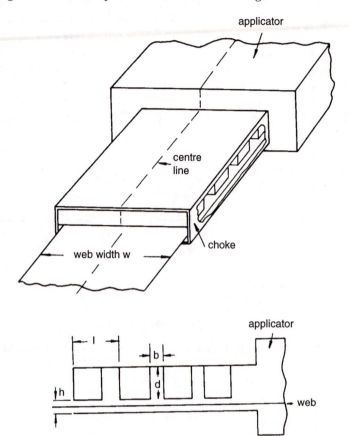

Figure 8.7   Idealised corrugated choke (Metaxas and Meredith, 1993)

irrespective of their direction because they can be resolved into components at right angles aligned with the axes of the slot patterns. The slot patterns can be formed from a set of square posts mounted on the wall of the tunnel, spaced apart to form the slots.

The analysis of Van Koughnett's choke follows conventional filter theory for periodic structures (Lines, Nicoll and Woodward, 1949). Using the symbols of Figure 8.7, a modified form of the equation of Lines *et al.* gives the attenuation coefficient α in the stop-band as

$$\alpha = 8.686 \, arc \, \cosh\left[\cos\left(\frac{2\pi l'}{\lambda_g}\right) - \frac{b}{2h}\left\{\tan\left(\frac{2\pi d'}{\lambda_g}\right)\right\}\left\{\sin\left(\frac{2\pi l'}{\lambda_g}\right)\right\} \right] \text{ dB per section}$$

$$(8.13)$$

where the term in square brackets must be greater than +1 in the stop band. To satisfy this condition in a practical choke, $1/4 < l'/\lambda < 1/2$, and the maximum theoretical attenuation (infinity) occurs when $d'/\lambda = 1/4$. Values of λ_g are appropriate to each mode in the overmoded tunnel, but in practice it is sufficient to use the free-space wavelength because most modes are well removed from cutoff. For this reason the few modes which are close to cutoff may suffer little attenuation, representing a possible limitation on performance. The primed dimensions refer to the effective electrical lengths; curves showing the adjustments from the mechanical lengths are given by Marcuvitz (1986)

Figure 8.8 shows curves plotted from eqn. 8.13 giving the attenuation per section as a function of normalised slot depth d/λ for selected ratios b/h, and a section length chosen as $l/\lambda = 3\lambda/8$. The slot depth becomes progressively more critical as the headroom height h increases. Figure 8.9 is a graph of attenuation as a function of b/h for 5% and 10% 'detuning' of the slot depth, showing that, for 5% detuning, the attenuation will be just over 20 dB per section when $h = b$, and 15 dB when $h = 2b$. If $b = \lambda/4$, there is a theoretical attenuation of 15 dB per section with a working height of 165 mm at

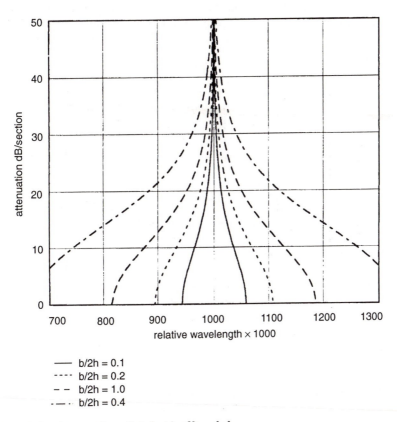

— b/2h = 0.1
- - - b/2h = 0.2
– – b/2h = 1.0
·—·· b/2h = 0.4

Figure 8.8 Attenuation of choke-pin filter choke

915 MHz, and 61 mm at 2450 MHz. A detuning of 5% from optimum is chosen to allow for tolerances, and to avoid spuriously high attenuation levels which in practice are unreliable.

High values of attenuation are achieved by cascaded sections, for example a four-section filter as described above would have an attenuation of 60 dB. In practice, it is desirable to insert a section of resistive attenuation before adding further filter sections.

Van Koughnett's choke, in principle, comprises two filters as above, with their axes orthogonal; the choke appears as a set of square blocks forming channels at right angles, a so-called 'double corrugated' choke. He further simplifies the mechanical design by replacing the square blocks by circular rods. Although the resulting 'choke-pin' design has departed far from the original model, it does still have a substantial attenuation and is more practical. Where a greater headroom height is required, a double-sided form is used; its performance must be established by experiment or numerical modelling.

A very simple choke for large apertures comprises a section of radial waveguide terminated in a circular reflector having an aperture equal to the desired headroom, as shown in Figure 5.14. and described in Section 5.7.3.

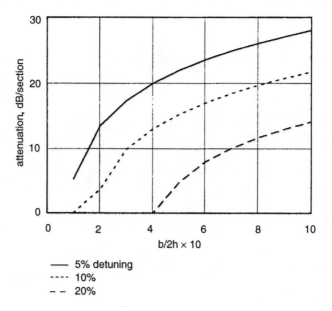

Figure 8.9   Effect of detuning of filter on attenuation

8.5 Active electronic systems

Active systems for preventing excess leakage are designed to trip out the microwave generator(s) under fault conditions, and take two forms:

(i)   those which detect the high microwave field directly; and
(ii) those which detect the absence of workload, where the workload forms a major part of the attenuation in the tunnel.

Both have a rapid-response capability. In conventional magnetron power supplies, the EHT to the magnetron may be tripped, or the applied magnetic field to the magnetron may be increased to raise the π-mode threshold voltage above the open-circuit EHT voltage of the power supply. In a switch-mode power supply, the EHT voltage can be reduced to zero in a few microseconds.

   The microwave field detector is usually a microwave diode because of its fast response. It is conveniently mounted in a coaxial line terminated in a small loop antenna inserted to sample the field in the choke tunnel near its open aperture. The loop antenna forms a DC return path for the diode, making it more suitable than a probe. Figure 8.10 shows the essential features. The DC output, typically in the range 0.5-1.0 V on alarm, feeds to a trip line in the power supply. Precautions must be taken to ensure that spurious signals do not reach the diode via the DC-output connection since they will be rectified and produce a false signal. Such spurious signals are

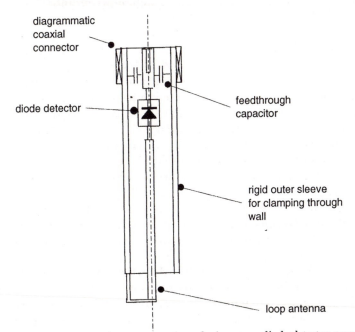

diagrammatic coaxial connector

diode detector

feedthrough capacitor

rigid outer sleeve for clamping through wall

loop antenna

Figure 8.10   Diagrammatic representation of microwave-diode detector assembly

mainly from stray microwave leakage, magnetron noise and circulating earth currents at power frequency.

Workload sensors may take many forms, but the most reliable are those based on a visible-light or infrared beam which is interrupted by the workload. They are not inherently fail safe because dirt and foreign bodies can also interrupt the beam. Automatic self checking on start-up, using a test lamp, reduces this problem to minimal risk; however a duplicated system is very advisable. Another system is to use infrared light reflected from the workload surface, which is slightly superior in the fail-safe role because its 'go' setting relies on the establishment of an optical link between its source and its receiver via the workload, unlike the direct interrupted-beam method. However, variation in optical reflectivity of the workload surface may make this system unusable. In both cases, modulated light beams are usually used with a correlation detector, which virtually eliminates false alarms from normal lighting.

Other sensors which have been used include capacitive proximity detectors, weighbridges and 'feeler' arms. The first is very prone to false alarms and requires close proximity between the sensor and target, which cannot usually be realised. Weighbridges are expensive, and feeler arms are prone to fail to danger unless very regularly maintained.

It must be recognised that all active-protection systems are likely to fail to danger in some circumstances, and it is essential for the design engineer to be aware of such possibilities and to audit their performance and probability of failure in the context of the particular system in its entirety.

8.6 Microwave fuses

Fuses are often used as the final 'end stop' to a microwave leakage-protection system. The principle is very simple: a fuse carries a pilot current which must be maintained for the microwave generator(s) to be operated. If the current is absent, it is impossible to switch on, and if it is interrupted the generator(s) trip. The fuse is connected in a microwave circuit coupled to sample the microwave power near the open aperture of the choke tunnel. If this power exceeds a preset limit, the fuse blows, irreversibly tripping the microwave power sources.

In practice, it is necessary to use a fuse of about 100 mA rating because of the difficulty of achieving adequate sensitivity with higher-rated fuses which have a low DC resistance. High sensitivity requires a good microwave-impedance match into the fuse wire which is possible with a low-current fuse of the glass-cartridge variety. As with the detector diode, the fuse is coupled via a loop antenna, but the power required is much greater and the loop is made resonant at $\lambda_0/2$ circumference as a square or circle. A DC break must be included in the microwave circuit to avoid the loop bypassing the fuse at the control frequency.

8.7 Screened rooms

Where it is not feasible to achieve the very low residual leakage demanded by a radio-interference specification by normal choking techniques, a fully screened room housing the entire equipment must be used. In practice, a screened room is only likely to succeed when the attenuation required is less than about 40 dB because of the difficulty of making and maintaining a screened room worthy of the name. Meticulous care in design and manufacture is necessary, and the temptation to reduce standards for the sake of cost saving must be resisted. A single leak will severely impair performance and can be extremely difficult and time consuming to locate.

The room itself must be a *totally enclosed* metal box, including the floor, and all the joints in the skin must have 100% integrity metal-to-metal contact along their entire length. A $\lambda/2$ gap (165 mm at 900 MHz) will radiate freely as a resonant-slot antenna, and should be reduced to less than 1% of this length for adequate screening; in practice this means no gaps. In this context, a gap may be a fine crack in a weld. Perforated sheets are not recommended because of the difficulty of making a satisfactory joint between them.

All electrical conductors entering the room must be individually filtered, and this applies to the individual conductors of a multicore cable. Any unfiltered conductor will transmit power through the wall with little attenuation in a TEM mode, and will operate as a receiving antenna inside the room and a transmitting antenna outside it, defeating completely the objective. A filter attenuation of >60 dB should be the aim. For low-power control lines up to 250 V, a pair of feed-through capacitors (1000 pF) in succession through a double skin is usually adequate. For high-power 3-phase systems, each phase, neutral and earth must pass through a microwave filter tuned to the operating frequency. Such a filter may comprise a coaxial structure of the low-high-low impedance type with quarter-wave sections (Section 4.8.3).

The screened room will form a multimode cavity, and to avoid mode resonances it is desirable to line at least one wall with a microwave-absorbent material to give an overall reflection coefficient < 0.2 from the wall.

Access doors are the most difficult to seal, as the same exacting requirements must be achieved. The door and its jamb must be of precise dimensions and stiff enough to ensure that the sealing gasket is correctly compressed at all points around the circumference. The hinges must carry the weight of the door, plus the forces resulting from compressing the gaskets, with minimal and predetermined deflection. The door stiffness must be such that the forces exerted by the hinges and door clamps do not cause deflection greater than allowable for correct compression of the gasket. Door contact with the floor is best secured by a threshold step; it is impossible to arrange a satisfactory contact to a plain level floor with a 'contact strip'.

The size of the room can be reduced by using fully screened generators mounted externally, feeding power via waveguide(s) sealed into the wall. This may remove the need for high-power cable filters.

8.8 Choking of drive shafts

It is frequently necessary to insert a moving rod or rotating shaft through the wall of an applicator. The problems encountered are leakage of microwave energy and local overheating or arcing between the shaft and the wall.

The simplest arrangement is to use a shaft of dielectric material of diameter small enough that it forms a cutoff circular waveguide in a stub tube mounted in the wall (see Section 4.6.1). Figures 4.11 and 4.12 are graphs showing the attenuation of cutoff waveguides when filled with a selection of common dielectric materials. The length of the choke tube should be at least two diameters more than the length required to give the desired attenuation, to eliminate the uncertainty of attenuation near the open ends. The end of the tube opening into the oven should be radiused (2–5 mm radius) to minimise field concentration. With a low-loss dielectric there should be little heating and a low risk of arcing. The main disadvantage is the restricted diameter, which limits the mechanical strength.

Metal drive shafts may be 'choked' by a contact ring through which the currents induced in the shaft can return to the wall. A segmented graphite ring, spring loaded into a tapered housing, has been a successful technique, but it must be kept clean of dust and lubricant. Ball bearings may also be used, the currents flowing from stator to rotor via the balls, but arcing is a risk and can rapidly destroy the bearing.

A noncontact choke may be used on a metal shaft of the low-high-low-impedance type (Section 4.8.3). The shaft forms a TEM coaxial transmission line in the housing, a journal bearing being incorporated in a low-impedance section, preferably remote from the oven. A further bearing must be provided external to the choke. Again, edges must be radiused to avoid arcing. Great care must be taken to ensure concentricity of the shaft in the choke housing; also to note any TE or TM coaxial modes which may propagate, negating the choke action (Marcuvitz, 1986).

Chapter 9

Microwave generators
1: microwave power tubes

9.1 Introduction to microwave power sources

Most industrial microwave heating systems demand power in excess of 10 kW, often extending into the range 100 kW to 1 MW. In addition, the following are essential requirements for an industrial production plant:

(i)   a high conversion efficiency from incoming power to usable microwave power output, especially important as the power rating rises;
(ii) operation within the prescribed frequency band at all times;
(iii) a low capital cost per kilowatt of output power; this is important because the generator often represents more than half the cost of the installation;
(iv) robustness in surviving incidents typical of industrial operation: hostile environments, electrical power surges and transients, arcing in the applicator and feed system, incorrect adjustment of operating conditions, vibration etc;
(v)   simple to operate, with the absolute minimum of user-adjustable controls;
(vi) simple to maintain, fault-find and remedy;
(vii) low running cost, not only in electrical power consumed but also in replacement of consumables, notably the microwave power tube itself. This implies a long magnetron life; and
(viii) comprehensive, accurate documentation to enable installation, service and operating teams to understand the equipment fully within their designated roles.

Currently (1997), the magnetron is overwhelmingly superior in satisfying the above needs compared with the klystron or with solid-state power sources. Although progress is being made in solid-state power supplies, they are far short of the power-output requirement and are very expensive, and will not be considered further.

Klystrons can generate adequate power and are widely used in radar, communications, television and scientific research. They are usually used as amplifiers, and their particular virtues are purity of spectrum (i.e. freedom from unwanted sidebands), tunability over a wide frequency band, and freedom from distortion of modulated signals. None of these is particularly relevant to industrial heating. Nevertheless klystrons are sometimes considered, especially when driving resonant applicators where their frequency stability is important, and so their principal features are reviewed in Section 9.2.

Magnetron generators are considered in Sec 9.3

9.2 Klystron generators

For over 20 years, a klystron (Maloney and Faillon, 1974) of 50 kW output at 2450 MHz has been available for industrial heating. Until recently, magnetrons at 2450 MHz had been limited to 6 kW, although this has now increased to 30 kW. It is for this reason that klystrons are sometimes considered for high-power heating installations at 2450 MHz.

9.2.1 Klystrons: principle of operation

A diagrammatic view of the amplifier klystron described by Maloney and Faillon is shown in Figure 9.1. The essential components are

(i)   an electron 'gun' which forms an intense parallel beam of electrons which are accelerated towards the resonator assembly by a DC extra-high voltage applied between the gun assembly and the resonator structure. For convenience, the resonator structure operates at earth potential, so the gun has an applied voltage of about −25 kV DC;

(ii)   a resonator assembly comprising two or more circular cavities (TM_{010} mode), having holes in the end plates to allow the electron beam to pass through. The cavity endplates have stub sleeves forming a gap across which a high-value microwave E field is developed which interacts with the electron beam. So-called 'drift tubes' separate the cavities; their length is critical in achieving optimum power output. The cavities are tuned to the operating frequency, and may be 'stagger tuned' across a prescribed operating band to achieve particular performance characteristics of uniformity of amplifier gain and linearity of phase;

(iii) an input waveguide (or coaxial line) coupled to the first cavity, and an output waveguide to the last;

(iv)   an applied axial DC magnetic field (about 0.15 T) which collimates the electron beam. It is absolutely essential that no, or in practice very few, electrons hit the cavity structure because their energy is so high that rapid melting would occur, destroying the klystron. This is provided by a solenoid requiring some 3 kW DC power;

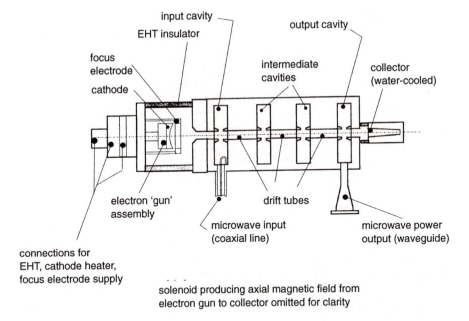

input cavity

EHT insulator

output cavity

focus electrode

intermediate cavities

collector (water-cooled)

cathode

electron 'gun' assembly

drift tubes

microwave input (coaxial line)

microwave power output (waveguide)

connections for EHT, cathode heater, focus electrode supply

solenoid producing axial magnetic field from electron gun to collector omitted for clarity

Figure 9.1   Schematic view of high-power klystron (Maloney and Faillon, 1974)

(v)   a water-cooled 'collector' which forms an anode for the electron beam. The collector is usually at the same DC potential as the cavity assembly, but higher overall efficiency can be obtained by applying a negative DC to slow the velocity of the electrons before impact, thereby reducing their kinetic energy (depressed-collector operation);

(vi)   being an electron valve, the klystron has a high vacuum, and in a structure of this size and complexity it is provided with an ion pump to maintain the requisite vacuum, and this also has a need for a high-voltage DC supply; and

(vii) water cooling is also provided for the cavity assembly and the solenoid: this must be of high quality to avoid scale deposition on the heat transfer surfaces.

In operation, electrons passing through the first cavity (driven from an external source) are, on alternate microwave half-cycles, accelerated and decelerated by interaction with the microwave E field. As they leave this first cavity, some are travelling faster than average, some slower. As the beam travels onward through the drift tubes, the faster electrons 'catch up' the slower, and bunches are formed at intervals in position along the beam, causing the density of the electron cloud to be high. Intermediate cavities are located at the positions of these bunches, and microwave energy is induced in them; the resultant E fields further intensify the bunching, and the pattern of further cavities is repeated until the final cavity, where the bunching is at its most intense, and the power coupled into it is transferred to the output waveguide. The bunching technique is described as 'velocity modulation'.

The Thomson-CSF TH2054 klystron has five cavities, and develops 50 kW with an anode efficiency of 60% optimum. The overall efficiency including subsidiary power supplies is about 3% less. As an amplifier, it requires a drive of about 700 mW, which can be supplied from a solid-state source. Alternatively, it is possible to make the klystron self-oscillate by abstracting a small fraction of the microwave output and feeding it, via a transmission cavity to determine the frequency, to the input.

This klystron, including its solenoid, occupies a space envelope 415 mm in diameter, 1050 mm. long; it weighs 280 kg, of which the solenoid assembly is 215 kg. Such an assembly necessarily requires in-built handling equipment.

From the above outline description it is evident that the klystron is inherently expensive, and moreover because of the many subsystem power supplies needed (filament, focus electrode, electron-beam EHT, solenoid, ion pump) the power-supply assembly to drive it is expensive too. Also, because the specification requires extensive monitoring of the operating parameters, partly for automatic protection and partly to establish conditions pertaining to a failure and possible warranty claim, the instrumentation is a significant further cost.

Although it is claimed that the life of the klystron exceeds 15 000 h, experience with industrial magnetrons is that premature failure is not uncommon due to application incidents. For magnetrons, these are less prevalent than formerly due to systematic elimination of their causes; klystron power supplies are likely to follow the same pattern of evolution.

It is for the above reasons that klystrons have not yet been used significantly in industrial applications; nearly all have been associated with scientific research.

9.3 Magnetrons

The magnetron was developed intensively during the Second World War as a source of high-power microwaves, crucial to the success of precision radar. Such was its importance that some would place the magnetron at the top of the list of contemporary scientific achievements which brought victory to the Allies. Since then it has been extensively developed to cover a very wide frequency range, (0.5–100 GHz) in a wide range of power outputs for radar and microwave heating. The greatest single application in terms of number in use is undoubtedly the domestic microwave cooker, where small magnetrons at 2450 MHz with power output ratings in the range 600–1500 W are made in great quantities by mass-production methods giving exceptionally low cost. Some industrial installations use these domestic magnetrons because they are cheap and their drive power supplies are very simple. There has been, and continues, a long debate on the relative merits of providing a power source in the form of a large number of these cheap 'domestic' magnetrons, or as a much smaller number of high-power magnetrons; the pattern is evolving towards the latter, and the relative merits of the two approaches are compared in Section 10.5.

9.3.1 Magnetrons: principle of operation

The principal components of a high-power industrial, continuous-wave (CW) magnetron are:

(i)   a cylindrical heated cathode emitting electrons;

(ii)   a circular anode concentric with and surrounding the cathode, having an array of radial slots forming resonators tuned to the desired operating frequency;

(iii)   a high-strength magnetic field aligned axially to the anode–cathode assembly. This may be provided by a permanent magnet in small magnetrons, or an electromagnet for higher-power magnetrons;

(iv)   a high-value DC EHT voltage (usually in the range 2–20 kV) applied between anode and cathode, with the anode earthed for convenience, and the cathode at negative potential;

(v)   the whole enclosed in a vacuum envelope and sealed with a high vacuum of about 10^{-6} mm Hg;

(vi)   a probe antenna or slot, coupled to the resonator and sealed by an output window of microwave transparent material; and

(vii)   a construction which provides for easy rebuilding after end-of-life failure or accidental damage.

The assembly is illustrated diagrammatically in Figure 9.2, typical of high-power magnetrons at 915 MHz and 2450 MHz. Figure 9.3 is a sectional drawing of a magnetron type CWM75L (California Tube Laboratory Inc.), rated at 75 kW output at 915/896 MHz. In operation, electrons are drawn from the heated cathode by the electric field created by the applied DC EHT. In the presence of the magnetic field, the electrons, instead of travelling radially outward to the anode, travel a spiral path in the 'interaction space' between the cathode and anode. They have, therefore, a circumferential component of velocity, which can be controlled by the combined adjustment of the EHT and applied DC magnetic field. If the magnetic-field strength is very high, the outermost radius of the trajectory is less than the internal radius of the anode, so that no electrons actually reach the anode, and no anode current flows. As the magnetic-field strength is reduced, the radius of the trajectory of the outermost electrons increases, and there is a critical point at which it equals the radius of the anode bore, whereupon anode current starts to flow.

    Now consider the microwave aspects of the anode. It comprises a ring of cavities which are in principle $\lambda_0/4$ transmission lines short-circuited at the outer end and open at the inner, where the microwave electric field is

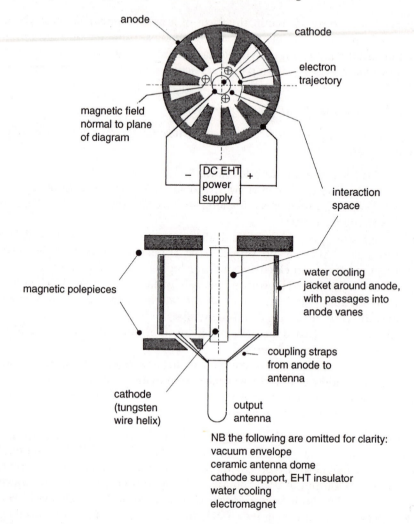

Figure 9.2   Diagrammatic representation of 10-cavity magnetron

therefore high. At this end, the electrons are at grazing incidence. It is well known that such an array of coupled cavities has several resonant frequencies with characteristc field patterns, the simplest being that in which alternate cavities are excited precisely in antiphase, i.e. π radians phase difference between them. This is called the *π-mode*, and is the desired pattern for the magnetron operation. It has a quasisinusoidal field distribution around the anode in which the number of complete cycles is $N/2$ where N is the number of cavities. Other field patterns at other resonant frequencies have $(N/2 - 1)$, $(N/2 - 2)$ etc. cycles: the first of these is called the '$(π - 1)$' mode and is undesirable, as are all the rest.

The microwave-field pattern around the anode appears as a standing wave of very high VSWR, and following Section 4.3 can be represented by the sum of a forward and a reverse wave of equal amplitude: specifically one wave travelling clockwise and one travelling anticlockwise. Clearly, if the peripheral velocity of, say, the clockwise wave is close to that of the electrons as they graze the anode, there will be an interaction. Those electrons which enter a positive microwave-electric-field gradient will accelerate, and those which enter a negative gradient will be retarded, and so the electron cloud becomes bunched and would appear as a rotating spoked wheel. If the speed of rotation of this 'spoked wheel' is increased by raising the applied EHT voltage, the average speed of the electrons at grazing incidence tries to exceed that of the microwave field, but it cannot because there is synchronism between the two. The high-velocity electrons are slowed by the microwave field, and their kinetic energy (KE) is dissipated in it; the microwave power generated is proportional to their change in kinetic energy and the time rate of arrival of electrons at the anode, which also determines the anode current. A few electrons are deflected back to the cathode and dissipate their kinetic energy as heat, a process called 'back bombardment'; to prevent the cathode overheating, the filament-heater current is programmed to reduce as the anode current rises.

The power transfer in the magnetron is therefore:

DC power input $\Rightarrow$
rate of increase in KE of electrons leaving the cathode $\Rightarrow$
rate of decay of KE of the electrons arriving at the anode $\Rightarrow$
microwave power in the anode resonator system $\Rightarrow$
microwave power coupled to the workload

The transfer of power is extraordinarily efficient, and high-power magnetrons at 900 MHz have typical efficiency greater than 88%, and at 2450 MHz greater than 70%. The lost power is mostly due to the residual energy of electrons hitting the anode having lost most of their velocity and dissipated most of their KE to the microwave field. This lost power is dissipated as heat, so the anode is liquid cooled except for low-power magnetrons which are air cooled. Some power is lost as I^2R heating of the anode by microwave currents flowing.

In high-power industrial magnetrons, the microwave-output configuration comprises an antenna enclosed in a high-purity alumina 'dome' forming one end of the magnetron, as illustrated in Figures. 1.1 and 9.3. The dome is inserted into a dedicated waveguide launcher having a short-circuit at one end. The distance of the short-circuit plate from the centre line of the antenna, the insertion depth of the antenna into the waveguide, and the design of the antenna itself are chosen and specified by the magnetron manufacturer for optimum performance. Optimum performance is a compromise of efficiency, power output and stability when operating with mismatched loads. Magnetrons are tested for stability by operating into a load

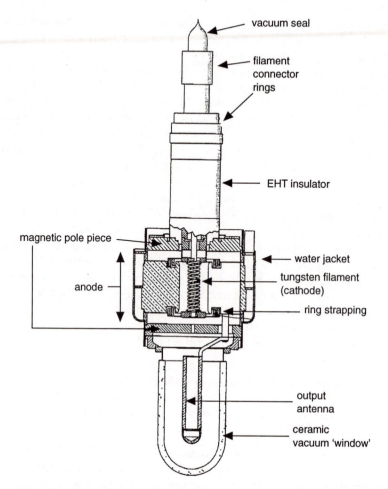

Figure 9.3   Sectioned view of 75 kW 915 MHz magnetron (Courtesy California Tube Laboratory Inc.)

with VSWR of 3.0 which is adjusted in reflection phase angle through 2π radians; the magnetron must continue to oscillate and display no adverse effects such as failure to oscillate, mode skipping [e.g. to the (π-1) mode] , 'flashing' or spectral deterioration.

9.3.2 Magnetrons: performance characteristics

Magnetron manufacturers give very precise specifications for the maximum ratings of their tubes and details of typical operating conditions. It is very important to ensure that the magnetron is operated at all times within the limits and conditions of these specifications, as serious damage can result from failing to comply, even momentarily in some cases. In the following sections the principal features are considered in detail.

9.3.2.1 Electrical inputs

(*a*) *Cathode:* In most high-power magnetrons for industrial heating, the cathode is a directly-heated robust tungsten filament wound into a spiral, operating at a temperature about 1300°C. The life of the magnetron is determined by the life of the filament, which slowly evaporates depositing tungsten on the anode. There is usually a 'hot spot' which is the site of eventual thermal run away. It has higher-than-average evaporation rate which accelerates the reduction in cross-section of the wire, increasing the I^2R heating locally, raising the temperature yet further; failure almost invariably occurs at switch on of the filament supply due to the surge current.

From the above, it is important to operate the cathode at as low a temperature as possible, consistent with providing a sufficient flow of electrons to sustain the anode current demanded, with a margin to allow for momentary surges. Excessively high temperature causes more rapid evaporation and hence shortens the life.

The filament has a very low cold resistance, around 10 mΩ, and the power supply must be limited in its initial current at switch-on in accordance with specification. The usual practice is to provide electronic control of filament current in a closed-loop system, with a controlled ramp-up in starting, and a further control of current derived from a signal proportional to anode current.

The integrity of the filament power supply is extremely important because incorrect filament current is a principal cause of premature failure through the magnetron having insufficient cathode emission, causing it to operate in a higher mode, usually (π - 1), and resulting in gross internal heating.

Lower-power magnetrons (<15 kW rating) have oxide-coated cathodes which operate at lower temperature, and the above criteria apply, although the control is less critical.

In all cases it is important to allow the manufacturer's specified time delay 'heater warm-up time' between energising the filament/cathode and applying anode EHT.

(*b*) *Anode EHT, Performance Chart:* The anode voltage/current relationship of a magnetron is unusual, and is shown typically in Figure 9.4. As the anode voltage is raised from zero there is very little anode current until a critical voltage, the π-mode voltage (V_π), is reached. Thereafter, the anode current rises very rapidly giving a dynamic slope resistance usually < 100 Ω.

The reason for this characteristic is that at low values of applied EHT (V_a) electrons circulating around the cathode do so with an outer orbit having radius less than the anode bore and so none reach it. When $V_a = V_\pi$, the outer electrons reach the anode and current starts to flow.

Such a V_a/I_a characteristic requires a substantially constant-current power supply for stability. A simple unregulated transformer–rectifier supply for the EHT will give wide variations in power output from the magnetron from

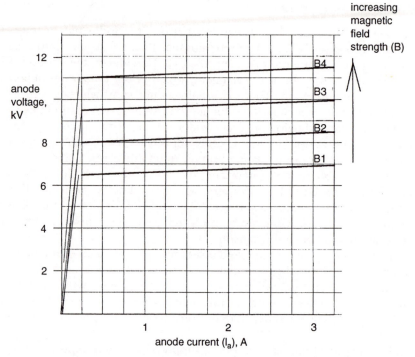

Figure 9.4   Magnetron V_a/I_a characteristic with magnetic-field strength as a variable parameter

small changes of mains supply voltage. Techniques for EHT power supply are discussed in Chapter 10.

Figure 9.5 shows the so-called 'performance chart' of a magnetron. It differs from Figure 9.4 in showing the contours of constant power output, and contours of constant efficiency, defined as

$$(power\ output)/(V_a \times I_a)$$

One widely used method of power control is by adjusting the magnetic field: paradoxically *lowering* the applied field *increases* the power output, as can be seen from the chart.

Note that it is possible to prevent a magnetron from oscillating by raising the magnetic field to give $V_\pi > V_a$, a technique commonly used to hold a magnetron temporarily in standby mode.

The performance chart is always plotted for the magnetron operating into a matched load. For mismatched loads, the general shape is similar but the efficiency may be very different depending on the phase and amplitude of the reflected wave.

(*c*) *Electromagnet supply:* In the high-power magnetrons (> 10 kW), the magnetic field is provided from a separate electromagnet into which the

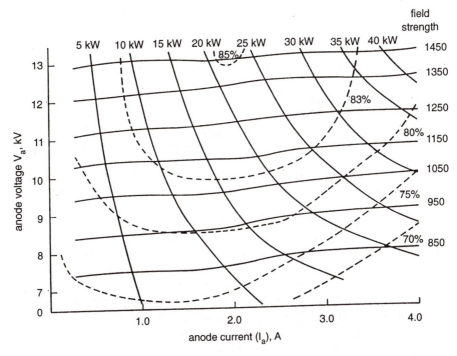

Figure 9.5   Performance chart of 915 MHz magnetron showing contours of power and efficiency (Twistleton, 1964)

magnetron is inserted. The design is specified in principle by the magnetron manufacturer in terms of ampere turns and the magnetic polepieces. A typical magnet for a 915 MHz tube has 2450 turns and operates in the range 2.8–5.0 A, with a resistance of 12 Ω. It is driven from a single-phase thyristor-controlled supply with full-wave bridge rectifier. The inductance of the magnet provides sufficient smoothing.

9.3.2.2 Mismatched loads: frequency and power 'pulling'

For a general understanding of magnetron performance when operated into a mismatched load, a simplified equivalent circuit shown in Figure 9.6 is a good approximation. A reference plane is chosen where the anode can be represented by a simple parallel-tuned circuit with components L_A , C_A etc.; the electron cloud presents a parallel admittance $Y_E = G_E + jB_E$, where G_E has a negative value representing a power source. The load $Y_L = G_L + jB_L$ is connected through a length of transmission line with the parameters shown. The magnetron will oscillate if the total conductance of the circuit is zero or less, and the frequency will be that value of ω giving zero overall susceptance.

Qualitatively, it is immediately apparent that the oscillation frequency is a function of the loading because it will present a susceptance B_L (positive or

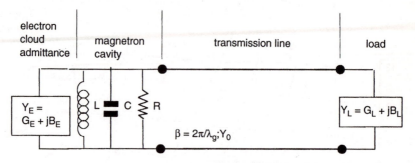

Figure 9.6   Simplified equivalent circuit of magnetron with transmission line and load

negative) at the magnetron reference plane dependent on Y_L and the length of the transmission line. This susceptance will add algebraically to that due to the cavity susceptance B_C. Changing the length of the line will vary the load susceptance 'seen' by the magnetron, causing the frequency to change to restore the zero-susceptance condition. Similarly, the conductance component of the load 'seen' by the magnetron changes with line length, resulting in a change in power output, and more especially efficiency. For a small reflection, changing the length of the line progressively through $\lambda_G/2$ results in a quasisinusoidal change in frequency and, displaced by $\lambda_G/4$, of power output.

Combining the work of Lythall (1946) and Twisteleton (1964), the operating frequency of a magnetron having a mismatched load of complex reflection coefficient $\rho e^{j\theta}$ is given by

$$\frac{(f_0 - f)}{f_0} = \frac{\sin\theta}{Q_x(\xi + \cos\theta)} \tag{9.1}$$

where

$$\xi = \frac{(1 + \rho^2)}{2\rho} = \frac{1 + S^2}{1 - S^2} \tag{9.2}$$

Q_x is the 'external Q-factor' associated with the cavity resonator and the conductance G_L' of a matched load. Q_x is determined by the coupling between the magnetron and its output waveguide and is chosen by the designer as a compromise between efficiency and stability (Twistleton, 1964). The requirement is principally for stable operation with loads of VSWR $\leq$ 3:1 in any phase without anomalous behaviour. Typically $120 < Q_x < 180$.

Figure 9.7 is a plot of eqns. 9.1 and 9.2 showing the fractional frequency change for a typical magnetron with $Q_x = 150$, with various values of load VSWR as shown, as the phase changes through two whole cycles. Note that the rate of change of frequency is a rapidly varying positive function of phase

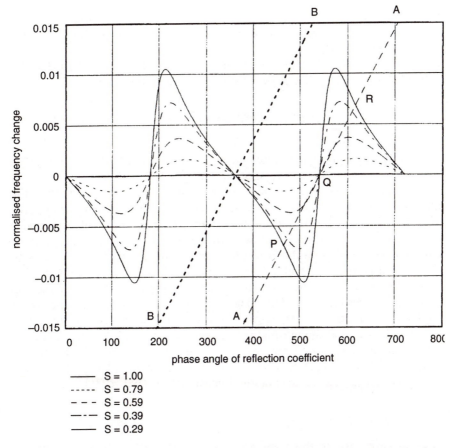

Figure 9.7   Frequency of magnetron against phase of reflection coefficient of load for various values of VSWR, also showing locus of frequency-sensitive loads

angle in the vicinity of θ = 180°, 540° etc., but for phase angles in the region of θ = 0°, 360° etc. it is negative, and much less rapid. This has important implications for the stability of the magnetron when working with mismatched loads, discussed below.

9.3.2.3 Frequency-pulling figure

The frequency pulling figure is a specification parameter for a magnetron and is derived from eqns. 9.1 and 9.2 as

$$f_p = \frac{f_0}{Q_x} \frac{2|\rho|}{1-|\rho|^2} = \frac{f_0}{Q_x}\left(\frac{1}{S} - S\right) \qquad (9.3)$$

where f_p is the total frequency change as the load phase angle is varied through a complete cycle, 2π radians, and f_o is the frequency when operating into a matched load.

The 'pulling figure' of a magnetron is the value of f_p when the load VSWR is 0.666 (or $\rho = 0.2$), whence by substitution in eqn. 9.1,

$$pulling\ figure = 0.417\frac{f_0}{Q_x} \tag{9.4}$$

Typical pulling figures for industrial magnetrons are 2.5 MHz at 915 MHz, and 7 MHz at 2450 MHz, corresponding to $Q_x \approx 150$ for both.

The above treatment of frequency-pulling effects assumes that the admittance of the electron cloud remains constant; in practice it varies partly as a function of the anode current, and partly as the load admittance changes, so that the measured values may differ slightly from those given by eqn. 9.3. There is also a thermal effect, that a change in efficiency brought about by a change in load will cause the anode to expand (or contract) thermally because of the change in anode dissipation and its effect on temperature distribution. The magnitude of the thermal effect may be comparable with the pulling figure, but its response time is slow, of the order of seconds. Thus a change in load reflection will result in an immediate frequency change by the mechanism of eqn. 9.1, followed by a thermal change which adds algebraically over the ensuing seconds. The pulling figure is defined and measured to exclude thermal effects (BSI, 1971).

9.3.2.4 Frequency-pushing figure

It is already noted that the electronic admittance Y_E contributes a susceptance in the equivalent circuit of Figure 9.6 which affects the frequency. It is also a function of the anode current because it is proportional to the number of electrons arriving at the anode per unit time, or the density of the electron cloud. The magnetron frequency therefore changes with anode current, and also again through thermal expansion of the anode because the anode dissipation increases with anode current.

The 'frequency-pushing figure' is defined as the instantaneous change in frequency resulting from a small change in anode current and is expressed in units of megahertz per ampere and is typically 1 MHz/A at 915 MHz, and 2 MHz/A at 2450 MHz.

9.3.2.5 Rieke diagram

The Rieke diagram is a Smith chart (Section 4.3.4) drawn to represent the admittance of the load on a magnetron at a convenient reference plane such as the launch-waveguide flange. Superimposed on the chart are contours of

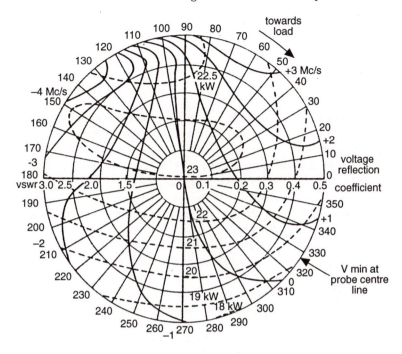

Figure 9.8   Magnetron Rieke diagram (Twistleton, 1964)

constant output frequency and of constant output power (or efficiency), as functions of load admittance as defined by the Smith chart. Using the Rieke diagram, it is thus possible to view the magnetron performance over the whole range of load admittance plotted.

The frequency and power contours are, in principle, mutually orthogonal, but in practice they are inclined due to the variation of electronic admittance (Thomson and Callick, 1959). A typical Rieke diagram is shown in Figure 9.8.

Note that the frequency contours crowd together on one side of the diagram where the power output is greatest, and the rate of change of frequency with load is at its greatest, corresponding to the upper region in Figure 9.8. This zone is called the 'sink', and is an area of potential instability, declared a forbidden zone in some magnetron specifications.

The Rieke diagram is only of value if the full operating conditions for the tests are given: specifically and primarily, the location of the reference plane, the anode voltage and current, and the magnetic field. If these parameters change during the test, this must be noted. The magnetron is usually operated in a constant-anode-current system, in which case the resulting changes in magnetic field or anode voltage should be recorded. Unfortunately, these data are often not adequately presented by the magnetron manufacturers. The Rieke diagram of Figure 9.8 is plotted with the magnetron operated in a series-field configuration, where the anode current passes through the electromagnet winding to give a self regulation of anode current as considered in Section 10.3.7.

9.4 Stability

In the overall system design the magnetron is a component which cannot be considered entirely in isolation. It usually forms an element in a closed-loop control system, and so, as a 'black box', its several transfer functions are important operational parameters to the designer in determining loop gain and response time. The anode-voltage/current characteristic, itself a transfer function, is a function of the microwave load impedance, and so transient changes in load in turn affect the loading on the EHT supply, so that the output impedance of the supply is an important factor.

Magnetrons may in extreme conditions be caused to 'mode skip', resulting in operation at the wrong frequency with very low efficiency; most of the input power becomes internally dissipated, resulting in rapid destruction. Arcing may also occur due to gas in the vacuum envelope, sometimes initiated by operation outside the maximum limit of a parameter such as anode current.

These topics are the subject of the following sections.

9.4.1 Transfer function within a closed loop

Because of its very low dynamic impedance, the magnetron is invariably operated with a constant-current or constant-power control system; otherwise the power output varies far beyond acceptable limits. Various control methods are used, discussed in detail in Chapter 10, but they all operate by control of either the applied EHT, or the applied magnetic field to the magnetron.

Low-power magnetrons (< 6 kW rated power output) at 2450 MHz have permanent magnets and the only available control method is by anode-voltage control. For these the V_A/I_A characteristic is provided by the manufacturer, from which the static (V_A/I_A) and dynamic $(\partial V_A/\partial I_A)$ impedances are readily derived. The response times of both parameters are virtually instantaneous, being determined by the Q_X value of the magnetron. A typical response time would be around 0.02 μs, which is negligible for a control loop. Typical values when operating at rated power output into a matched load are given in Table 9.1.

In Table 9.1, the first two magnetrons are low-power 'packaged' types, i.e. having built-in permanent magnets. The last three have electromagnets.

Note that the incremental impedance varies widely, and varies over the operating range, especially at low power output less than 20% of rated power. The incremental impedance may be negative locally, giving rise to spurious, low-level oscillations at a frequency determined by the stray inductance and

Table 9.1 Typical values at rated power output into a matched load

Magnetron type	Frequency	Rated power output	V_A	I_A	V_A/I_A	$\partial V_A/\partial I_A$
	(MHz)	(kW)	(kV)	(A)	(kΩ)	(Ω)
YJ1540	2450	1.26	4.5	0.4	11	1400
2M53A	2450	0.83	4.0	0.3	13	1000
YJ1600	2450	6.0	7.2	1.15	6.3	<30 to 600
CWM60L	915	60	18	3.8	4.7	<10 to 150
CWM30S	2450	30	16	2.7	5.9	<20 to 200

capacitance of the EHT conductors adjacent to the magnetron cathode.

Although switch-mode power supplies allowing direct anode-voltage control are being introduced, most high-power magnetrons are controlled by adjustment of their magnetic field to maintain constant anode current or, less frequently, constant power output. The relevant parameter is the incremental change in anode current for a small change in the electromagnet current providing the magnetic field ($\partial I_A/\partial I_{MAG}$) at constant EHT. Note that an *increase* in electromagnet current *reduces* the anode current so that $\partial I_A/\partial I_{MAG}$ has a negative value. The response time has two components: that of the inductance and circuit resistance of the electromagnet, and that due to the 'shorted-turn' effect of the magnetron anode, which is effectively a copper ring. It is the magnetic field inside the ring which is important, but any change in applied magnetic field induces an EMF in the ring, causing a current to flow which has a magnetic field opposing the change inside the ring. Steady-state conditions do not occur until this current has decayed to a negligible amount.

Typical electromagnets have inductance in the range 0.2–0.8 H and resistance about 10–15 Ω, giving a time constant ($T_C = L/R$) of around 40 ms. It is desirable to measure the electromagnet parameters to establish specific figures.

The time delay due to the copper-ring effect can be estimated roughly by assuming an equivalent copper ring for the magnetron anode and performing a classic calculation of its self inductance and resistance, which yields a time constant (T_M) of about 10 ms.

For a 60 kW 915 MHz magnetron of type CWM60L, the value of $\partial I_A/\partial I_{MAG}$ is about 15–20 with a standard electromagnet of 2450 turns, giving 0.056 T/A. For control of anode current from the electromagnet current, the transfer function of the CWM60L magnetron–electromagnet combination is therefore

$$\frac{\partial I_A}{\partial I_{MAG}}\left(\frac{1}{1+pT_C}\right)\left(\frac{1}{1+pT_M}\right) \approx -17.5\left(\frac{1}{1+0.04p}\right)\left(\frac{1}{1+0.01p}\right) \quad (9.5)$$

where p is the Heaviside operator (Pipes,1946; James *et al.*, 1947; Distefano *et al.*, 1990). For the customary Nyquist or Bode analysis of control-loop performance in the frequency domain, insert $p = j\omega$, which then gives the frequency response of eqn. 9.5.

Similar expressions can readily be derived for other magnetrons having variable magnetic fields controlled from their electromagnets.

Because of its very high efficiency, the input impedance (V_A/I_A) of the magnetron is affected by the RF load impedance, specifically the real part of the complex RF load impedance measured at the reference plane corresponding to the equivalent circuit of Figure 9.1. In the limit condition, the magnetron may operate into a load of VSWR = 3.0 in any phase, which means that the real part of the complex load impedance varies over the range 3 to $1/3$ (= 0.333), i.e. a load-impedance variation of 9:1. If the anode voltage and magnetic field are held constant while such a load is varied in phase angle through a complete cycle, the anode current will vary through the ratio 9:1, attenuated by the efficiency of the magnetron. The magnetron efficiency itself varies with this procedure so precise data can only be obtained by experiment, but the typical variation is likely to be around 6:1.

It will be seen from the above that the overall performance is affected by several parameters, including the load. If the load impedance is frequency sensitive, then there is the added effect of the magnetron frequency-modulation effects (pulling and pushing) becoming a feature of control-loop gain and stability. Because of the complexity of the system and the interactions between the parameters involved, the use of a microwave circulator to isolate the magnetron from its load has become almost universal for high-power installations. However, circulators are expensive and have been a source of unreliability; in particular they are vulnerable to faults such as waveguide arcing. With the advent of switch-mode power supplies with their very high-speed control characteristics, it is probable that circulators can be avoided in many installations where the load is substantially broadband in frequency response. In such cases it then becomes essential to address the above effects quantitatively.

It should be noted that a constant-power control system may cause the anode current to rise above the permitted maximum if the load impedance varies adversely to reduce the magnetron efficiency, and that an electronic limit stop must be used to limit the anode current; otherwise arcing or mode skipping may occur.

9.4.2 Moding, (π-1) mode, spurious emissions

In Section 9.3.1, the mode patterns of a magnetron are discussed and it is explained that the desired fundamental mode is that in which adjacent segments of the magnetron are in opposite RF phase. The magnetron is designed to operate in this 'mode' (called the 'π-mode') at high efficiency and with good stability. Other modes of oscillation are possible but they are all undesirable because they oscillate at the wrong frequency, and have very low efficiency. Their low efficiency results in most of the input power to the anode being dissipated internally, causing rapid heating and severe damage to the magnetron in a very short time (1–10 s typical).

Part of the magnetron designer's skill is in separating the π-mode from the others to minimise the risk of their excitation (Twistleton, 1964). One technique is 'double ring strapping' where alternate segments of the anode are connected together by a pair of separate copper rings, one for the even-numbered segments, the other for the odd numbers. It is a simple method of ensuring that the segments have the correct phase relationship for the π-mode: in principle no current flows in the rings in the π-mode because the connected segments are all in phase and of equal instantaneous voltage. For other modes, a heavy current flows because there is a phase difference between alternate segments. In a magnetron which has operated in a wrong mode the ring straps have usually melted, but other damage may also have occurred.

Wrong modes are characterised by an abnormally high value of EHT to the anode, or abnormally low magnetic field. Limitation of the operational range of these parameters within the power supply is an effective safeguard against moding. Moding is caused mainly by insufficient emission of electrons from the cathode to support the desired anode current when the anode current is controlled in a closed loop. Under these conditions, the control system will, depending on its type, raise the anode EHT or reduce the electromagnet current, and if not limited will reach the condition for oscillation in the next $(\pi\text{-}1)$ mode. In a magnetron with a pure tungsten filament this happens if the filament temperature is too low due to insufficient filament current, so accurate control and monitoring of filament current is an essential need. In magnetrons with oxide cathodes, emission may fall due to aging, which does not happen with the tungsten filament.

Magnetrons emit various spurious signals beside the main power output. These are emitted from the cathode EHT insulator which is transparent to all RF and microwave frequencies. The signals comprise a small leakage at the operating frequency and a wide spectrum of noise at low level extending from about 30 kHz upwards. These emissions are dependent not only on the magnetron but also on the power supply, especially the detailed wiring arrangements of the EHT circuits. The magnetron specification does not usually set a limit on these emissions, except for some small low-power packaged magnetrons fitted with external filters and absorbers. Although the magnetron has in-built choking to minimise the emission of energy at the operating frequency, the attenuation is finite, of the order of 50 dB. Thus the cathode-stem leakage from a 50 kW magnetron is of the order of 0.5 W at the operating frequency. Other frequencies are at much lower level, but they can be a source of radio interference unless precautions are taken. The implications of these effects are discussed in Chapter 10.

9.4.3 Long-line effect and resonant high-Q-factor loads

In Section 9.3.2.2 it is shown that the oscillation frequency of the magnetron is affected by the RF-load impedance, an effect quantified by eqn. 9.1. and illustrated in Figure 9.5. If the load impedance is itself a function of frequency, an interaction must occur between the magnetron and its load so that the frequency adjusts to a value at which the load impedance corresponds to a point on the magnetron Rieke diagram of that same frequency. Such a point may not exist, or it may be unstable, resulting in frequency jumping, and most probably anode-overcurrent tripping.

Lythall (1946) analysed this effect for long transmission lines in pulsed magnetron systems for radar, but the same argument applies for CW systems. Consider, for example, a magnetron coupled to a load of VSWR 0.3 of variable phase. The frequency curve of the magnetron as a function of phase angle is shown in Figure 9.7. Note that the slope of the curve $df/d\theta$ has regions of positive and negative values. Superimposed on the graph is a straight line AA which represents the change of phase angle with frequency of the reflection coefficient of a load of VSWR 0.29, located at the end of a long transmission line. The load itself is assumed to be insensitive to frequency over the range of interest, and the phase change 'seen' by the magnetron is purely that due to the change of electrical length of a long feeder with frequency. Lythall shows that the slope $d\theta_A/df$ of this line is given by

$$\frac{d\theta_A}{df} = \left(\frac{4\pi l}{c}\right)\left(\frac{\lambda_g}{\lambda}\right) \tag{9.6}$$

where $c = 3.10^8$ m/s, and l is the length of the feeder.

Possible operating points are where the line AA intercepts the magnetron curve at P,Q,R. The slope of the line AA ($d\theta_A/df$) is positive: stable operation occurs at the point(s) where it intercepts the magnetron curve having negative slope, e.g. P or R. Intercepts where the slopes have the same sign are unstable, because a small change in frequency results in a change of phase angle in the same direction for both magnetron and load, resulting in a cumulative departure from the intersection point. Whether P or R is the final operating point depends on the transient-starting conditions, discussed in detail by Lythall. In a CW system, jumping from one point to the other is a possibility. However, if the transmission-line length is changed by $\lambda_g/4$ the line AA moves to BB, where it will be seen that there is now only one intersection at a stable operating point. Moreover, there is a range of values of phase angle for the line BB before it intercepts the magnetron frequency-pulling curve corresponding to $s = 0.29$, where instability might reappear.

In an elementary case such as above with just one fixed load at the end of the feeder, the simple expedient of changing the line length by a nominal $\lambda_g/4$ in principle stabilises the system, although a greater or lesser change may represent an optimum. However, the load is usually frequency sensitive, with distributed reflections from bends etc. along the feeder, and so in practice stabilisation may be difficult. The use of a circulator usually overcomes these effects completely.

Where the load is a high-Q-factor resonator, a similar stability condition exists, but, in contrast to the long-line-effect condition above, both the amplitude and phase of the reflection coefficient are now variables with frequency. The admittance–frequency locus of a resonator plotted on the Smith chart is a circle (Metaxas and Meredith, 1993), ideally with the resonant point lying along the $(g + j0)$ line, and the circle tangential to the $(0 + jb)$ circle. If the magnetron Rieke diagram is superimposed on the chart, the magnetron frequency contours will intercept the cavity locus.

From the Rieke diagram, the magnetron's frequency sensitivity to load can be directly read in terms of an incremental change in susceptance $(df/db)_{MAG}$. Similarly, the sensitivity of the resonator to frequency $(db/df)_{RES}$ can be estimated from the diagram. Instability occurs if, at any point on the *whole* Rieke diagram (i.e. covering a full Smith chart extending to $|\rho| = 0$),

$$(df/db)_{MAG} \times (db/df)_{RES} = -1 \qquad (9.7)$$

There are two cases of interest: first with an assumed lossless and zero-length line between the magnetron and resonator, and secondly where there is significant loss.

For the first, at frequencies far removed from resonance, the resonator's admittance locus on the Rieke diagram tends to the $\pm \infty$ point. At all points of intersection the LHS of eqn. 9.7 is positive, so that there is unconditional stability. The ultimate example of this is the 'coaxial magnetron', where the usual magnetron anode cavities are coupled to a surrounding resonator which has a high value of Q_0 and determines the operating frequency. Coaxial magnetrons have excellent frequency stability and life but are, unfortunately, very expensive. If, however, the transmission line has finite length but remains lossless, the admittance locus 'seen' by the magnetron extends beyond the $\pm \infty$ point and the LHS of eqn. 9.7 becomes negative and, depending on the exact conditions, may result in instability.

If the line is $n\lambda_g/2$ long and has finite attenuation (A decibels), the admittance locus remains a circle but crosses the $(g + j0)$ axis of the chart at two finite values of conductance g. (It is assumed that the Q_L of the cavity is sufficiently high that the phase length of the transmission line remains constant over the frequency range of interest.) These two intersections correspond to:

(*a*) the cavity completely de-tuned, and
(*b*) then tuned to resonance.

For case (*a*) the cavity input admittance (as seen at the cavity end of the lossy line) is

$$y_i = \infty + j0 \qquad (9.8)$$

After traversing the lossy line, the reflection coefficient is reduced in magnitude by the factor $10^{(-A/20)}$. Using eqn. 4.8, the input conductance to the line is then, with $g_0 = \infty$,

$$g_{i,a} = \frac{1 + 10^{(-A/20)}}{1 - 10^{(-A/20)}} \tag{9.9}$$

Similarly, for case (b) the input conductance is given by eqn. 7.7; substituting $g_0 = g_C + g_L$, putting $\delta\omega = 0$ and using the admittance form of eqn. 4.8 with $\phi = 0$, whence

$$g_{i,b} = \frac{1 - \left(\dfrac{1 - g_0}{1 + g_0}\right) \times 10^{(-A/20)}}{1 + \left(\dfrac{1 - g_0}{1 + g_0}\right) \times 10^{(-A/20)}} \tag{9.10}$$

It will be seen that the effect of the line attenuation is to reduce the diameter of the circular admittance locus of the cavity as 'seen' by the magnetron, quantified by eqns. 9.9 and 9.10.

Figure 9.9 illustrates an idealised chart showing a cavity-admittance locus plotted on a Rieke diagram. The cavity is shown to be overcoupled, the $(1 + j0)$ point lying *within* the admittance circle. Note that the cavity-susceptance differential $(db/df)_{RES}$ changes sign when the detuning frequency exceeds that at which the circle is tangential to the constant-susceptance contours of the Smith chart. Further detuning increases the magnitude of $(db/df)_{RES}$, and the condition of eqn. 9.7 may be reached, resulting in unstable operation, manifest in operation with the cavity substantially off tune, and/or frequency jumping.

If the load has a low value of Q_L, stable operation is possible with direct coupling of magnetron to load. However, with higher Q_L these problems are usually overcome by using a circulator to isolate the magnetron from reflected power from the cavity. Most circulators have an isolation around −16 to −23 dB, reducing the value of $(db/df)_{RES}$ to about 10% as 'seen' by the magnetron. This may not be sufficient to avoid the condition of eqn. 9.7, and further circulator(s) may be required to secure unconditional stability.

9.4.4 Anode-current surges and arcing

Magnetron specifications always set an absolute maximum value for anode current to avoid operation beyond the rated output, and to limit damage under fault conditions. If not specifically stated, the current limit is interpreted as an instantaneous value.

It is usual practice to fit a fast-trip system to switch off the EHT if the anode current reaches the absolute maximum limit, the response time of which is

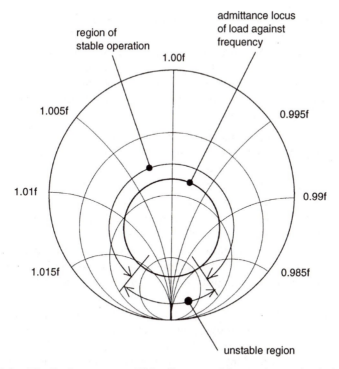

region of
stable operation

admittance locus
of load against
frequency

1.00f

1.005f

0.995f

1.01f

0.99f

1.015f

0.985f

unstable region

Figure 9.9   Idealised magnetron Rieke diagram with superimposed admittance locus
of resonant-cavity load

about 40 ms, being, mostly, the time required to open the main power
contactor of a conventional (i.e. not switch-mode) power supply.

There are several reasons for the anode current reaching its limit:

(i)   a transient disturbance in the power supply to the equipment. If this is
        of sufficient magnitude, the anode-current regulator may momentarily
        lose control through saturation due to the large error signals which
        result. The anode EHT supply may rise faster than the electromagnet
        field, forcing an anode-current surge. An extreme case is the
        momentary interruption of the incoming power supply. A well designed
        power supply will have an in-built routine to restart the generator
        subsystems in the correct order following an interruption of supply;

(ii)   a transient disturbance in the load impedance 'seen' by the magnetron.
        This may be a microwave arc originating in the applicator or elsewhere
        in the waveguide system between the generator and load. Such arcs
        usually have a large reflection coefficient, causing a sudden jump to
        another operating point on the Rieke diagram;

(iii)   with resonant loads instability due to the mechanisms outlined in
        Section 9.4.3 may result in an anode current surge;

(iv)   the magnetron may have residual gas of sufficient pressure that ionisation occurs at or below the required operating electric field within the magnetron. The ionisation voltage may be less than the anode EHT, in which case the magnetron cannot be switched on because the discharge forms a near short-circuit between anode and cathode. In some cases the RF voltage at the anode-vane tips adds to the EHT so that the effective voltage is higher; this results in a limit on power output before ionisation occurs.

Residual gas may obviously be the result of a vacuum leak through damage, or occasionally faulty manufacture, in which case the magnetron must be reprocessed. However, gas pressure can build up due to faulty operation of the magnetron beyond its specification limits, when overheating causes gas occluded into the vacuum envelope to be liberated. If the amount of gas liberated is small, it can often be 'cleaned up' in a magnetron with a tungsten filament by running the filament alone at its standby current. The filament acts as a vacuum pump (of limited capacity) and may restore the vacuum in about 30 min running. A small amount of gas may collect in a magnetron which has not operated for a long period, and similar treatment often clears the problem.

Anode-overcurrent trips due to ionisation result in intense local heating within the magnetron envelope due to arc formation, particularly with conventional power supplies comprising a large power-frequency EHT transformer with rectifiers and smoothing choke. Much of the energy stored therein is dissipated in the arc; further energy is dissipated due to the delay in disconnecting the EHT power supply from its feeder. The local heating may be sufficient to cause local pitting of the surface inside the magnetron due to vaporisation of the metal: gas is released and the pressure rises. It is imperative to delay restarting a magnetron in this condition until it has operated on filament alone for at least 5 min in an attempt to restore the vacuum. Repeated application of EHT immediately causes repeated arcs resulting in a progressive increase in gas pressure, beyond which recovery becomes impossible without reprocessing the magnetron. Many magnetrons have been irretrievably damaged by operators unaware of the importance of following the correct procedure.

The introduction of high-frequency switch-mode power supplies is an important advance in the technology because they have much less stored energy, and can be switched off electronically in about 50 µs. The fault energy dissipated within the magnetron is less than 0.1% of that with a conventional power supply. Stored energy has been a constraint on the development of higher-power magnetrons at both 915 and 2450 MHz, because it is proportional to the rated power output of the magnetron. For a 100 kW magnetron, a conventional power supply has so much stored energy that an arc is likely to result in an irrecoverable rise in gas pressure.

A significant reduction in the fault energy of a conventional power supply can be achieved using an ignitron to 'crowbar' the supply as a short circuit.

The ignitron is connected directly in parallel with the magnetron and is triggered by an impulse derived from the short-circuit current of the initial fault. A response time of the order of 20 µs can be achieved. However, there remains a total short circuit on the supply, placing a heavy duty on the EHT rectifiers, and there is a risk of the ignitron firing prematurely.

9.5 Cooling

Cooling is a very important feature of high-power-magnetron operation, and careful design and operation of the cooling system is essential for reliability and long life. Liquid cooling is applied to the anode, and air cooling to the output window (usually a ceramic dome) and to the anode connector. The integrity of the cooling systems cannot be overemphasised, requiring the highest standards in engineering, in both hardware and software.

9.5.1 Liquid cooling

Water cooling must be applied at all times in accordance with the magnetron specification, which usually gives an absolute limit on the maximum outflow temperature of the coolant, as well as a minimum flow rate. Nearly all the internally generated heat is abstracted by the liquid cooling. In magnetrons of high efficiency the internal heat dissipation under normal operating conditions is a relatively small proportion of the power input; however, a small reduction in efficiency due to a mismatched load, for example, causes a large increase in internal heating. For this reason it is prudent to provide an excess of cooling capacity beyond the minimum stated in the specification.

The specification usually requires cooling to be applied for several minutes after all power supplies have been removed to ensure that residual heat is conducted away to avoid excessive temperature rise. Failure to comply with this requirement often results in cracking of the ceramic dome of the magnetron, because the small volume of water in the cooling jacket is quickly boiled away by the residual heat, whence the temperature rises rapidly resulting in thermal stressing of the metal–ceramic seal.

The best practice is to use distilled or demineralised water for cooling in a pumped closed circuit, using a water–water heat exchanger. The flow passages within the magnetron are small and can easily block, not only from foreign particles but also from scale due to hard water. Such a system has a high thermal capacity isolating the magnetron from variations in supply pressure common in industrial cooling systems.

9.5.2 Air cooling

Air cooling requires the same care in design and operation as liquid cooling, inadequate cooling being a prime cause of magnetron failure. Although the heat removed by air cooling is small, it is critical in limiting the temperature of sensitive zones, notably ceramic parts and ceramic/metal seals. The manufacturer's specification must be followed strictly, and suitable allowance must be made for reduction of flow due to blockage of air filters etc. Dust must not be allowed to enter the waveguide system as a build-up will eventually cause an arc, and filters should be fitted to cooling-air intakes, particularly where the process is dusty, e.g. carbon dust in a rubber factory, or flour in food manufacture. Generous overcooling alleviates uncertainty about the actual measurement of flow rate.

9.6 Operation practice

Although modern magnetrons are robust and reliable, having inherently long lives, they are sophisticated structures and can be damaged through operation under conditions beyond their specification limits. The engineer must take great care in system design that the magnetron is correctly operated at all times, not only to preserve its life, but also to avoid invalidating the manufacturer's warranty.

In common with engineering systems generally, start up and shut down (colloquially, takeoff and landing) are the most critical phases when failure may occur in a microwave generator, because thermal stresses and electrical transients are at their greatest. Steady-state operation, in which all the parameters are substantially constant, especially thermal gradients, gives virtually fault-free operation unless an external disturbance occurs.

For maximum reliability it is therefore desirable to operate a system continuously with as few stoppages as possible. This is not to say that a system with frequent interruption is unreliable, for there are many in this category giving very satisfactory performance, the reduction in magnetron life (typically 30–50%) being more than compensated by the convenience of operation in this role.

9.6.1 Start-up procedures

Whilst low-power magnetrons (<1 kW output) may be energised simply by switching on the cooling air, cathode and EHT simultaneously, power being generated when the cathode reaches operating temperature, higher-power magnetrons require their auxiliary supplies energising in a predetermined order. For a conventional power supply with power control by regulation of applied magnetic field, the sequence is:

(i)   water cooling and air cooling;
(ii) electromagnet supply, set to such a field intensity that the magnetron π-mode voltage exceeds the open-circuit EHT by about 5%;
(iii) filament/cathode supply;

(iv) delay, typically 3 min to allow cathode/filament to reach operating temperature;

(v) anode EHT supply;

(vi) delay, > 1 s; and

(vii) reduce electromagnet current to give required anode-current/power output.

The above is a mandatory sequence of applying the supplies to the magnetron and must be used in its entirety following a total shut down. However, it may be re-entered at the appropriate stage following a partial shut down.

Obvious variations are for switch-mode power supplies where power control is by regulating the EHT anode voltage at fixed magnetic field, but all the first four steps remain mandatory. Note especially that the filament/cathode must never be energised without all cooling systems operating, because the heat radiated from the cathode would melt the anode.

In contemporary generators the above sequence, and many other operating functions including interlocking and fault annunciation, are controlled by a progammable-logic controller (PLC) of proprietary manufacture.

9.6.2 Automatic protection systems

To minimise risk to the generator and magnetron, the following interlocks are essential:

(i) *Cooling systems*

Water:

(*a*) flow switch from the primary supply main; this is best fitted at the discharge pipe to check the integrity of the pipework;

(*b*) flow switch of secondary coolant through the magnetron, similarly, best located downstream of the magnetron;

(*c*) thermostat switch located to measure temperature of outflow from magnetron.

Air cooling:

(*a*) airflow switch mounted in the air duct leading to the magnetron; a pressure switch is not recommended because it fails to danger if there is a partial blockage downstream.

(ii) *Electromagnet current*

This interlock is set to 'drop out' at about 75% of the lowest envisaged. Its purpose is solely to determine the integrity of the supply rather than monitor its magnitude; in practice the most probable fault is a total absence of current.

(iii) *Filament current*

This current is preferably measured in the secondary of the EHT insulated filament transformer, and is again set to about 75% of the lowest operating current by the argument in (ii) above.

(iv) *Anode current*

This interlock is the most critical, and is set to 'drop out' the EHT anode supply if the anode current exceeds the absolute maximum specified. The energy dissipated within the magnetron under overcurrent fault conditions is in most part proportional to the reaction time of the interlock, which is the time taken for the main contactor (in a conventional power supply) to clear its contacts (about 40 ms.). In a switch-mode power supply this interlock is entirely electronic and operates in about 50 μs, which is a major advantage, greatly reducing the energy dissipated in the magnetron under fault conditions.

(v) *Reflected power*

A magnetron may be damaged due to excess reflected power, and protection must be provided. A ferrite circulator is often used, diverting reflected power away from the magnetron into a waterload. Alternatively, a microwave sensor coupled to a directional coupler is used, the sensor often being a microwave diode detector. If the reflected power exceeds a preset limit, the magnetron is switched off.

(vi) *Arc detector*

Microwave arcs may occur in the waveguide system from a variety of causes, and they are generally unstable and progress towards the generator where they can cause serious damage from burning/overheating. An optical detector using a photodiode is an effective protection. The light from the arc is intense in an otherwise substantially dark environment. However, some light is emitted from the cathode of some magnetrons, visible through the RF output window; care must be taken to ensure that this light does not create false alarms.

9.6.3 Restarting after a fault trip

Restarting a generator after a fault trip requires care, the procedure being dependent on the nature of the fault. In all cases, it is imperative to have understood the cause of the fault and to have taken the necessary action to ensure that these are no repeat faults. This is especially necessary after anode-overcurrent or arc-detector trips, events which are potentially damaging; arc damage must be totally cleaned up before restarting, otherwise a repeat is almost certain.

Table 9.2 lists starting procedures following fault trips.

9.6.4 Shut-down procedure

Shut-down is usually to one of two levels: standby or total shut down. In the first, the magnetron is in a nonoscillating state of readiness with all its auxiliary supplies operating, i.e. at the end of stage 4 in Section 9.6.1. In a conventional power supply the anode EHT is energised but the magnetic field is set to a high value so that the π-mode voltage is greater than the

Table 9.2   Starting procedures following fault trips

Item	Fault	Procedure	Note
1	Incoming-supply trip	Re-enter full starting sequence	
2	Cooling-water trip	Re-enter full starting sequence	
3	Air-cooling trip	Re-enter full starting sequence	
4	Electromagnet trip	Re-enter at 2	
5	Filament trip	Re-enter at 3	
6	Anode-current trip	Re-enter at 4	1
7	Reflected-power trip	Re-enter at 5	2
8	Arc trip	Re-enter at 5	3

Notes

1 Following an anode-current trip, the magnetron should be allowed to remain on standby filament current for at least 5 min before re-application of anode EHT. See Section 9.4.4. Strenuous efforts must be made to establish the cause of the trip, with remedial action if necessary.

2 A reflected-power trip may have several causes, some such as disturbance or loss of workload feed being obvious. Damage to the waveguide feeder or applicator is a possible but unusual cause. Possible causes include an accretion of debris in the applicator, arc damage (particularly to a circulator) and water ingress from condensation or leak (e.g. from a waterload). A reflected-power trip may be accompanied by an arc trip; see below.

3 An arc trip must be investigated in depth and the cause remedied before resuming production. Its cause may be any of those listed in 2 above. Occasionally, it is caused by careless fitting of the magnetron into its socket, the microwave-sealing gasket being out of position. If this has happened, it is likely that the magnetron has been damaged through a cracked ceramic window. Note that this may not be obvious, as a fine hairline crack, not visible, will spoil the vacuum irreversibly.

supply, resulting in zero anode current. This condition can be maintained indefinitely. In a switch-mode power supply, the anode EHT supply is off, but at immediate readiness. Reverting to full power output from standby is achieved in < 1 s.

The full shut-down sequence begins with switching off the anode EHT supply, then the filament followed by the electromagnet supplies. A 3 min delay then elapses before the cooling systems are shut down. The electromagnet must remain energised until the anode EHT is zero to prevent the magnetron drawing a heavy anode current as a simple diode, resulting in an overcurrent trip.

It is very important to avoid switching off all supplies simultaneously as this results in severe thermal stressing of the magnetron, with risk of failure.

9.7 Magnetron life

The design and manufacture of the modern magnetron has evolved to a high state of perfection, yielding long life, *provided that the power supply driving it is of similarly high quality and is maintained and serviced regularly.* In the past, many magnetrons have failed early in life because of imperfections in the design or maintenance of the power supply, and this point cannot be overemphasised.

The normal end of life is due to cathode failure. In large-power magnetrons with directly heated tungsten filaments, this is a 'sudden-death' event without warning, when the filament 'open circuits' through melting at a hot spot. Full rated power output is obtainable until this happens, typically at 8000–12 000 hours at 915 MHz, and 4000–6000 hours at 2450 MHz, depending on operating conditions. The life is dependent on the compromise between running the filament hot enough to achieve the desired emission to sustain the anode current, and not shortening its life by making it so hot that evaporation of tungsten proceeds unnecessarily fast. Usually, the filament fails at the moment of switch on due to the surge current and transient temperature changes.

Lower-power magnetrons at 2450 MHz normally fail through loss of cathode emission, but this is not usually so dramatic, and is indicated by a reduction of power output, loss of stability and a tendency to mode skipping.

Magnetron warranties are usually based on a full allowance against the purchase price for cathode operation within a short time, followed by a linear fall in allowance to zero to a longer time commensurate with the predicted life. The initial time period is typically 50 h cathode/filament life, with the zero point at 2000 to 5000 h depending on manufacturer and magnetron type. There is also an overriding 'shelf-life' limitation of two years from date of purchase, after which all warranty ends. For a successful warranty claim, the user must satisfy the manufacturer that the magnetron has been operated within specification limits at all times.

9.7.1 Magnetron rebuilding

High-power magnetrons at 915 MHz and 2450 MHz can generally be rebuilt following failure, at a cost around 50–60% of new price. Most parts are renewable except for the anode itself. A magnetron with severe melting of the anode due to loss of coolant is not usually recoverable, although melted ring straps (due to a mode-skip incident) can be repaired.

After renewal of the faulty parts, the magnetron has to be reprocessed to restore the vacuum and retested to specification. The magnetron is then in 'as-new' condition and carries the same warranty as originally. There seems no limit to the number of times a magnetron can be rebuilt, and there are many in service which have been rebuilt more than five times.

Chapter 10

Microwave generators
2: electrical power sources

10.1 General requirements

This chapter reviews the power supplies required to drive microwave generators, and it is devoted especially to supplies for magnetrons as they are overwhelmingly the most widely used microwave power sources in industrial heating.

The particular anode current and voltage characteristics of magnetrons discussed in Chapter 9, having an exceptionally low dynamic impedance, demand that the power supply must have a constant-current characteristic for stable operation. Such a constant-current characteristic implies that the power source must have a very high dynamic output impedance.

There are several ways of achieving such a power source, the choice depending on, inter alia, the factors listed in Table 10.1.

Table 10.1   Factors involved in power source selection

1	Power rating
2	Ease of adjustment of output power
3	Tolerable ripple of the output
4	Efficiency
5	Specific cost (per kilowatt of output)
6	Size
7	Weight
8	Cooling
9	Radio-frequency interference (RFI)
10	Stored energy
11	Switch-off time in fault conditions
12	Harmonics of power-line frequency imposed on the supply current
13	Power factor

In the following sections various power supplies are considered ranging from the simplest to the most technically advanced, and the lowest to the highest power output.

10.2 Variable-voltage supplies

Many industrial microwave ovens use a plurality of low-power packaged (i.e. permanent-magnet) magnetrons where the anode current is stabilised by anode-voltage control. The comparative merits of many low-power magnetrons against a few at high power giving the same total power is considered in Section 10.5. For lowest cost these small power supplies are single-phase, and are arranged in groups of three to give symmetrical loading on a three-phase supply.

These supplies comprise a line-frequency (50 or 60 Hz) EHT transformer driving a bridge rectifier in conventional format to provide the anode EHT supply, sometimes via a smoothing inductor. The anode-current waveform, without smoothing (Figure 10.1), is a succession of quasirectangular pulses, because of the V_A/I_A characteristic of the magnetron. Sometimes the EHT transformer has an additional secondary winding supplying the magnetron filament.

10.2.1 Variable transformer ratio

The variable voltage may be supplied by a variable-ratio transformer (VRT) (e.g. Variac® or Regavolt®) motor driven from a controller sampling the anode current and comparing it with a set reference, the error signal feeding to drive the motor. It is a very simple system and, being a second-order

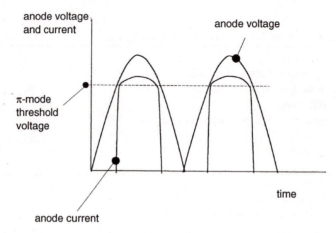

Figure 10.1   Anode voltage and current waveforms for fullwave unsmoothed bridge rectifier

servomechanism (zero position error), gives very precise control of *average* anode current. However, the response time is slow, being dominated by the inertia of the drive motor, and large errors in anode current follow a transient disturbance until the steady-state status returns. In its simplest form (Figure 10.2) the VRT rating equals that of the EHT transformer, but the use of an auxiliary transformer, as in Figure 10.3, reduces the rating required of the VRT. In the latter circuit, the minimum output voltage is designed to be about 20% less than the magnetron π-mode voltage with the VRT set to minimum, and to give maximum anode current with the VRT set to a high output limited by an adjustable electrical endstop

10.2.2 Thyristor-phase-angle control

Alternatively, the variable supply voltage may be provided by a pair of thyristors switched to conduct over part of the halfcycle each, by control of the phase angle of the firing angle, in a conventional system. This forms a first-order servomechanism having a small steady-state error, but has a fast response limited by the line-power frequency which sets the switching frequency of the thyristors. It is less expensive than a variable-ratio transformer. Special care must be taken in choosing the EHT transformer rating because of the relatively high RMS line current due to the chopped waveform, and of the voltage rating of the thyristors because of the transformer inductance. No significant load current flows until the output

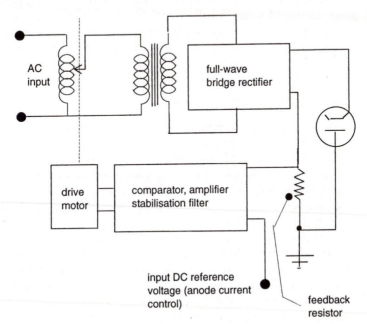

Figure 10.2 Variable-ratio-transformer anode-current controller

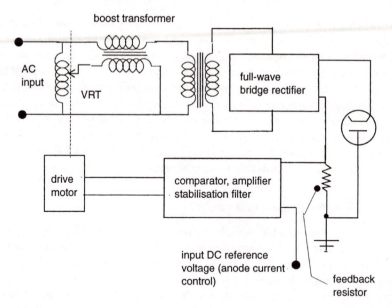

Figure 10.3   Modified variable-ratio-transformer (VRT) anode-current controller with reduced rating on VRT

voltage reaches the π-mode voltage of the magnetron, and until this point is reached the transformer appears as an inductor with a near-quadrature lagging power factor. This may result in unstable firing of the thyristors with transient high-voltage surges; a 'snubber' circuit to control such transients should be considered (Williams, 1987). A typical embryo circuit is shown in Figure 10.4.

10.2.3 Saturating-transformer control

The most widely used magnetron power supply for low-power magnetrons is based on the resonant characteristics of a tuned circuit in which a capacitor resonates with the leakage inductance of a specially designed EHT transformer having a saturating core. Such power supplies are universal in domestic and catering microwave ovens, and are also used in industry where the power source is a multiplicity of small magnetrons. Optimally designed, they are very cheap and reliable.

Figure 10.5 shows an equivalent circuit of the system in which, to highlight the important aspects, the leakage inductor is separated from the EHT transformer, which is assumed to be an ideal component. The leakage inductor has an inductance which is a function of the current passing through it, resulting from partial saturation of the iron core; as the current increases the core becomes more saturated and the inductance value falls. The capacitor C_1 resonates with the inductor at a frequency some 10% higher

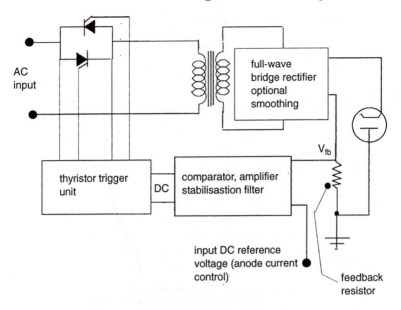

Figure 10.4   Basic phase-controlled thyristor magnetron power supply

than the line power frequency (50 or 60 Hz.), and the bandwidth of the resonance curve is such as to place the power-line frequency at about the half-power point on the resonance curve, Figure 10.6.

Thus, as the load current rises, the inductance value falls causing the resonant frequency to rise, so the resonance curve of Figure 10.4 shifts to the right as shown. The intercept with the power-line frequency then reduces in amplitude, causing the load current to reduce. These two opposed actions result in a stabilisation of the load current.

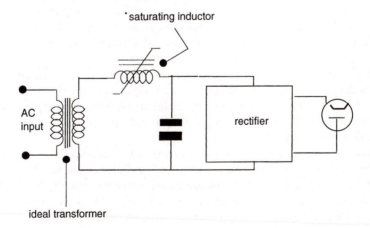

Figure 10.5   Equivalent circuit of saturating choke/transformer anode-current regulator

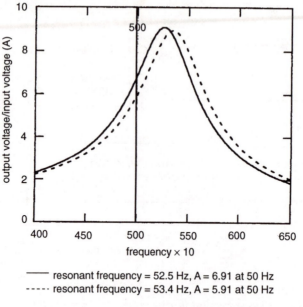

Figure 10.6   Resonance characteristic of saturating-current anode-current regulator

The design of the circuit is empirical, being particularly focused on the transformer design, to secure the desired leakage inductance and rate of change of inductance with load current at the chosen anode current in the magnetron.

Figure 10.7 shows a typical stabilisation of anode current with variation of incoming supply voltage. It will be seen that, over a supply-voltage range of ± 10%, the anode current change is only ± 0.5%, a stabilisation ratio of 20. However, the anode current changes rapidly outside these limits, and so it is very important to ensure that in the initial setting-up the nominal supply voltage is well centred within the control range.

Obviously the stability of the supply main frequency is very important, and in practice the anode-current change for a 1% frequency change is typically 9%. Changing the power supply from 50 to 60 Hz generally requires an adjustment to both the transformer design and the value of the capacitor C_1 to retain optimum performance.

Correctly designed, these power supplies give excellent performance and, having no active components, are extremely reliable. Care must particularly be taken in choosing the rating of the transformer because the RMS values of the currents in primary and secondary are high, the waveforms being far from sinusoidal and changing markedly through the control range. Similarly, the capacitor C_1 must be chosen carefully not only for its DC working voltage but also its high current AC duty.

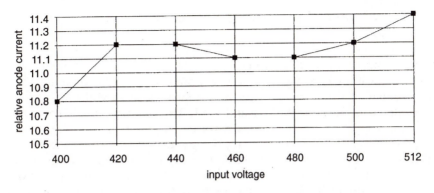

Figure 10.7   Stabilisation of resonant anode-current regulator

Power supplies for domestic ovens are designed for a low duty cycle, often less than 50%, with 'on' time short compared with the warm-up time of the transformer. The transformer is then much reduced in size and cost. For industrial use 100% duty cycle is essential, operating in a high ambient temperature, e.g. 55°C: this results in a transformer of much greater size and weight than the domestic version. Failure to observe these points has, in the past, led to premature transformer failure through overheating.

Referring to Table 10.2 these power supplies compare as follows:

Table 10.2 Features of variable-voltage supplies

1	Limited power rating, typically up to 1.5 kW microwave power output
2	Essentially a fixed-power device, not adjustable without diminishing the stabilisation performance
3	High ripple on microwave output: up to 100% amplitude modulation of microwave power output; high ratio of peak to mean power
4	Efficiency around 40% overall
5	Low cost
6	Large size
7	Heavy
8	Air cooled
9	Low RFI
10	High stored energy for its rating, but not a significant factor because the power rating is low
11	Switch-off time about 30 ms, principally the time to open the supply contactór
12	Harmonic content of load current high
13	Power factor lagging, poor

10.3 Conventional power-frequency supplies for high-power magnetrons

These power supplies are for magnetrons of more than 15 kW rated power output with electromagnets, where power regulation is achieved by adjustment of the magnetic field of the magnetron (Twistleton, 1964), as considered in Sections 9.3.2.1*b* and 9.4.1. This is a particularly convenient control system because the control involves low-power circuits only, the bulk DC EHT power for the anode supply being provided by a simple unregulated three-phase EHT transformer with full-wave bridge rectifier and smoothing inductor, of conventional format as shown in Figure 10.8. The alternative of controlling the EHT voltage by the methods described in Section 10.2 would be relatively expensive at these power levels, and with thyristor control would create serious harmonic problems in the load current drawn from the supply. However, recent important developments in high-frequency switch-mode power conversion enable control of EHT to be accomplished effectively and economically, having particular relevance to the series-field configuration considered below.

Two methods of control are possible. In one the electromagnet is connected in series with the magnetron so that a spurious increase in anode current increases the magnetic field in the magnetron, thus opposing and correcting the initial increase (Twistleton, 1964). Although attractive for its simplicity, at low current the magnetron efficiency is very low because of the low magnetic field, a regime in which the magnetron is not designed to operate. This problem can be overcome by an auxiliary power supply providing a biasing magnetic field, but this is additional equipment, partly negating the simplicity. Figure 10.9*a* shows the essential features.

The other method of control is to have a completely separate, controllable DC power supply for the electromagnet. The power supply forms part of a closed-loop controller of the anode current. By measuring the anode current and comparing it against a preset reference (the 'set-power' control), the error signal is amplified and used to control the current in the electromagnet, as shown in Figure 10.9*b*. Many power supplies for high-power

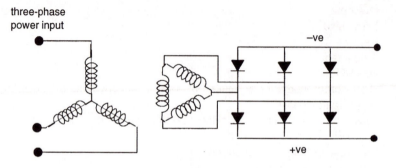

three-phase
power input

−ve

+ve

Figure 10.8   DC EHT power conversion with three-phase full-wave rectifier

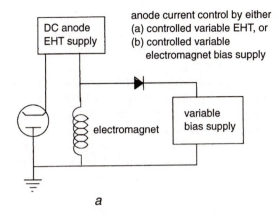

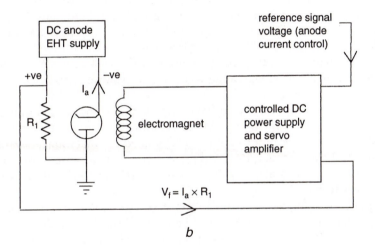

Figure 10.9   *Block diagram of methods of anode-current control*

     a Control by series-field anode-current regulation
     b Control by variable electromagnet-bias supply

magnetrons operate in this way, and the salient design features are considered below.

10.3.1 EHT DC anode supply

This supply basically comprises a three-phase delta/star transformer driving a full-wave bridge rectifier, with a smoothing inductor. The output is shunted by a safety resistor bleed chain and a series capacitor and resistor to reduce the rate of rise of voltage in transient conditions. The essential circuit is shown in Figure 10.8.

The minimum transformer rating should be 1.047 $P_{dc\ max}$ (Ryder, 1960) and the overall efficiency of power conversion to DC EHT is typically > 96%. The phase voltage (i.e. line to star point of the star secondary) of the transformer is 0.606 V_{DC}.

For a three-phase full-wave bridge rectifier, the lowest ripple frequency is $6f_L$ Hz where f_L is the supply frequency. However, there may be a small ripple at frequency f_L if the supply voltage has a phase imbalance or if the leakage inductances of the three phases of the transformer are unequal.

It is desirable to limit the current flowing under anode/cathode short-circuit conditions. This is provided by a high leakage reactance in the transformer of 5–10%, which limits the anode current in a 70 kVA system to about 50 A.

The ripple voltage at the rectifier output is 4% of the rectifier DC output, corresponding, for example, to 360 V peak at 18 kV DC. Since both the power supply and the magnetron dynamic impedances are small, this ripple voltage would create a high ripple current, causing high amplitude and frequency modulation of the microwave output. An inductor of 0.5 H has an inductive reactance of 940 Ω at 300 Hz (i.e. 6 f_L), which, in series with the load, limits the ripple current to 0.38 A peak, or about 10% peak of the direct anode current of a magnetron operating at 60 kW microwave output (67 kW DC EHT input at 18 kV DC).

Although the steady-state voltage across the smoothing inductor is relatively low, under switch-on and fault conditions it can momentarily equal the full DC EHT. The inductor must be insulated for this voltage across the winding, and also between the winding and earth. It is common practice to fit a spark gap (set to about 4 kV) across the inductor to limit the voltage across the winding.

The rectifier duty is for a peak inverse voltage (PIV) of 1.045 V_{DC}, an average current per diode of $I_{DC}/6$ and a peak current per diode of 1.05 I_{DC}. Silicon avalanche diodes have proved extremely reliable, having inherent self protection against reverse high-voltage transients. Note that these data are for normal steady-state operation, and that the rectifiers must be able to withstand the transient fault conditions for a short-circuit.

A spark-gap is often fitted directly across the magnetron set to strike at about 30% higher voltage than the DC EHT . This is to protect the EHT components against excessive voltage surges

10.3.2 Stored energy and fault energy

If an anode/cathode short-circuit occurs in the magnetron, energy is dissipated internally which may cause damage, and the amount of energy must be limited. This energy is from various sources:

(i)   magnetic energy stored in the core, and the leakage flux, of the EHT transformer. For a 70 kVA transformer probably in the order of 50 J;

(ii)   stored energy in the smoothing choke, typically 0.5 H at 4 A, so stored energy is $^1/_2\,LI^2 = 4$ J;

(iii)   energy from the supply main continuing until the circuit-breaker/contactor is opened, typically in 30 ms. This is the greatest energy, in the region of 20 kJ for a 70 kVA system, and is proportional to the kVA rating of the power unit; and

(iv)   energy stored in a smoothing capacitor, if used; a capacitor of 0.1 µF charged to 18 kV stores 16 J.

From the above it is clear that almost all the energy dissipated is from the supply main in the period until the circuit is isolated.

In practice magnetrons up to 75 kW output at 915/896 MHz, and 25 kW at 2450 MHz operate satisfactorily in this regime, but for higher ratings the risk of serious damage under fault conditions rises significantly. These ratings are probably the maximum which can be achieved without resorting to a specific form of energy management.

One method of energy control is to employ an electronic short-circuit in parallel with the magnetron, so the short-circuit current is diverted away from it. This so-called 'electronic crowbar', well known as a protection for high-power travelling-wave tubes and klystrons, comprises a thyratron or ignitron directly in parallel with the magnetron. An ignitron is usually used because it is a cold-cathode device, requiring few additional components; its firing electrode is connected to a current-monitoring resistor in series with the magnetron and fires if the voltage rises above a threshold limit, exceeded if the magnetron becomes short-circuit. The firing time is a few microseconds, reducing the energy dissipated in the magnetron by some three orders of magnitude. However, the short-circuit current continues to flow until the system is isolated from the supply, placing a heavy burden on the rectifiers. A further problem with electronic crowbars is that the fault current they draw from the supply may so depress the supply voltage as to cause malfunction of other equipment connected to the supply, especially other microwave generators.

Switch-mode power supplies have very fast reaction times and inherently store little energy. They are therefore ideally suited to driving magnetrons, and have other major advantages too; they are considered in Section 10.4.

10.3.3 Power control by electromagnet

10.3.3.1 Separately excited electromagnet

This is the most common form of conventional high-power magnetron controller, using the electromagnet current as the control parameter for the anode current, as considered in Section 9.3.2.1 *b* and *c*. As the power for the electromagnet is small, around 200 W (typically 50 V at 4 A), it is easy to control with an inexpensive thyristor regulator (Williams, 1987), which

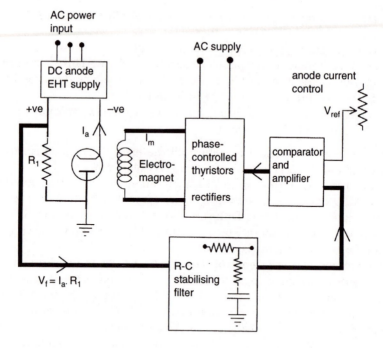

Figure 10.10   Closed-loop control system, separately excited electromagnet

usually comprises a pair of phase-controlled thyristors in a single phase system, driving a transformer with a full-wave bridge rectifier.

A typical schematic circuit is shown in Figure 10.10. A closed loop is formed around the circuit printed heavily in the diagram. The anode current passes through the current-monitoring resistor R_1 (value typically in the range 1–10 Ω), giving a proportional positive DC signal V_f. The signal feeds into a passive loop-stabilising RC network before being compared with an adjustable reference (effectively the power control) V_{ref}. The difference $(V_{ref} - V_f)$ is an error voltage which is then processed to control the firing angle of the AC thyristors feeding the current I_m to the electromagnet. This then controls the anode current via the magnetron coefficient $\partial i_a / \partial i_m$ (see Sections 9.3.2.1 *b*, and 9.4.1), thus completing the loop.

There is an inherent time delay in the electromagnet and magnetron due to the inductance of the former and shorted-turn effect of the latter, expressed by eqn. 9.5. There is also a secondary delay involving the source impedance of the EHT anode supply, the smoothing choke and the dynamic resistance $\partial V_a / \partial I_a$. Referring to the performance chart (Figure 10.11), the steady-state load line of the EHT power supply AB is superimposed, from which it can be seen that the operation of the magnetron as the magnetic field is changed is along the line AB, a change in magnetic field affecting both the anode voltage and current.

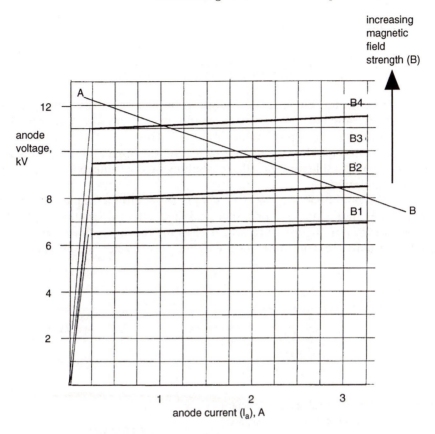

Figure 10.11   Typical magnetron anode-voltage/current characteristic with magnetic-field strength as a variable parameter, showing load line of EHT supply

It is also clear that if there is a step change in magnetic field causing a change in anode current, the smoothing inductor will develop a voltage initially to oppose the change in anode current, and there will follow an exponential change towards the new steady-state condition with a time constant $T_C = L/(R_S + R_D)$. The smoothing inductor and associated circuit are therefore another inherent time delay in the loop which must be considered in the design of the loop-stabilising filter, with a transfer function

$$\frac{1}{1+pT_C} \qquad (10.1)$$

Typically, for a 70 kW DC supply at 17 kV, with 5% voltage drop from no load to full load, the source impedance is 250 Ω, and the magnetron dynamic impedance is about 100 Ω. Using a smoothing inductor of 0.5 H, this gives the time constant $T_C = 1.5$ ms. This effect can be incorporated into eqn. 9.5

to give the overall transfer function of the magnetron assembly as

$$\frac{\partial I_A}{\partial I_M} = -17.5\left(\frac{1}{1+0.04p}\right)\left(\frac{1}{1+0.01p}\right)\left(\frac{1}{1+0.0015p}\right) \tag{10.2}$$

The third term representing the smoothing inductor is small compared with the others, and so the inductor would not affect the performance significantly for the typical value indicated. However, if the inductor is increased to, say, 5 H in an attempt to secure less ripple, the time constant becomes comparable with the others, resulting in a possible loop instability if not compensated.

Conventional Bode analysis (James *et al.*, 1947) gives a simple view of the loop-stabilisation criteria. Loop stabilisation is effected by the simple RC network shown in Figure 10.12, which gives the response shown, where the time constants are determined by the design aim of achieving an open-loop

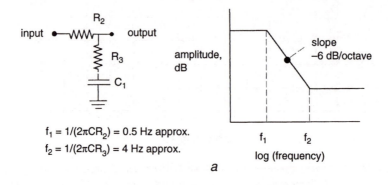

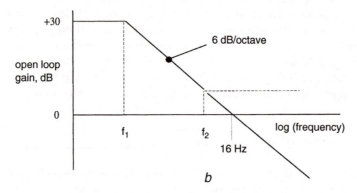

Figure 10.12   Typical loop-stabilising network and overall response of loop

> *a* Loop-stabilising filter and frequency response
> *b* Open-loop frequency response with typical values

response of –6 dB/octave from the lower frequency upwards until the open-loop gain is less than 0 dB at the frequency corresponding to the centre term of eqn. 10.2, namely 16 Hz. This then gives an unconditionally stable control loop. The lower frequency is determined by the desired DC loop gain: for example five octaves would yield a loop gain of $(5 \times 6) = 30$ dB, and the lower frequency for the network would be 0.5 Hz. Note that a loop gain of 24 dB corresponds to a voltage amplification of $\mathrm{antilog}_{10}(30/20) = 32$; thus an external disturbance is reduced in its effect on anode current by 1/32, or 3%, of the disturbance without the closed-loop control. Higher loop gain can be obtained by reducing the lower frequency.

The location of the stabilising filter is important, because it serves the additional purpose of attenuating the low-frequency 'noise' (including ripple) present in the magnetron anode current, which otherwise can cause saturation of subsequent amplifier circuits. In addition, it is good practice to incorporate into the stabilising filter a zener diode to limit its output voltage under fault conditions. Under anode short-circuit conditions the anode current may momentarily approach 100 A, and the corresponding voltage transient may damage subsequent semiconductors.

10.3.3.2 Series-magnet control

The principal advantage of the series-field-circuit arrangement (Fig.10.9a) is the potential elimination of the auxiliary power supply for the magnet, particularly when the EHT supply is from a switch-mode system, as considered in Section 10.4.

The anode current is controlled by adjustment either of the EHT anode supply, or of the magnet-current-bias magnitude. The former is applicable to switch-mode power supplies, and the latter to a conventional EHT supply which is essentially of fixed voltage. Note that with a variable-voltage EHT supply (V_a) the power output is no longer proportional to the anode current, because the power input is $V_a \times I_a$.

Twistleton (1964) gives the relation of incremental supply voltage and anode current as

$$\frac{dV_S}{dI_a} = \left(\frac{dV_A}{dI_A}\right)_{fixed\ field} + \left(\frac{dV_t}{dB} \times \frac{dB}{dI_m}\right) + R_m \qquad (10.3)$$

where $V_S = V_A + V_m$, and V_m is the voltage across the magnet winding. B is the magnetic-field strength, I_m is the magnet current (equal to the anode current I_A in this circuit), and V_S and V_A are the supply voltage and anode voltage, respectively.

In practice, dV_S/dI_A calculated from eqn. 10.3 is about 5 kV/A (i.e 5000 Ω), giving 10% change in power for 5% change in supply voltage. Also, this dynamic impedance is very much higher than the magnet-winding resistance of about 15 Ω, which means that the increase in winding resistance with temperature has negligible effect on the anode current.

It is necessary to protect the magnet winding from high-voltage inductive surges due to transient changes in anode current using a 'snubber' comprising a nonlinear resistor (back-to-back zener diodes, or a varistor). In the absence of the snubber, the time constant of the magnet due to winding inductance is L/R_{eff} where R_{eff} is the effective dynamic-source resistance. If the EHT supply is from a constant-current switch-mode power supply, R_{eff} is very high, giving a very short time constant of a few microseconds. The limit condition with the snubber fitted is for all the anode current to flow through it momentarily, the time constant then being L/R_m. This effect has implications for control-loop stability, the response time being similar to that of the 'conventional' power supply described in Section 10.3.3.1.

10.3.4 Control-loop stability: effect of RF load

The considerations of the previous sections have assumed that the microwave-load admittance is constant over the small range of frequencies over which the magnetron output may vary as the anode current is modulated by the control loop. In most cases, the RF load is sufficiently broadband for this to be a reasonable assumption, especially if a circulator is interposed between generator and load. However, if the load admittance changes rapidly with microwave frequency due to resonance, the assumption may well be invalid, in extreme cases even if a circulator is used. Such loads are lightly loaded multimode ovens with a low spectral density of modes, or resonant applicators. When the latter are used to create a low-pressure plasma discharge, there is also the possibility of step changes in admittance as the power amplitude changes, an effect usually displaying a marked hysteresis.

These effects are generally instantaneous in the context of the criteria for loop stability, but this may not always be so. If the workload properties vary sufficiently with temperature, the rate of rise of temperature may affect the loop stability by adding another time delay; similarly the body of a high-Q factor cavity may heat and slowly change its resonant frequency through expansion.

In the absence of significant time delays, these effects simply change the magnitude of the transfer function of the magnetron assembly given by eqn. 10.2, and the design of the control loop must take account of these possible variations. If time delays, and especially hysteresis, are present, the system becomes very complicated and stability design may have to be empirical. If there is, in addition, a long-line effect or similar instability of the form considered in Section 9.4.3, stablility may be impossible to achieve without resorting to several circulators in cascade to achieve sufficient isolation from the load.

Note that the important parameter in magnetron–load interaction is dY_L/df, not Y_L. It is not effective to tune Y_L to a perfect match at a spot frequency within the magnetron frequency range, because the matching device will be relatively broadband. It is for this reason that several circulators may be required to counteract these forms of instability. Note that a typical circulator has an isolation of 20 dB, which corresponds to a reduction in amplitude of reflection coefficient of only ×10.

10.3.5 Cathode supplies

The cathode of the magnetron has to be heated to a high temperature to provide a source of electrons. In oscillation, a few of the electrons are returned to the cathode at high velocity and their kinetic energy is dissipated as heat, an effect proportional to the anode current and known as 'back bombardment'. An auxiliary supply is required to provide the primary heating and, for high-power magnetrons with pure tungsten cathodes, includes a circuit to reduce the cathode-heater current proportionally to the anode current, so as to maintain substantially constant temperature and preserve the life of the cathode. Low-power magnetrons usually have a simple step change downwards in filament current at a predetermined anode current.

The tungsten filament is usually supplied with AC power as this is the simplest system, avoiding high-current, low-voltage rectifiers. Typical filament supply requirements are given in Table 10.3.

Table 10.3 Typical filament-supply requirements

Magnetron frequency	V_f standby	I_f standby	V_f at rated power output	I_f at rated power output
(MHz)	(V rms)	(A rms)	(V rms)	(A rms)
896/915	11.0	112	8.5	80
2450	11.5	50	0	0

As the filament operates at EHT potential to earth, a filament transformer is used with the dual purpose of impedance matching the low-resistance filament to its supply, and insulating the secondary to provide for the EHT voltage. It is good practice to fit a current transformer for filament-current monitoring in the direct filament lead, rather than in the primary of the filament transformer, thereby avoiding errors due to the magnetising current of the filament transformer.

Specifications of high-power tungsten-filament magnetrons require the filament current to be set to < ± 2% of nominal to ensure the optimum operating temperature. It is very important to note that this is the RMS current, and care must be taken to ensure that the measurement instrument responds inherently to the RMS value of the waveform, which is not sinusoidal if phase-controlled thyristors are used to regulate the current. Suitable instruments are moving-iron analogue ammeters, or certain digital instruments which continuously calculate the RMS value by integrating the mean-square value of a large number of points in the cycle. DC moving-coil instruments with rectifiers, calibrated to read RMS values of a sinusoidal waveform, give large errors when the waveform becomes chopped, and should not be used for filament-current monitoring.

Some manufacturers of power supplies for large magnetrons arrange for the reduction in filament power, rather than current, to be proportional to anode current.

Another technique, based on the resistance–temperature characteristic, is to regulate the filament current to maintain constant resistance, and therefore mean temperature. Although perhaps academically superior, these circuits add complication, accompanied by an extra risk of breakdown. In practice the magnetron specification states the RMS filament current at both standby and as a function of anode current, and these data make allowance for the variation of filament resistance with temperature. Extensive experience is that the lifetimes of magnetrons operating with a filament-current regulator are entirely adequate.

10.3.6 Electromagnet supplies

Electromagnets for large-power magnetrons typically operate at 3–5 A mean at 50 V DC for 896/915 MHz, and 5.0 A mean at 40 V DC for 2450 MHz magnetrons. The inductance of the electromagnet, typically 0.4 H (Section 9.4.3.1), provides inherent smoothing, and operation from a single-phase full-wave bridge rectifier at power frequency (50/60 Hz) is satisfactory, with a residual magnetic-field ripple of about 3% peak-to-peak at 100/120 Hz. The electromagnet supply can be controlled by phase-controlled thyristors, providing a very simple low-power means of regulating the power output of the magnetron.

10.4 Switch-mode power supplies

Recent developments in power electronics have made possible high-frequency converters capable of supplying the anode EHT of high-power magnetrons directly. With switching frequencies around 20 kHz, these systems permit very rapid control of EHT, with the ability to switch off the supply in less than 100 µs. This is particularly important in limiting the energy dissipated within the magnetron under fault conditions of an internal arc, and represents a three-orders-of-magnitude reduction in fault energy. These techniques provide a viable supply for magnetrons of mean power output in excess of 100 kW.

Two types of switch-mode power supply are considered here. Both have, inherently, very high dynamic-output impedances, ideal for regulating the anode current of a magnetron with its very low dynamic impedance (Section 9.4.3.1 and Table 9.1). Indeed, the anode current is regulated to virtually constant-current conditions from the normal operating condition down to a short circuit of the magnetron.

As constant-current supplies, these power converters can be considered current sources, enabling them to be connected in parallel so that the total current is the sum of the individuals. Thus ten 12 kW supplies can be connected in parallel to provide 120 kW to drive a 100 kW-output magnetron.

The first power supply (pulsed resonant inverter) is attributed to Chambers (Chambers and Scapellati, 1994). The second, the Boucherot inverter, is based on the inherent constant-current characteristics of the resonant circuit shown in Figure 10.15 (Boucherot, 1919) developed as a self-oscillating inverter (Gurwitz and Morris, 1991). Their principles of operation are very different, and are considered below.

10.4.1 Pulsed resonant inverter

Chambers (1982) described an inverter system in which a classic series-resonant circuit L, C, R executes a series of single-cycle, damped free oscillations following repetitive closure and opening of switches to a DC source. The switches are opened after the completion of each single cycle in readiness for the next. Isolated gate bipolar transistor (IGBTs) are used as the switches. The bridge circuit shown in Figure 10.13 will be seen to accomplish the polarity reversal. The resulting current passes through the primary of an EHT stepup transformer, giving an EHT-secondary voltage which is rectified and supplied to the magnetron.

The well known transient analysis of the series L, C, R circuit (Weller, 1946) shows that the current flowing after connection to the DC supply is oscillatory provided that

$$R \langle 2 \sqrt{\frac{L}{C}} \tag{10.4}$$

In practice, there is an optimum value for R for maximum dissipation therein of

$$R = \frac{1}{2} \sqrt{\frac{L}{C}} \tag{10.5}$$

Assuming that the circuit losses are negligible, R represents the equivalent resistance of the load.

With R having a value satisfying eqn. 10.4, the current in the resonant circuit is (Weller, 1946).

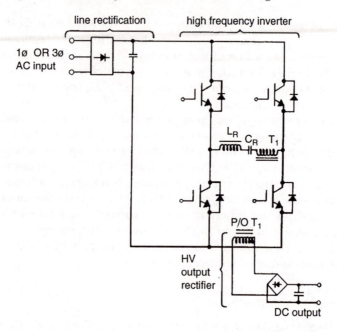

Figure 10.13   Damped resonant-inverter circuit (Chambers and Scapellati, 1994)

$$I = V_{DC} \sqrt{\frac{C}{L}} \, \exp\left(-\frac{R}{2L}t\right) \sin\left\{ \sqrt{\left(\frac{1}{LC} - \frac{R^2}{4L^2}\right)} \right\} t \qquad (10.6)$$

Note that if the load becomes short-circuited ($R = 0$), the maximum possible peak instantaneous current is

$$I_{MAX} = V_{DC} \sqrt{\frac{C}{L}} \qquad (10.7)$$

I_{MAX} (transformed through the EHT transformer and rectifiers) is set to correspond to about 25% more than the maximum rated anode current of the magnetron, or the proportional share if more than one power supply is feeding it, by choice of the parameters of eqn. 10.6. The energy dissipated in the load resistor R is then

$$\zeta = \int_0^{T_1} I^2 R \, dt \quad \text{joules per cycle} \qquad (10.8)$$

where T_1 is the time period (seconds) between the start of this cycle and the next, and I is given by eqn. 10.6. The power dissipated in the load is

$$P = \frac{\zeta}{T_1} \quad \text{watts} \qquad (10.9)$$

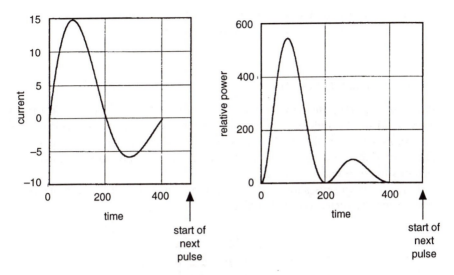

Figure 10.14   Typical current and power waveforms in pulsed resonant inverter

One method of controlling the power output is by varying the time interval T_1, which enables the power to be varied over a wide range, but results, at the lower powers, in a pulsating output which may not be desirable. Figure 10.14 shows a typical current waveform computed with $C = 1.3$ μF, $L = 26.7$ μH and a load resistance of 2.5 Ω. A pause is shown between the end of the first cycle and the beginning of the next (inverted) cycle to indicate operation at 80% of full power output. The natural frequency is 20 kHz.

A further development (Chambers and Scapellati, 1994) of the control system, which overcomes the pulsating output at low power, comprises two identical power supplies in which the relative phase angle of their respective current waveforms (Figure 10.14) is infinitely adjustable through 180°. The resultants of the two waveforms therefore add vectorially, resulting in an amplitude variable between zero (antiphase) and twice the current of one supply when the two are in phase.

The bulk, power DC supply is derived from a simple polyphase bridge rectifier with smoothing choke; usually by direct rectification of the three-phase incoming supply without an intermediate transformer. It is unregulated because the magnetron output is controlled in a high-gain closed-loop system in which a sample of the anode current is compared with a reference, and the difference (error) signal is used to adjust the triggering of the IGBTs to control the anode current, as described above.

The control loop has a very fast response because the main inherent delay is of the switching frequency of the IGBTs. Moreover, the time taken to switch off in emergency is the time duration T_1 of one cycle.

Note that the ratio of peak to RMS current of the waveform of Figure 10.14 is significantly higher than for a continuous undamped sinusoidal waveform of $\sqrt{2}:1$. This results in a lower power output for a given IGBT than could be obtained from a true sinusoid, where peak current is a principal limiting parameter.

10.4.2 Boucherot inverter

Boucherot (1919) described a series-resonant circuit in which the load is connected in parallel with either the capacitor or the inductor, and showed that the current in the load is constant, regardless of the load impedance, provided that the circuit is excited at its resonant frequency. Consider the circuit shown in Figure 10.15a.

Using the symbols shown, the generator current I_G can immediately be written for the series–parallel configuration

$$I_G = \frac{V_G}{j\omega L + \left\{ \dfrac{R_L \dfrac{1}{j\omega C}}{R_L + \dfrac{1}{j\omega C}} \right\}} \tag{10.10}$$

Also,

$$V_C = V_L = I_G \left\{ \frac{R_L \dfrac{1}{j\omega C}}{R_L + \dfrac{1}{j\omega C}} \right\} \tag{10.11}$$

Substituting eqn. 10.10 into eqn 10.11 to eliminate I_G and then rearranging, the load voltage is

$$V_L = \frac{R_L}{R_L \left(1 - \omega^2 LC\right) + j\omega L} \tag{10.12}$$

If the circuit is driven at its resonant frequency so that, $\omega = \omega_0 = 1/\sqrt{(LC)}$, eqn. 10.12 becomes

$$V_L = V_G \frac{R_L}{j\omega L} = I_L R_L \tag{10.13}$$

and

$$I_L = V_G \frac{1}{j\omega_0 L} \tag{10.14}$$

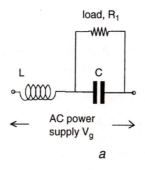

load, R_1

L   C

AC power
supply V_g

a

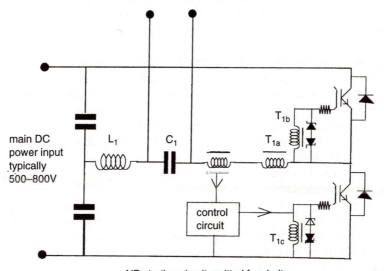

load (EHT transformer primary)

main DC
power input
typically
500–800V

L_1 C_1

T_{1b}

T_{1a}

control
circuit

T_{1c}

NB starting circuit omitted for clarity

b

Figure 10.15   Schematic diagram of Boucherot inverter

a Basic Boucherot resonant circuit
b Self-oscillating Boucherot inverter

The load resistance R_L is eliminated in eqn. 10.14, showing that the circuit behaves as a constant-current source independent of the load impedance.

Gurwitz and Morris (1991) submitted a patent application for a self-oscillating Boucherot circuit in which semiconductor switches (e.g. IGBTs) connect the Boucherot circuit of Figure 10.15*a* to a DC supply in alternating polarity, at a frequency equal to the resonant frequency due to the inductor and capacitor. The IGBTs are triggered from the load-current waveform, using a current transformer. Further, by providing a variable delay in the triggering of the IGBTs the effective generator voltage is variable, permitting smooth control of load current form zero to rated output.

A specially important feature is that the Boucherot circuit is in continuous oscillation and the current is sinusoidal at full rated output. The ratio $I_{PEAK} : I_{RMS}$ has the optimum of $1 : \sqrt{2}$, allowing the maximum possible output from the IGBTs for their rated peak and RMS-current specifications.

10.4.3 EHT circuits for resonant inverters

Both the inverters described in Sections 10.4.2 and 10.4.3 operate at relatively low voltage and high current, e.g. a 500 V DC supply giving 60 A RMS at 20–50 kHz. A transformer and rectifier system provides the EHT voltage for the magnetron, typically in the range 10–20 kV. The system is single-phase using a double-wound transformer, with a rectification system which may be a full-wave diode bridge, or voltage multiplier, usually a doubler or quadrupler (Ryder, 1960). The choice of system is a compromise of transformer-insulation specification, regulation (i.e. voltage drop as a function of load current) and cost. A smoothing choke is used to reduce ripple current.

The transformer requires specialist design as a power component operating at 20–50 kHz. It is much smaller than its 50/60 Hz counterpart, one of the secondary advantages of switch-mode systems. The magnetic-flux density, usually limited to 1000 G to minimise core losses, a core area of 0.01 m², and a frequency of 25 kHz gives about 100 V/turn (Say, 1948; and Chambers, 1982). Skin depth in copper at 25 kHz is about 0.02 mm, and multistranded, individually insulated 'Litz' wire must be used for the windings to avoid excessive I^2R eddy-current heating at these frequencies. Likewise, the transformer core is usually of ferrite with high resistivity to minimise eddy currents. Although these transformers have high efficiency, c. 93%, their small volume relative to their rating results in very high power-dissipation density, necessitating oil cooling. A 40 kVA transformer dissipates about 3 kW internally, and an oil–water heat exchanger is used to remove this waste heat. Again, the high voltages and small size create very high-voltage stresses, necessitating oil insulation. These same considerations apply to the smoothing choke (typically 5 mH) which is a high-voltage component, both from winding to earth and across the winding itself where the full ripple voltage is developed.

Rectifiers must have very fast recovery t_{rr} (around 40 ns) to minimise reverse current at these high frequencies, in addition to consideration of the usual parameters of peak inverse voltage (PIV) or V_{RRM} and forward RMS current I_{FRMS}. Cooling is very important, and adequate heat sinking and/or oil immersion is necessary. It is convenient to mount the EHT transformer and rectifiers in the same oil tank, which incorporates a simple water–oil heat exchanger.

10.4.4 Resonating inductors and capacitors

Inductors have values of about 10 μH, carrying a current of about 200 A and can, in principle, be air-cored. However, the surrounding magnetic field at

20–30 kHz would cause serious induction-heating losses in neighbouring materials, and so a ferrite core is essential. A toroidal core is usually used, with Litz wire for the winding.

Capacitors also require careful selection because of the high current, and to minimise losses. Capacitors with polypropylene dielectric and metal-foil electrodes are suitable (Chambers, 1982). Connection terminals and insulation bushes must be appropriate to the current and voltage demands.

10.4.5 Switching devices

In early systems silicon-controlled rectifiers (SCRs) were used for the semiconductor switches, but their performance has been surpassed by IGBTs. The important parameters in a switch-mode power supply are:

(i)   forward current ratings, peak and RMS;
(ii)   turn-on time: the device must be fully conducting (i.e. minimum voltage drop) within a nominal 10° of the start of the current waveform (about 1 µs at 25 kHz);
(iii) peak reverse voltage, normal working and instantaneous;
(iv)   turn-off time: this is affected by the circuit conditions, particularly dV/dt which must be controlled by a 'snubbing circuit' comprising a series resistor and capacitor shunting the device.

10.5 Power supplies: a large number of low-power units, or vice versa?

Through three decades of the history of industrial microwave engineering there has been constant debate over whether a few high-power generators are preferable to many at low power. That the debate has continued for so long is perhaps indicative that there is no clear-cut answer. Certainly a noticeable trend is for large plants, based on initial development in academic establishments, to be scaled-up using a multitude of low-power generators, probably because the designer feels more familiar with such generators. Such subjective aspects as familiarity or personal preference should not be allowed to influence a decision which can be assessed logically against a set of objective criteria.

The decision is a very important one, not only affecting the initial cost of the equipment, but also radically affecting the logistics and operating cost through its lifetime.

Some of the parameters of the decision are:

(i)   overall efficiency: ratio of microwave power usefully dissipated to electrical power consumed;
(ii)   lifetime of individual generators;

(iii)   there is a statistical probability that some low-power generators will be inoperative at any moment, and so extra generators should be provided to allow for this if the full power rating must be provided at all times;

(iv)   the actual power required may not allocate conveniently between a few high-power generators, so that extra capacity may need to be provided inherently;

(v)   low power magnetrons are 'throw-away' items. Although they are cheap there are many of them;

(vi)   high-power magnetrons can be rebuilt at about 50% of the price of new units;

(vii)   the time taken to change a low-power magnetron is about the same as for a high-power one;

(viii)   the man-hours spent, and 'downtime', in changing magnetrons may become very significant for the low-power-magnetron route;

(ix)   the electrical wiring (power feed, control and monitoring) involves much labour cost with many generators;

(x)   small magnetrons will be distributed inconveniently for access unless a waveguide feed system is provided for each, which greatly increases cost and complexity;

(xi)   small magnetrons are less efficient than high-power units; moreover, they are usually air cooled. In a large installation, the total heat dissipated in air may be hundreds of kilowatts, representing an environmental problem in the workplace. If the magnetrons are water cooled there is a very complex and expensive pipework system, and the problems of changing magnetrons are greatly magnified;

(xii)   power control with small magnetrons is usually by switching on/off. This results in possible degradation of heating uniformity in the oven as the power is progressively reduced;

(xiii)   small magnetron power supplies are, in this context, of the saturating-transformer type, as in Section 10.2.3. These have a supply-current waveform high in harmonics, and a poor power factor. A large installation may cause problems to the supply authority;

(xiv)   the choice may be dependent on operating frequency, as there are at the time of writing no commercially available low-power magnetrons at 896/915 MHz;

(xv)   installations with a large number of low-power magnetrons are limited to multimode ovens. Sophisticated applicators of other types tend to be incompatible with that approach;

(xvi)   in a multimode oven there may be phase locking of magnetrons if there is appropriate cross-coupling and their frequencies are sufficiently close. This is a statistical and uncontrolled event which may affect uniformity of heating. If it does not occur, the peak instantaneous microwave-field stress will increase by $\sqrt{2}$ from its value if the two magnetrons are phase locked. This may be significant in multimode ovens operating under vacuum, in reducing the margin to ionisation.

The above list is not exhaustive, but from it there is obviously a wide range of parameters to be considered in making a decision. In the past there has been the question of the state of development of the power supplies, but both low- and high-power generators have now evolved to a high state of performance and reliability. In particular, high-power generators based on switch-mode technology are likely to advance to yet higher powers, making their advantage even greater.

Chapter 11

Outline of microwave measurements on components and materials

11.1 Introduction

All equipment manufacturers and research institutions must be well equipped to perform microwave measurements on waveguide and other components, applicators and workload materials. The main purpose of such measurements is to obtain quantified data for the design of applicators and waveguide feeders for optimum performance (impedance matching), and, through data on ε' and ε'', to assess the suitability of materials for particular applicator designs.

Formerly, these measurements were carried out with a microwave test bench comprising an assortment of instruments including microwave oscillators, standing-wave detectors (slotted lines), diode detectors and sensitive measuring amplifiers (Barlow and Cullen, 1948). Much experience and considerable patience was required to achieve correct results. Fortunately, this equipment has been superseded by the network analyser (NA), but nonetheless it is instructive to read the early literature (e.g. Harvey, 1963) because of the fundamental principles exploited in its use.

In summary, the NA comprises a precision microwave generator which can be swept in output frequency over a preset range and a sensitive measurement receiver capable of accurate analysis of phase and amplitude of a reflected and/or transmitted wave relative to the forward wave from the generator. Received data are processed internally and displayed in any desired format, e.g. VSWR, complex reflection or transmission coefficients against frequency, or Smith chart. Some instruments can display two or more parameters simultaneously. The NA gives a virtually continuous and instant plot of the chosen parameter over the selected frequency band, avoiding the previous tedium of point-by-point measurements at spot frequencies; it also ensures that an undesired high-Q-factor resonance is detected, where in the past it could have been missed between the chosen spot frequencies, leaving a doubt in the designer's mind.

Bryant (1993) give a detailed presentation of the theory and practice of NAs to which the reader is referred. That subject is beyond the scope of this Chapter, in which only the specific use of the instrument in relevant measurements is considered.

These measurements are essentially low power, with the NA delivering a power of some tens of milliwatts only. Heating is therefore negligible, and it must constantly be remembered that the high-power performance of a component may differ from that measured, e.g. because of temperature variations of ε' and ε''. Chapter 12 reviews testing of equipment at high power.

In addition to the NA and its accompanying transmission/reflection test-set unit (in coaxial cables for convenience), the following tests require:

(i)   two-off coaxial-to-waveguide transitions for each waveguide size in use. These should be broadband components with a good impedance match (VSWR > 0.85 over a 5% frequency band centred on the nominal operating frequency of the equipment). The residual reflection can be cancelled electronically in the NA by an established setting-up procedure (given in the NA manufacturer's handbook);

(ii)   waveguide-matched low-power terminations for each waveguide size in use, at least two of each size. These may be proprietary items, but good low-cost terminations can be made from a wooden taper fitting the waveguide with a running clearance allowing it to slide easily. The taper is about 15°, and is followed by a parallel section of at least the same length as the taper. The taper allows a gradual change in wave impedance in the waveguide, giving minimum reflection, and the full-width section is of sufficient length to absorb 90% of the power. Beech is usually used. By sliding the load, the residual reflection can be seen on the NA Smith chart as a small circle described as the wooden taper is moved. An excellent match can be obtained by cut-and-try methods. The disadvantage of the sliding wooden load is its length and weight. The most compact load is a coaxial/waveguide transition, terminated in a proprietary precision-matched coaxial load.

(iii) a plain short-circuit plate (aluminium) drilled to mate with the waveguide flange in use, one being needed for each waveguide size.

11.2 Measurements on waveguide components

This section is devoted to measurements on passive waveguide components such as bends, directional couplers, matched terminations, waterloads, circulators and magnetron launch sections.

11.2.1 Measurement of VSWR, reflection and transmission coefficients, the Smith chart

Figure 11.1 shows the set-up for measuring the input impedance of a 2-port waveguide component. The device under test (DUT) is connected to a

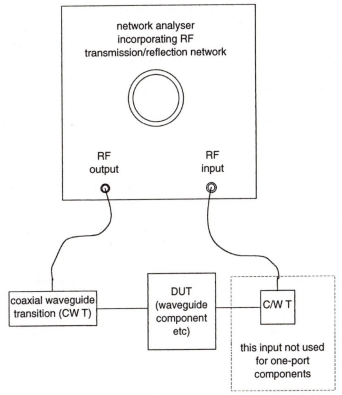

Figure 11.1   Test set-up for reflection and transmission measurements on 1- or 2-port components

coaxial–waveguide transition at the input, and to a matched termination. The NA is set to sweep the desired frequency band, and the display is set to 'Smith chart'. The impedance locus as a function of frequency will be displayed on the Smith chart. By terminating the input coaxial waveguide transition in a short-circuit plate, it is possible, by following the NA manufacturer's routine, to refer the impedance plot on the Smith Chart to the input flange of the DUT.

By appropriately setting the NA, the following can also be displayed:

(i)   VSWR on simple cartesian co-ordinates;
(ii)   reflection-coefficient amplitude (as $|\rho|$);
(iii)   reflection coefficient phase angle; and
(iv)   complex reflection coefficient on real/imaginary Cartesian axes, or as a polar display.

A frequent procedure is setting-up the match of a water load. This is a case where the VSWR is affected by the power dissipated because of the wide variation in ε' and ε'' with temperature for water. If a very low reflection is sought, it is necessary to estimate the likely water temperature and to perform the matching with the load filled with water at that temperature. This will not truly represent the on-load conditions, where there is an inevitable temperature gradient which cannot be simulated in a static low-power test. However, the error resulting in practice is usually insignificant; it is easy to check the sensitivity to temperature by observing the match as a function of temperature.

Sometimes water loads are required to operate with a water/glycol mixture; the same mixture must be used for setting up, again at a representative temperature.

The integrity of a magnetron-launching waveguide can be assessed by treating it as a 1-port structure, with a special load fitted to represent the magnetron. Comparison with the magnetron manufacturer's data of impedance match (if available) validates the launcher. Similarly, a suspect magnetron can be fitted and the 'cold' input impedance to the launcher checked against the manufacturer's data: there should be a resonance (due to the anode structure) at a frequency removed from the 'hot' operating frequency by a specific amount Δf. If there is no resonance, or it occurs at a different value of Δf, then the magnetron is probably faulty through a damaged anode.

Where the DUT has more than two ports (e.g. a directional coupler or circulator) it is usual to fit matched terminations to all the outputs, in which case the reflection coefficient is the scattering coefficient S_{11}.

11.2.2 Measurement of attenuation and coupling

Attenuation is most easily measured by the insertion method, in which the attenuation of the direct transmission path between the output of the NA and its receiver is measured, and then remeasured when the DUT is inserted in the path, the difference then being the attenuation of the DUT. Where the DUT is a 2-port waveguide component, the path must include two coaxial–waveguide transitions for making connection to the DUT. The attenuation of these components can be compensated by the setting-up procedure of the NA. Figure 11.2 shows the set-up.

In practice the attenuation of the path without the DUT inserted can be memorised and 'zeroed' by the NA over the frequency band, so that when the DUT is inserted the result displayed is the actual attenuation value. The display may be in decibels or expressed as $|\rho|$ or as a complex transmission coefficient in polar or Cartesian coordinates.

In the above it is assumed that the DUT has negligible reflection loss compared with its insertion loss, i.e. it has a good impedance match. If this is not the case, the reflection loss must be measured by the method of Section 11.2.1. Again, if the insertion loss is small (< 1 dB) the true insertion loss is simply the difference (in decibels) between the measured insertion loss and the reflection loss.

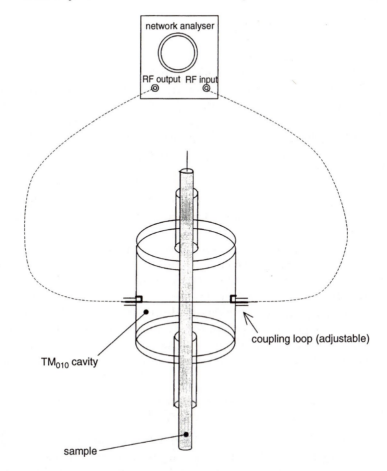

Figure 11.2 *TM_{010} resonant cavity for dielectric measurements*

However, if the insertion loss is large (> 1 dB), and the DUT is not well matched, the reflection coefficient may be the vector sum of two or more reflections within the DUT. For example, the principal loss may be a workload in a slotted waveguide: a reflection will be generated at the leading and trailing edges of the load, and the vector sum will be determined by the distance apart of the edges. An accurate result for the attenuation can only then be obtained by measuring the *S*-parameters of the DUT and performing calculations using the lossy-transmission-line equations (Chapter 4).

Where the DUT is a multiport structure, e.g. a directional coupler, the procedure for measuring coupling between any two ports is identical to the above, except that matched terminations must be fitted to all other ports.

11.3 Measurements on dielectric materials

This section is devoted to an outline of methods of measuring ε' and ε'' of materials at microwave frequency. Workload materials vary in dielectric parameters over an extremely wide range from the nearly lossless to those with a skin depth rendering them unsuitable for microwave heating. No single method is satisfactory over the whole range. Fortunately, most industrial-heating applicators are quite tolerant of variations in ε' and ε'', and so the accuracy usually required is not high, ± 30 % generally being adequate for system design. However, as design techniques improve, particularly in computational electromagnetics, the need for higher accuracy grows. The network analyser has much simplified the procedures for dielectric measurements, allowing swept-frequency measurements to be made very quickly.

Many techniques are used for dielectric measurements, and this section considers only those most commonly used; the reader is referred to the list of references for a more detailed review (Roberts and Von Hippel, 1946; Von Hippel, 1954; Tinga and Xi, 1993).

As the temperature of the workload rises during the process, the values of ε' and ε'' change. Moreover, the composition may change through drying or by chemical reaction, features which are time dependent, and often irreversible. Considerable effort has been made to measure the parameters under these high-temperature dynamic conditions.

11.3.1 Insertion-loss measurement of dielectric properties

The designer is often interested primarily in the attenuation rate of a workload dielectric rather than the values of ε' and ε''. This is particularly the case when a travelling-wave or meander-line applicator is planned, for the treatment of a thin web or a conveyor band loaded with a powder.

A very simple insertion-loss measurement can be made with a plain waveguide arranged to support a sample of the workload about the centre line of the broad face so that the plane of the workload web is aligned with the electric field at its position across the waveguide of maximum electric-field strength. The transverse-slotted-waveguide applicator shown in Figure 6.5 forms a suitable measurement jig. The insertion loss of the assembly is then measured by the method of Section 11.2.2, first without the workload sample present and then with it in position, to obtain the attenuation of the load in decibels per metre, knowing the length of the load. The test may be repeated for various thicknesses of workload, and the data obtained may be sufficient for the applicator design to be developed.

This method is only suitable for materials of medium to high loss (> 2 dB/m). If the loss is less, the material is unlikely to be suitable for a travelling-wave applicator, and a resonant system or multimode applicator should be considered. If ε'' is required, it can be calculated directly using eqn. 6.1 or 6.2.

11.3.2 Resonant-cavity techniques

For materials with low value of ε'' (< 3), the relative permittivity ε' can be measured by observing the change in resonant frequency of a resonant cavity when a sample is introduced, and the loss factor ε'' by the change in Q-factor (Horner *et al.*, 1946). Various cavity shapes and modes may be used, but the E_{010} circular cylindrical cavity is the most convenient in practice, and it is capable of analytic solution for ε' and ε''. The test sample, in circular rod form and of small diameter compared with the cavity diameter, is inserted concentric with the cavity and occupies the full height. The set-up is shown in Figure 11.2, where the cavity is operated as a transmission cavity, loosely coupled to avoid reducing the Q-factor by external loading. The NA is set to sweep a band of frequencies centred on the cavity resonant frequency, and the receiver is set to display the transmission coefficient on a decibel scale; the display is of a classic resonance curve, and a frequency marker is set at the peak of the curve. The sample is then introduced and the cavity resonant frequency decreases, a new frequency marker being set on the maximum point. The change in resonant frequency Δf gives the value of ε' as

$$\varepsilon' \approx 1 + 0.539 \frac{a^2}{b^2} \frac{\Delta f}{f_0} \tag{11.1}$$

where

a = cavity diameter
b = sample diameter
Δf = frequency shift with sample inserted
f_0 = resonant frequency of unloaded cavity

Eqn. 11.1 is a simplification of the exact relation, but for $\varepsilon' = 3$ the error is 1% for $b/a = 0.05$, and 25% for $b/a = 1/3$.

The loss tangent ($\tan \delta$) is obtained from the change in cavity Q-factor between the unloaded and loaded conditions. Q-factor is measured by first adjusting the NA so that the display amplitude is on a decibel scale, setting the peak of the resonance curve to 0 dB as a reference point. Frequency markers (f_1 and f_2) are then set at the two -3 dB points on the flanks of the resonance curve (i.e. the two halfpower points, one on each side of the resonance peak).

The Q-factor of the system is given by

$$Q = \frac{f_0}{\delta f} \tag{11.2}$$

where $\delta f = f_1 - f_2$ is the frequency difference between the frequency markers at the -3dB points on the resonance curve.

Two measurements of Q-factor are required, one for the empty cavity (Q_1), the second with the sample inserted (Q_2). The loss factor ε'' for the sample is then given by

$$\varepsilon'' \approx 0.269 \frac{a^2}{b^2}\left(\frac{1}{Q_2}-\frac{1}{Q_1}\right) \tag{11.3}$$

The error in using the approximation of eqn. 11.3 is several per cent when $b/a \leq 1/20$ and $\varepsilon' = 3$. For most industrial-heating purposes, this is adequate, but for greater precision it is necessary to use the exact analysis of Horner's paper. Typical cavity dimensions are as in Table 11.1.

Table 11.1   *Typical cavity dimensions*

Frequency (MHz)	896/915	2450
Diameter (mm)	250	93
Axial length (mm)	125	46

It is important to secure the highest unloaded Q-factor for the cavity, which should be of high-conductivity copper or aluminium. Wall currents flow axially in the barrel and radially across the end plates, the current paths being across the joint between the barrel and end plates. These joints must have electrical contact of high integrity around their circumference at the inside surfaces to avoid degrading the Q-factor. The simplest joint comprises a precision-machined thick flat end plate, with the end of the barrel machined to have a taper of about 3° less than a right-angle end. When the two surfaces mate, the contact is at the peak of the taper on the inside surface of the barrel. Several (at least eight) high-tensile bolts secure the end plate to the barrel; when they are tightened the mating surfaces deform to create an excellent contact.

Coupling of power into and from the cavity is best achieved by means of small magnetic loops inserted through the barrel. The loops, terminating the ends of coaxial cables, are inserted the minimum possible distance and are arranged so they can be rotated to adjust the coupling. The coupling should be as light as possible consistent with an adequate signal-to-noise ratio at the NA's receiver. Loop size is easily adjusted, and typical areas are 15 mm^2 at 896/915 MHz and 2 mm^2 at 2450 MHz.

It is important that the sample fills the full height of the cavity, especially if ε' has a high value, otherwise the effective impedance of the resulting air gap appears in series with that of the sample, reducing the apparent values of ε' and ε''. Powders and liquids are often measured by placing them in a test tube. If the value of Q_1 is measured including the empty test tube, the resulting measurement gives a result of sufficient accuracy for most purposes.

Resonant-cavity techniques are particularly suited to materials of low loss factor such as oils, waxes and insulating materials. Many workload materials have such high loss that the halfpower points of the resonance curve are so far separated in frequency that the frequency response of other components in the measuring system may vary and affect the result. Reducing the diameter of the test piece is an obvious procedure in this case, but there are always practical limits.

11.3.3 Surface-contact measurements

A very convenient technique (Gabriel and Grant, 1989; Stuchly and Stuchly, 1980) for dielectric measurements involves the measurement of the admittance between a pair of electrodes placed on the surface of the test piece. The real and imaginary parts of the admittance correspond to ε'' and ε', the values of which are computed directly from the measurement (Hewlett–Packard). This is not an 'absolute' method and requires prior calibration, but it gives results of sufficient accuracy quickly and easily. In its fully developed form it comprises a hand-held probe connected to a network analyser with built-in computer and software permitting a direct readout of ε' and ε'' to be presented over a chosen frequency range.

In practice, the probe is a rigid coaxial line terminated in an abrupt open circuit, with the outer terminated in a plain flange machined flat, and coplanar with the end of the inner line and solid insulation. The probe is connected by a precision flexible coaxial cable to the NA, for convenience. The assembly is sealed to prevent ingress of liquids, and is capable of making measurements over the temperature range $-40°$ to $+200°C$.

The limitations of this method are for thin materials where the thickness approaches the penetration depth, and for low-loss materials.

Chapter 12

Procedures for testing high-power installations

12.1 Introduction

Large microwave-heating installations can rarely be tested under full-scale production conditions in-house at the equipment manufacturer's premises. This may be because of the logistics or cost of arranging for provision of the appropriate workload, or available space, or limitation of available electrical power. Testing must therefore be a two-stage procedure in which the manufacturer tests all the items prior to shipment, followed by testing to the operational specification on site.

Testing presupposes that a formal specification exists against which the tests can be made. The existence of the equipment-performance specification is a vital contractual necessity for both purchaser and vendor, and its preparation must be done with the most extreme care to ensure that all relevant expectations of the customer are properly expressed, and are within the realisable performance of the manufacturer's equipment. Appendix 2 is devoted to preparation of specifications.

12.2 Microwave generators

Although the equipment manufacturer will have a detailed in-house test specification for all the subsystems of the microwave generator, the customer specification will concentrate on power output, frequency, efficiency and spurious emissions. It will also be concerned with external control and monitoring, interfacing with other equipment, and safety.

12.2.1 Power input, output and efficiency

The measurement of power output and efficiency of the generator requires accurate measurement of power input and of microwave power output, both parameters requiring great care and understanding of the principles involved. Unexpectedly, the power input is more difficult to measure because the waveform of the current input is usually nonsinusoidal. The precise definition of efficiency is important because some manufacturers, more correctly, include the power consumed by auxiliary systems (electromagnet, filament supplies, cooling-water pump etc.), while others ignore it and consider only the principal load of the magnetron-anode power. In the latter case the three-phase load current can usually be assumed balanced, while in the former some of the supplies may be single phase, introducing an imbalance to the phase currents, and a current flow in the neutral conductor if the supply is a four-wire system..

Efficiency is often a crutial specification parameter since it affects the operating cost of the equipment. It is therefore important to measure both power input and output accurately, and to know the likely error.

12.2.1.1 AC input-power measurement

From elementary AC theory, the power input in the general case of an unbalanced three-phase load is the sum of the powers transmitted in each phase:

$$P = V_{P1} I_{L1} \cos\phi_1 + V_{P2} I_{L2} \cos\phi_2 + V_{P3} I_{L3} \cos\phi_3 \qquad (12.1)$$

where $P =$ is the total power input (watts); V_{P1} etc are the RMS phase voltages between line and neutral (assumed sinusoidal); I_{L1} etc. are the RMS line currents *of the fundamental of the harmonic series representing the nonsinusoidal current waveform*; and cos ϕ_1 etc. are the power factors of each phase, where ϕ_1 is the phase difference between the fundamental components of the voltage and the current waveforms.

I_{L1} etc., are not the RMS values of the overall current waveforms, and are not the values that would be given by ammeters inserted in the line feeds, unless both voltage and current waveforms are true sinusoids. Most instruments measure true RMS values only of sinusoidal waveforms, and can be grossly in error if harmonics are present. Because the average power flow due to a sinusoidal voltage at fundamental frequency, associated with a sinusoidal current at harmonic frequency, is zero*, it is clear that the instrumentation must exclude harmonics from the measurement, unless it computes it precisely.

The total power can obviously be measured with three wattmeters, moving-coil dynamometer instruments in which signals proportional to the voltage

---

* This is because $\displaystyle\int_0^{2\pi} \sin\theta \sin(n\theta + \phi_n)\, d\theta = 0$

for all integral values of n except $n = 1$. ϕ_n is an arbitrary constant.

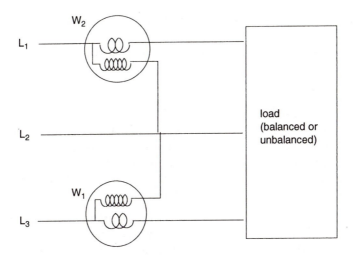

L₁

L₂

L₃

W₂

W₁

load
(balanced or
unbalanced)

W₁ and W₂: dynamometer wattmeters

total power = W₁ + W₂, irrespective of balance,
power factor or harmonics

NB no neutral conductor

Figure 12.1   Classic two-wattmeter method of measuring three-phase power

and current respectively are applied to the moving-coil movement, and an electromagnet to provide the magnetic field of the instrument. The deflection of the instrument is proportional to the *average* value of the product ($V \times I$), which is the power transmitted, irrespective of waveform or power factor.

If the power supply is a three-wire system (i.e. with no neutral), the classical 'two-wattmeter method' may be used (Kemp, 1949; Cotton, 1946). The total power is readily shown to be the sum of the readings of the two instruments, irrespective of harmonics and balance of the load between the phases. Figure 12.1 shows a typical circuit arrangement.

Another seemingly accurate method is to use a three-phase induction-disc integrating kilowatt-hour meter, measuring the total energy used over an accurately timed period, from which the average power is immediately calculable. Unfortunately these instruments have been found to have substantial errors when harmonics are present in the current waveform, and the method cannot be recommended.

Recently, specialised digital instruments have become available which measure three-phase power directly, and inherently make allowance for the above difficulties. They measure the instantaneous power $V \times I$ at a high sampling rate (i.e. at many points through a cycle), and then, by integration, compute the average power. They also compute the active power and reactive power of the load.

The accuracy of power measurement with correctly chosen instruments is <2% error.

12.2.1.2 Microwave-power-output measurement

Power output of high-power microwave generators is invariably measured calorimetrically using a water load to absorb the power. The power is deduced from measurement of flow rate and temperature rise of the water. Several precautions must be taken to ensure accuracy, which can usually be within <2%:

(i)   The water load must have a good admittance match, i.e minimum reflection coefficient. If the generator incorporates a circulator, reflected power is mostly deflected away from the magnetron so that the magnetron 'sees' a matched load. However, if the circulator has less than 20 dB isolation, the residual reflection reaching the magnetron may be of sufficient amplitude to affect the power output. The extent of this effect can be estimated from measurements of the water-load reflection, the measured isolation of the circulator and reference to the magnetron Rieke diagram. If there is no circulator, this is particularly important. It is important to ensure that the water-load match is optimised at the operating frequency of the generator.

(ii)   The water-load admittance will vary with water temperature, and so is dependent on the temperature rise of the water, and consequently on the power input. The effect is small but must be assessed for accurate measurement.

(iii)   Heat loss occurs from the waterload, reducing the apparent temperature rise. Heat losses occur from convection from the water-load body, which should be insulated. They also occur due to conduction to adjacent waveguides.

(iv)   The water should be from a constant-head supply to ensure a constant flow rate; direct connection to a mains water supply is usually unsatisfactory because of fluctuation of pressure. A large-volume constant-head supply also ensures a constant temperature for the input water. A needle valve is recommended for flow regulation, as it permits fine control.

(v)   Water-flow rate requires accurate measurement. A calibrated Rotameter® flowmeter is excellent because it is possible to observe fluctuations in flow, if any. A 'bucket-and-stopwatch' measurement is accurate provided that the flowrate remains constant during the measurement, which should be made over at least 2 min for accuracy. Flowrate-measurement accuracy should have less than 1% error.

(vi)   The water load has a high heat capacity, so at least 3 min must be devoted to steady-state running before a stable, reliable measurement of temperature rise of the water is recorded.

(vii) Temperature rise can be measured by mercury-in-glass thermometers with a suitably wide scale to read temperature to ± 0.2 deg C. Thermocouple or platinum resistance thermometers are suitable, but care must be taken that any electronic instrument is unaffected by stray microwave leakage.

(viii) To minimise the error due to conduction and convection heat losses, the temperature rise should be minimised by adjustment of water flowrate. However, the temperature rise must be high enough to permit its accurate measurement, and a temperature rise of about 20 deg C is a good compromise.

(ix) If the temperature rise is unavoidably high (e.g. > 40 deg C) because of high power, local boiling may occur in the waterload. This will cause the reflected power to increase with consequent reading error, and may damage the water load.

The relation between power, flowrate and temperature rise is

$$P = 4.18 \times Q \times \Delta T / 60 = 0.697 \times Q \times \Delta T \ kilowatts \tag{12.2}$$

where Q = flowrate of water (litre/min)
ΔT = temperature rise of water deg C
4.18 = mechanical equivalent of heat (Ws/cal)

A convenient setting often used is a flowrate of 1.43 litres/min, when the power is 1.00 kW per deg C.

Table 12.1 gives power in kilowatts for various values of flowrate and temperature rise, based on eqn. 12.2.

It is sometimes necessary to measure the power transmitted to an applicator, which can be done using a directional coupler (Section 4.8) and thermistor. The thermistor is a resistor with a high temperature coefficient of resistivity. It is mounted in a coaxial or a waveguide structure which is designed to have a good impedance match, with all the microwave power incident being absorbed in the thermistor. The coupling factor of the directional coupler is chosen to feed 1–10 mW microwave power to the thermistor, which is sufficient to raise its temperature to give an accurately measurable change in resistance. The thermistor is previously calibrated by feeding an accurately known power at low frequency to give a similar resistance change. Several proprietary instruments are available which give a direct readout of power.

This method of measurement relies on an accurate knowledge of the coupling factor of the directional coupler. This is typically in the range 55–65 dB; the exact value would need to be known to ± 0.05 dB for ± 1% accuracy for an absolute measurement. This is very difficult to achieve, and it is more practical to perform an overall calibration of the combined system beforehand, using a water load as above.

Table 12.1   Water-load power against flowrate and temperature rise

Temp rise	Flowrate (l/min)				
	1	2	3	4	5
(deg C)	(kW)	(kW)	(kW)	(kW)	(kW)
1	0.70	1.39	2.09	2.79	3.49
2	1.39	2.79	4.18	5.58	6.97
3	2.09	4.18	6.27	8.36	10.46
4	2.79	5.58	8.36	11.15	13.94
5	3.49	6.97	10.46	13.94	17.43
6	4.18	8.36	12.55	16.73	20.91
7	4.88	9.76	14.64	19.52	24.40
8	5.58	11.15	16.73	22.30	27.88
9	6.27	12.55	18.82	25.09	31.37
10	6.97	13.94	20.91	27.88	34.85
11	7.67	15.33	23.00	30.67	38.34
12	8.36	16.73	25.09	33.46	41.82
13	9.06	18.12	27.18	36.24	45.31
14	9.76	19.52	29.27	39.03	48.79
15	10.46	20.91	31.37	41.82	52.28
16	11.15	22.30	33.46	44.61	55.76
17	11.85	23.70	35.55	47.40	59.25
18	12.55	25.09	37.64	50.18	62.73
19	13.24	26.49	39.73	52.97	66.22
20	13.94	27.88	41.82	55.76	69.70
21	14.64	29.27	43.91	58.55	73.19
22	15.33	30.67	46.00	61.34	76.67
23	16.03	32.06	48.09	64.12	80.16
24	16.73	33.46	50.18	66.91	83.64
25	17.43	34.85	52.28	69.70	87.13
26	18.12	36.24	54.37	72.49	90.61
27	18.82	37.64	56.46	75.28	94.10
28	19.52	39.03	58.55	78.06	97.58
29	20.21	40.43	60.64	80.85	101.07
30	20.91	41.82	62.73	83.64	104.55

It is very important that the terminating load fitted to the 'reverse' arm of the directional coupler has a good match (> 40 dB return loss), and also that the directivity of the coupler[†] is high (> 40 dB), to avoid contamination of the wanted signal to the thermistor. If a water-load calibration is performed as above, this effect will be included and eliminated in the calibration.

[†]40 dB return loss represents a reflection coefficient of 0.01. This will add vectorially to the apparent forward wave to give a resultant voltage amplitude in the range 0.99–1.01, or apparent power (by squaring) in the range 0.98–1.02, or ± 2 % error.

12.2.1.3 Efficiency

Efficiency is simply the microwave power output divided by the electrical power input as measured by the above methods, taking care to ensure that the power consumed by auxiliary systems is included, or not, as required by the specification.

The efficiency may be required to be measured over a range of operating conditions (e.g. a range of power output, mismatch loading, range of magnetic field applied to magnetron).

Chapter 13

Equipment safety

13.1 Review of hazards

Industrial microwave equipment comprises many systems which must be assessed for safety. Apart from the high profile of microwave exposure, there are high voltages (medium and EHT), hot surfaces and steam, and mechanical power drives with both slow- and fast-moving machinery. All these represent a hazard, but no more so than for other more conventional machinery; and the microwave component, correctly handled, is probably the least risk. It often raises anxiety and fear in people who are unfamiliar with it; they cannot see it, and because it is sometimes classified as radiation there is a confused feeling of risk of a radioactive nature.

13.2 Microwave leakage

Radiation is an emotive word with connotations of dangerously harmful radioactive emissions, classed as 'ionising' radiations. Microwave radiation is not an ionising radiation and, in spite of various claims otherwise over many years, has no other substantiated and significant effect on the body than heating. Unlike ionising radiation, microwave is noncumulative: doses received on successive occasions are not additive and do not accumulate to an eventually harmful level. The only 'nonthermal' effect of microwaves on the human body, reliably reported, is an aural sensation in the vicinity of very high-power pulsed radar-installations (Puranen and Jokela, 1996), where a few people have reported that they can hear a note at the pulse-repetition frequency although they were too far away to hear the electronic equipment directly. The peak power of such pulses is many orders of magnitude greater than the levels encountered as leakage from industrial equipment.

Because of the hazard perceived in some people's minds, it is good management practice to address the safety issue well before the planned installation of major equipment. At an early date, the equipment supplier should meet all persons concerned with purchase, installation, operation and maintenance to explain the nature of microwave energy and the steps taken to eliminate any risk which might arise. There should be present representatives of all levels of management, operators, shop stewards, and safety personnel; such a meeting should allow plenty of time for questions and discussion, and should be allowed to consider, too, the other potential hazards. A representative of the local statutory safety authority should be invited to attend as an impartial commentator.

Upon installation and commissioning, it is good practice to record daily readings of microwave leakage from the equipment and to retain securely the data obtained. Such data could be extremely valuable evidence if a person at some time in the future claimed injury resulting, allegedly, from exposure to microwave power.

All microwave equipment must operate within the safety limits prescribed locally in the country of installation. Although there are variations, the generally accepted limit on safe stray leakage of microwave power density is 10 mW/cm measured at a distance of 50 mm from the equipment. The level 10 mW/cm is believed to have been chosen in the 1940s, based on the power flux just discernible as a sensation of heat. Experimentally, this was determined as about 100 mW/cm and the limit was chosen as an order-of-magnitude less. This specification was originally introduced for domestic and catering ovens, where the 50 mm dimension is readily identified. It was arbitrarily adopted for industrial equipment, but where there are open apertures the interpretation is subject to debate. Sometimes it is taken as the power flux at the inflow/exit aperture planes of the tunnel; on others a boundary around the equipment is considered outside which operators would normally be present for extended periods, where the leakage must not exceed 10 mW/cm.

It has long been recognised that the above limits applied to a large industrial plant were unrealistic, on account of the various interpretations of defining an appropriate boundary. In the United Kingdom the National Radiological Protection Board (NRPB) has studied this problem and issued a restriction specification (NRPB, 1993) based on heating of the body, taking into account the body's thermo–regulatory system and the penetration depth of microwave energy into the body at the principal ISM frequencies. In effect, this specification avoids reference to any boundary around an installation, being directed at the power flux incident on the body. Moreover, it gives an average maximum exposure within a 15 min time period which allows brief excursions to higher power flux, as long as the prescribed average is not exceeded.

The NRPB Basic-Restriction limits are expressed in terms of the specific energy-absorption rate (SAR) in units of watts per kilogramme, from which 'investigation levels' are established. These levels can be compared with measured levels to demonstrate compliance with the Basic Restriction levels.

If the investigation levels are exceeded, the exposure situation should be investigated further and compliance with the Basic Restriction assessed.

The NRPB Basic Restriction and power-flux investigation levels are listed in Table 13.1.

Table 13.1   NRPB Basic Restriction and power-flux investigation levels

Frequency	Basic Restriction, SAR	Power-flux density	
(MHz)	(W/kg)	(W/m^2)	(mW/cm^2)
2450	0.4	100	10
896 or 915	0.4	50	5.0

The last column of Table 13.1 can be compared with direct readings of a commercial instrument. In practice, the power flux varies between wide limits over potentially occupied zones near the equipment, and engineering judgement must be used if levels approach those of the last column.

There is an extensive literature on microwave hazards, including some recommendations of much lower safety levels (1 mW/cm^2) in some countries, notably the former Eastern Bloc in the 1970s, the need for which has never been substantiated.

Some relevant safety literature is included in the list of References.

(IEC, 1982; AMSI, 1992; NRPB, 1989; NRPB, 1993; CENELEC, 1995; IEEE, 1992; IRPA/INIRC, 1988; IMPI, 1975; Mumford, 1961; Stuchly, 1977).

13.3 Measurement of microwave leakage

Most microwave-leakage meters comprise a thermal sensor (thermistor, bolometer or thermocouple) which is heated by the microwave flux, the temperature rise being proportional to the intensity of the microwave flux. Electronic processing measures the temperature rise, or thermoelectric EMF, and gives a readout of power flux density in milliwatts per square centimetre. Usually, two or more sensors are used in a single assembly in the sensing head, coupled to separate antennas responding to mutually orthogonal planes of polarisation of the microwave field. Thus the instrument measures all polarisations, plane and elliptical, of microwave field.

The instruments are hand-held, and usually have a dielectric foam cover (dome) surrounding the antenna, dimensioned to place the antenna 50 mm from the outer contour. If the dome is placed against the wall of the equipment, the 50 mm dimension specified in the microwave-leakage specification is automatically set.

The instruments are tuned to the operating frequency; some having a switch to select the frequency band.

Accuracy is high in the best instruments: ±0.5 dB over a range 1–100 mW/cm^2. It is very important that instruments be recalibrated annually or as recommended by the supplier. Some instruments have interchangeable measuring heads to cover the full range of sensitivity. As it is easy to burn out the sensor, particularly the most sensitive one, it is good practice when first measuring a machine to begin cautiously, using the least sensitive, most robust, head.

Measurements on equipment must be planned to ensure that they are representative of all conditions under which the equipment might operate: fully loaded, part loaded or with no workload present. While the prime attention is directed to leakage emanating from open apertures or access doors, other zones should be examined to verify that there is no leakage from damaged or worn parts in important high-field zones. Note that machines with large open apertures may rely on the presence of workload to provide attenuation of leakage in the choke tunnels, and are provided with a variety of sensors to detect the presence of workload. These sensors switch the microwave power off in the absence, or partial absence, of workload. Some engineers have, in the past, insisted on making leakage measurements with the sensors made inoperative, arguing that the sensors might fail. Great care must be taken that such sensors are of the highest integrity, and that if they do fail the microwave power is disabled; checking this feature is more realistic than making measurements with them inoperative.

13.4 High-voltage protection

The voltages used in all microwave power equipment are lethal; in the range 1–25 kV DC, from high-current sources. It is therefore imperative to ensure beyond all doubt that access to those parts at high voltage is barred by mandatory, mechanical isolation of the electrical supply. The safety requirement is the same as in any other high-voltage equipment, not unique to microwave equipment, and the same methods of security apply.

There are usually several zones of EHT voltages, with a plurality of doors. Obviously the number of access doors should be minimised, but they must all be included in the safety system. In industrial microwave systems it is also the practice to include within the system, all doors giving access to 'hot' microwave zones such as oven doors for loading or maintenance.

Door switches which operate on opening the door are unreliable in the context of the needs of this safety system, because there is the possibility of them jamming in an unsafe position, and it is easy to override an electrical interlock system.

The most effective, and well proven, system is a mechanical door lock with dedicated keys. Each door has a lock with a unique key, which can only be removed from the door lock if the door is firmly closed and locked: the key must be used to unlock and open the door, and is then retained in the door lock. The door lock cannot be operated with the door in the open position and so the key is retained. To energise the microwave generators, all the door keys must be inserted into a dedicated key bank, which holds a master key for operating the main electrical isolator switch supplying the generators. Only if all the keys are inserted into the key bank can the master key be withdrawn; thus it is not possible to operate the equipment unless all the doors are positively closed and locked. When the master key is withdrawn from the key bank all the door keys are retained in it. Further, the master key is retained in the isolator when it is in the closed position. It follows that the master key cannot be returned to the key bank unless the isolator is opened and the equipment made safe. Several companies manufacture these key systems, prominent names being 'Castell'[®] and 'Fortress'[®].

13.5 Radio-frequency interference (RFI)

There are many users of the radio-frequency spectrum including the microwave range of frequencies, and it is important to avoid mutual interference which might impair their quality of performance. Although spurious emissions at frequencies out of the allocated ISM frequency bands must be limited, the main focus of interest is emission from industrial equipment operating at 896 or 915 MHz frequencies.

This is because in Europe (except the United Kingdom) and the Pacific rim countries there is no allocated ISM frequency band at 896 or 915 MHz. In the Americas the 915 MHz band is allocated, but this is threatened. The reason is probably historical: at the time these allocations were made, in the early 1950s, the potential for industrial use of microwave heating around 900 MHz was foreseen only in the United Kingdom and the USA, where appropriate reservations were made. Since that time, modest but very important use has been made of this ISM band.

However, the growth of the cellular telephone systems, both analogue and digital, has created a problem, not least in the UK where administrative blunders resulted in the 896 MHz ISM band being allocated for telephone use. The problem is compounded because the telecommunication authorities, unilaterally and without consultation with ISM users, set a limit on interference which is almost unattainable for industrial installations, corresponding to little more than residual noise in a sensitive receiver placed 30 m from the industrial equipment.

In practice, there has been very little interference between industrial equipment in the UK and the cellular-telephone system: to the author's knowledge only two cases in the mid-1980s, which were effectively resolved. This is in spite of the fact that the typical leakage of microwave energy at the 896 MHz nominal frequency is some 40–50 dB greater than the limit demanded by the cellular-telephone operators.

The reason for the low incidence of interference is principally that the industrial-microwave-generator output has substantial frequency modulation, mainly, but not entirely, at modulation rates at power-line frequency and harmonics thereof. Effectively, the frequency of the magnetron is swept over a frequency band of about 0.1%, i.e. approaching 1 MHz. The telephone receiver has a narrow bandwidth sufficient to accommodate the frequency bandwidth of speech, about 2 kHz. The effective interference power is thereby reduced in the ratio of these bandwidths, by about × 1/500 or –27 dB for this illustration.

Frequency modulation does not account entirely for the low interference experienced in practice. A further substantial effect is that the low limit set corresponds to the extreme conditions of the industrial installation being located either on the fringe of the service area of a telephone base station, or alternatively very close to a base station serving a large area, where the signal-to-noise ratios are both limiting. In the first case the cellular-telephone user would have difficulty with reception (hearing); in the second of transmission (being heard). Generally, only one of these conditions applies because the base-station transmit and receive frequencies are well separated such that if one falls within the ISM band, the other will be outside it and therefore will be unaffected by interference. It would appear from practical experience that these limiting conditions are rare, to the extent that it is unreasonable to impose 'blanket' limits based on them.

13.5.1 Elementary calculations of radio interference

The specification of permitted interference at microwave frequency is set by the Comité International Special des Perturbations Radio Electriques (CISPR), in conjunction with CENELEC and other bodies, and is stated in their Documents CISPR 11 and EN 55011 (CISPR, 1990; CENELEC, 1995). The limit is given as + 40 dB (μV/m), measured at a distance of 30 m from the outside wall of the building housing the equipment. The method of measurement is given by CISPR Document 16 (CISPR, 1993), which specifies the receiving equipment, with particular reference to the receiver bandwidth and the post-detection response.

Many assumptions have to be made in making calculations of field intensities because the propagation characteristics are affected by so many factors, such as the building materials used, reflections, shielding and diffraction from the neighbouring buildings, vegetation, overhead cables etc. It is assumed that the microwave equipment is a point source of leakage power, which it certainly is not.

The unit dB(μV/m) requires explanation because it effectively defines a power-flux density in free space, and implies the characteristic impedance of 377 Ω (Section 1.5). The reference level is 1 μV/m, which is an electric field intensity of 10^{-6} V/m.

In free space this reference power-flux density p is given by

$$p = \frac{E^2}{Z_0} = \frac{\left(10^{-6}\right)^2}{377} = 2.65 \times 10^{-15} \quad W/m^2$$

The specification limit allows + 40 dB relative to this reference level, i.e. $\times 10^{+4}$, so that the power-flux limit at 30 m from the building housing the equipment is

$$2.65 \times 10^{-15} \times 10^{+4} = 2.65 \times 10^{-11} \text{ W/m}^{-2}$$

To illustrate the magnitude of the problem this low intensity represents, assume that the ISM equipment behaves as a point source radiating isotropically in free space. The equipment is at the centre of a sphere of radius 30 m. Following the inverse-square law the total power radiated P watts is the above power flux multiplied by the surface area of the sphere:

$$P = 2.65 \times 10^{-11} \times 4\pi \times 30^2 = 3.0 \times 10^{-7} \text{ W} = 0.3 \text{ μW}$$

A typical equipment meeting the personal-safety limits (< 10 mW/cm) would radiate a total power in the order of 10 W, so the additional screening required would be 10 log $10/(3 \times 10^{-7})$ = 75 dB. This is a very demanding requirement which can only be satisfied at great expense and in many cases would render a proposed installation uneconomic. However, as the following additional effects should be included, the practicality becomes more realisable:

Frequency modulation: approx 27 dB
Building attenuation (metal-clad): 20 dB
Distance from equipment to outside wall of building: 10 dB
Total: 57 dB

This still leaves a requirement for a further 20 dB of screening, but modest additional choking would readily achieve the desired total.

Considerable study, both theoretical and practical, has been made of this problem by Kennington and Bennett (1993), and Last (1987).

The argument for raising the allowable interference power flux to a realistic level is persuasive from the studies already done, and from the minimal extent of interference suffered in practice.

Appendix 1

Outline of the economics of industrial microwave processing

The range of applications of microwave heating in industry is so wide that it is not possible to make a universal assessment of the overall financial implications. This appendix is intended to serve as an 'aide memoire' listing the parameters which have been found important in affecting economic viability. Many are obvious; some are irrelevant in many cases; others were unforeseen factors which emerged later as significant. No attempt is made to quantify them because of the diverse range of cases; it is for individual management to translate each into money. It is very important that an overall view is taken embracing not only accountancy but marketing, production and engineering aspects.

1 The installation

1.1 Purchase price of capital equipment
1.1.1 Microwave equipment including generators, oven(s) and accessories
1.1.2 Associated machinery before the microwave oven.
1.1.3 Associated machinery after the microwave oven
1.1.4 Conveyor systems
1.1.5 Control equipment
1.1.6 Process monitoring equipment
1.1.7 Computer
1.1.8 Air ducting and exhaust fans; scrubbing equipment
1.1.9 CIP system
1.1.10 No-break electrical power supply
1.1.11 Electrical substation and feeder
1.1.12 Loss of production during site construction and commissioning
1.1.13 Waste of material during setting-to-work and proving.

1.2 Site services
1.2.1 Planning of site and access for plant

1.2.2 Preparation of site (building and civil-engineering work)
1.2.3 Repositioning other equipment
1.2.4 Floors and foundations
1.2.5 Electrical supply to equipment
1.2.6 Cooling-water supply (may require glycol)
1.2.7 Closed-circuit cooling system with cooler (e.g. cooling tower)
1.2.8 Drains and associated pipework
1.2.9 Water supply: hose down or CIP use
1.2.10 Steam supply with water traps and fittings
1.2.11 Compressed-air supply (possibly oil-free)
1.2.12 Cranes and lifting equipment
1.2.13 Ventilation
1.2.14 Purchase of special test equipment, e.g. microwave-leakage meter

1.3 Spare parts: consumable spares
1.3.1 Magnetrons
1.3.2 Conveyor belts
1.3.3 Suppliers' recommended list of spares

1.4 Training of personnel
1.4.1 Operator training
1.4.2 In-house maintenance-staff training
1.4.3 Management training
1.4.4 Safety training

1.5 Process-development costs
1.5.1 Measurement/assessment of workload physical properties (ε', ε'', penetration depth, specific heat, density, thermal conductivity etc. over the range of conditions pertaining to the process)
1.5.2 Determining the preferred operating frequency
1.5.3 Planning, construction of experimental hardware, and conducting experimental trials
1.5.4 Development and design to full manufacturing drawings of the process plant
1.5.5 Hire of specialised consultants
1.5.6 Assessment of national and local regulations and laws affecting the operation and safety of the equipment, and verification of compliance

2 Direct operating costs

2.1 Electrical energy
    (including maximum-demand charge if affected)

2.2 Steam supply

2.3 Water supply and drain

2.4 Glycol

2.5 Demineralised water

2.6 Compressed air

2.7 Gas

2.8 Replacements:
2.8.1 Magnetrons
2.8.2 Conveyor bands
2.8.3 Suppliers' list

2.9 Operators' wages

3 Indirect operating cost

3.1 Factory overheads applicable
(i.e. rents, building provision and maintenance, lighting and heating etc.)

3.2 Interest on capital raised

3.3 Depreciation of plant

4 Maintenance and Cleaning

4.1 In-house cleaning

4.2 In-house maintenance and servicing

4.3 Contract servicing

5 The product

5.1 Throughput value using microwave

5.2 Throughput value not using microwave

5.3 Value of enhanced quality of microwave-treated product:
Improved turnover?
Improved quality?
Reduced scrap?
Greater profit?
Time saving in production?
Reduced value of work in progress?
Faster response to changed demand?
Unique product: no other method?

5.4 Production cost using conventional processing:
This requires a detailed assessment

6 Bonuses (often unforeseen)

6.1 Reduced maintenance on subsequent machinery

6.2 Floor-space saving

6.3 Cleaner, pleasanter environment

6.4 Improved professional image
Impresses customers

6.5 Saving management time in resolving problems of conventional plant

6.6 Improved workforce morale

6.7 Reduced sensitivity to quality of incoming product

6.8 Better quality control

6.9 Faster-response quality adjustment

6.10 Reduced energy wasted on stand-by or part-load conditions

6.11 Reduced labour force

6.12 Reduced product handling, e.g. fork-lift trucks

6.13 Reduced down time from breakdowns

7 Handbooks and supporting literature

7.1 Operator's handbook
This must be concisely and clearly written and illustrated to enable the operator to control the plant. It must be short, usable as a ready-reference to show what action to take in any foreseeable event. It requires great care in preparation: if the operator is unable to understand it quickly, this is a serious deficiency.

7.2 Plant engineer's handbook
This is a more detailed account of the equipment, with illustrated descriptions of each subsystem, what it does and how, with complete and accurate electrical circuit diagrams, and diagrams of cooling systems, drains, steam systems, CIP systems and air ducting.
This handbook must have detailed accounts of maintenance procedures, with particular reference to safety systems and procedures.

7.3 Translation
All the literature and handbooks *must* be presented in the language of the customer and user. Such translations must be done by professional technical translators conversant with the contemporary idiom of the language.

Appendix 2

Specification of industrial microwave equipment

A full and carefully considered specification is an essential prerequisite to a contract for the supply of industrial plant, especially so for microwave equipment where the technology is probably unfamiliar to the intending purchaser. It must be carefully set out and presented without specialised technical jargon so that both parties have a clear understanding of all the aspects of the equipment supply; there must be no surprises at any stage.

The following is a list of parameters which, from experience, may require specification. Not all will be relevant, but there should be a considered decision. It is assumed that a preliminary assessment of the project's financial viability is favourable.

1 The workload

1.1 What is it, by composition? Is there a variety of workloads?

1.2 What is the required process (drying, sterilising, pasteurising, vulcanising etc.)?

1.3 Is it toxic (*a*) before processing, (*b*) during or (*c*) after processing?

1.4 Are toxic gases released during processing?

1.5 Is it corrosive: (*a*) acidic (*b*) caustic?

1.6 Is it flammable or explosive?

1.7 Are there any incompatible materials?

1.8 Are bacteria present?

1.9 What form is the workload: liquid, fine powder, coarse powder, small irregular particles, large irregular pieces, blocks of preformed geometry?

1.10 Is the process batch or continuous?

1.11 What is the input mass flow?

1.12 What is the charge mass if a batch process?

1.13 What is the required process time?

1.14 What is the output mass flow?

1.15 What is the output-charge mass (batch process)?

1.16 What is the bulk density (input and output)?

1.17 What is the specific heat?

1.18 What is the starting temperature?

1.19 What is the input moisture content? This requires careful definition, as a wet-weight basis, or dry-weight basis, or percentage of solids. It is least ambiguously defined by stating the mass flow of dry matter, and the mass flow of water.

1.20 What is the output moisture content?

1.21 What is the thermal conductivity?

1.22 What are the final desired temperature and required temperature uniformity?

1.23 What are the the final desired moisture content and required moisture-content unformity?

1.24 Is there any other criterion by which the output quality is judged?

2 The process plant

2.1 Present detailed calculation of the heat balance, as in Section 3.5.

2.2 Determine the microwave properties of the workload over the range of operating conditions envisaged in the process (e.g. as a function of temperature, moisture content or other variable parameter).

2.3 Determine the total power required, choose the frequency and determine the optimum power density in the workload.

2.4 Estimate the E field within the workload and externally. Is there a risk of arcing?

2.5 Consider the options on applicator type. Determine the principal dimensions.

2.6 Decide on a batch or continuous-flow process.

2.7 Assess the probable uniformity of heating.

2.8 Determine the number of generators.

2.9 Decide on the microwave feed system.

2.10 Is on-load tuning required?

2.11 Is conventional heating required (*a*) in the applicator, (*b*) as a dwell tunnel?

2.12 Determine the type of auxiliary heating: (*a*) hot air, (*b*) infra-red, (*c*) steam

2.13 If hot air is used, can it be partially recirculated to save energy?

2.14 If steam is used, or if steam is generated by a drying process, consider the risk of condensation droplets, and design features to render them harmless; they must never be allowed to fall on to the workload.

2.15 If a batch process, determine the cycle time

2.16 If a continuous process, determine the nominal conveyor width, the bed depth and the conveyor speed.

2.17 For a continuous process, consider the loading arrangements: a uniform, or predetermined and controlled bed depth transversely is desired.

2.18 For a batch system is loading to be automatic or manual? Is handling gear required?

2.19 What are the local requirements on microwave leakage? Is 896/915 MHz an allocated frequency for ISM?

2.20 Consider the choke-tunnel design on a continuous system, as Chapter 8.

2.21 Pressure and vacuum vessels must be built to comply with national standards.

3 Construction materials and fabrication

3.1 Assess materials acceptable for the process: food customers may insist on stainless steel for all metal parts in contact with food.

3.2 Are there any corrosion problems foreseen arising from the composition of the workload?

3.3 Are there any prohibited materials associated with the workload or the customer's activities, e.g. PTFE is not permissible in contact with tobacco.

3.4 Advanced fabrication techniques may be required for the elimination of 'bug-traps' in equipment for the food and pharmaceutical industries.

4. The process

This section is concerned with assessing the performance against the specification expectations. It is important to agree a protocol by which the performance can be measured.

4.1 Is the workload feed to the machine in full accordance with the specification statement? What are the differences?

4.2 How is the mass flow to be measured? This can be a difficult measurement because of short-term fluctuations. The integration time as a percentage of the dwell time in the applicator is an important parameter: if it exceeds 10% there could be significant errors affecting the reconciliation of the heat balance.

4.3 How is the conveyor speed to be set and measured?

4.4 How is steady-state operation to be defined, before a set of measurements is taken?

4.3 How is the input temperature to be measured? Does it fluctuate and if so over what period?

4.4 How is the net microwave power to be measured? Facilities for this measurement are probably built-in, but they need coupling to a data-collection system.

4.5 How is the output temperature to be measured? This needs careful definition because of the imperfections in temperature rise. If an accurate temperature rise is sought for reconciling the heat balance, it is necessary to take a representative sample of the output workload under steady-state conditions and place it in a prepared thermally insulated box of low heat capacity, and to allow the temperature to equilibrate before measuring the final temperature. The temperature should have equilibrated to a residual variation less than 5% of the temperature rise, or some other agreed figure.

An alternative is to measure the temperature at a number of points, computing the mean and the standard deviation, and the percentage difference between maximum and mean and between minimum and mean.

4.6 How is moisture content to be measured: (*a*) at the input, (*b*) at the output? What are the variations? Is the measurement time compatible with the speed of the process?

4.7 How is the electrical-power input to be measured? The precise equipment needs careful consideration, and the cost of its hire agreed. The possibility of surges and momentary outages arising from switching operations, lightning strikes etc. must be considered.

4.8 Is it considered necessary to monitor any of the other service supplies? Experience suggests 'yes' because these supplies are often unreliable, having surges and dips in pressure which may cause the equipment to trip.

5. Standards

5.1 All national and local standards and regulations affecting the equipment must be studied to ensure compliance. These include:
Electrical wiring and equipment practice
Electrical safety
Electrical interference
Allocation of ISM frequency bands
Guarding of hazardous machinery
Pressure vessels
Toxic emissions
Fire hazards.

Appendix 3

Glossary of terms and symbols

The following is a list of words and phrases commonly used in the industrial RF and microwave vocabulary, for the assistance of those new to the technology.

Item	Symbol Units	Description
$(\pi - 1)$ mode		A specific mode in the magnetron in which there is one less cycle of variation of field intensity around the anode than for the π mode. Operation in this mode is to be avoided as damage may result through overheating.
π-mode		The fundamental oscillation mode of a magnetron, in which adjacent vane tips of the anode are of opposite polarity (i.e. 180° or π radians phase shift) of RF voltage.
π-mode voltage	$V_{a\pi}$ kV	The anode voltage threshold at which anode current just flows in a magnetron. It increases as the the magnetic field is raised.
AM		Amplitude modulation.
Admittance	Y mho	Reciprocal of impedance: $Y = 1/Z$.
Anode strap		One of a pair of copper rings joining alternate vanes of a magnetron anode together electrically, to ensure oscillation in the π-mode preferentially to other modes.
Antenna		A structure comprising a single feed with a reflector, or an array of radiating slots for directing microwave power into a cavity containing a workload.

Item	Symbol Units	Description
Arc		An intense local, destructive burning sustained by RF or micowave power.
Arc detector		A sensor, usually optical, for detecting an arc in an oven or feeder, arranged to switch power off instantly.
Attenuation	α Np/m or total neper	(i) The decay of an EM wave as it propagates along a transmission path. (ii) The attenuation of a device.
Attenuation coefficient	α Np/m	A measure of the rate of decay of an EM wave as it propagates through a medium.
Cavity		A metal box, closed except for feed point(s), forming a resonant structure.
Circulator		A 3-port waveguide junction having nonreciprocal properties, such that power couples between adjacent ports for one sequence direction, but not the other, i.e. $1 \rightarrow 2 \rightarrow 3$ but not $1 \leftarrow 2 \leftarrow 3$. It is used to divert away from the generator power reflected from a mismatched load. The active element is a ferrite material having gyromagnetic properties.
Conductance	G mho	The ratio of current to voltage in a resistor.
Cutoff wavelength	λ_c	In a waveguide, the critical wavelength greater than which an EM wave will not propagate in a designated mode without high attenuation. It is proportional to the dimensions of the waveguide.
E-bend		A bend in rectangular waveguide in which the broad faces bend and the narrow faces remain plane.
E-mode		A propagation mode in a waveguide, in which the axial-field component is an electric-field vector. The axial-magnetic-field vector is zero. Suffixes $E_{m,n}$ show the number of halfcycles of periodicity in space along the co-ordinate directions.
Electric field	E V/m	The electric-field stress, or voltage gradient, in a space or within a material resulting from a charge.
EM wave		Electromagnetic wave.

Item	Symbol Units	Description
FM		Frequency modulation.
Ferrite		A highly resistive ferromagnetic ceramic (formula MFe_2O_4) used for high-frequency transformer cores in inverter power supplies where M is a metal, typically nickel. For use in microwave circulators and isolators, where the ferrite displays gyromagnetic nonreciprocal properties. M is typically yttrium or gadolinium.
Free-space propagation velocity	c m/s	Velocity of propagation of an EM wave in free-space (velocity of light), 2.99×10^8 m/s; $c = \sqrt{(\mu_0/\varepsilon_0)}$.
Gassy		A condition of a magnetron in which the internal vacuum pressure is impaired, resulting in an arc between anode and cathode when the anode voltage is applied.
H-bend		A bend in rectangular waveguide in which the narrow faces bend and the broad faces remain plane.
H-mode		A propagation mode in a waveguide, in which the axial field component is a magnetic-field vector. The axial-electric-field vector is zero. Suffixes $H_{m,n}$ show the number of halfcycles of periodicity in space along the co-ordinate directions.
Impedance	Z Ω	A property of an electrical structure which determines the load current for a given applied voltage, equivalent to resistance in elementary Ohm's law. Impedance generally has a real and an imaginary part $Z = R + jX$, corresponding to the resistance and reactance of the circuit. X may be capacitive or inductive.
Klystron		An electronic valve for generating microwave power, usually by amplification from a low-power source. It has an electron beam which passes through a succession of tuned cavities. It is more sophisticated than a magnetron and very much more expensive. It has little industrial application except where its specialised features, e.g. frequency stability, are particularly desired.

Item	Symbol Units	Description
Lossless		Adjective applied to a dielectric material which does not significantly absorb power from an EM wave.
Lossy		Adjective describing a material which dissipates EM power.
Magic-T		A 4-port waveguide junction comprising an E-plane and an H-plane junction at the same position on a main waveguide. It forms a 3 dB directional coupler, and has other important electrical properties enabling it to be used as an impedance-matching device, in combination with adjustable short-circuit pistons.
Magnetic field	H A/m	The magnetic-field intensity resulting from a current flowing in a conductor, or from the displacement current resulting from the time rate of change of charge.
Magnetron		An electronic valve generating microwave power from a high-voltage source of DC power. It has a very high efficiency and the lowest cost of microwave power generators, and is the preferred generator for industrial microwave power.
Matching		The procedure of adjusting the admittance of a load to equal that of a transmission line feeding it to ensure maximum transfer of power.
Mismatch		A condition in which the admittance of a load does not equal the characteristic admittance of a transmission line feeding it.
Mode (cavity)		A defined pattern of distribution of the electric- and magnetic-field components of an EM wave excited in a closed cavity. The electric and magnetic-field components vary in a cyclical manner in the cavity space, such as to satisfy the boundary conditions at the cavity walls.
Mode (waveguide)		A defined pattern of distribution of the electric- and magnetic-field components of an EM wave propagating along a waveguide. The

Item	Symbol Units	Description
		TE_{01} (alias H_{01}) mode is the lowest-order mode in a rectangular waveguide.
Moding		An undesirable change of oscillation mode of a magnetron from the desired π-mode. The oscillation is at the incorrect frequency at low efficiency, and may lead to damage to the magnetron through overheating.
Multimode cavity		A cavity in which the dimensions are several half wavelengths long in at least two orthogonal directions. Several modes are resonant close enough to the generator frequency to be excited significantly.
Neper	α m^{-1}	Measure of attenuation based on the natural logarithm to base e. 1 Neper = 8.686 dB
Normalised admittance	y	Reciprocal of normalised impedance, $y = Y/Y_0$. Dimensionless.
Normalised conductance	g	$g = G/Y_0$. See normalised admittance.
Normalised reactance	x	$x = X/Z_0$. See normalised impedance.
Normalised resistance	r	$r = R/Z_0$. See normalised impedance.
Normalised susceptance	b	$b = B/Y_0$. See normalised admittance.
Normalised impedance	z	In a transmission line, a convenient method of evaluating impedance relative to the characteristic impedance of the the line Z_0 such that $z = Z/Z_0$. Dimensionless.
Packaged magnetron		A magnetron with integral permanent magnet; usually practical only for small-power magnetrons.
Permeability (relative)	$\mu = \mu' + j\mu''$	For a ferromagnetic material, the increase in permeability relative to free-space. μ' is the lossless component, and μ'' represents the dissipation in the material.
Permeability (free-space or absolute)	μ_0 H/m	The permeability of free-space. The quotient B/H where H is the applied magnetic field and B is the resulting magnetic flux. $\mu_0 = 1.257 \times 10^{-6}$ H/m.

Item	Symbol Units	Description
Permittivity (relative)	$\varepsilon = \varepsilon' + j\varepsilon''$	For a dielectric material, the increase in permittivity relative to free-space. ε' is the lossless component, and ε'' represents the dissipation in the material. The absolute permittivity of the material is $\varepsilon\,\varepsilon_0$. Dimensionless.
Permittivity (free-space or absolute)	ε_0 F/m	The permittivity of free space. The quotient D/E where E is the applied electric field and D is the resulting displacement current. $\varepsilon_0 = 8.854 \times 10^{-12}$ F/m.
Propagationt coefficient	γ m^{-1}	A constant defining the propagation of an EM wave through a medium along an axis (e.g. z axis) as $E = E_0\, e^{-\gamma z}$. It is a complex number.
Reactance	X Ω	The quotient of voltage/current of an inductor or capacitor. The current is in phase-quadrature with the voltage, lagging for an inductor, leading for a capacitor.
Reflection coefficient	$\rho e^{j\phi}$	A complex number representing the amplitude ρ and phase angle ϕ of a reflected wave from a mismatched load. Dimensionless.
Resistance	R Ω	The quotient of voltage/current in a resistor (Ohm's law).
Resonant		Adjective describing a structure, e.g. a cavity, whose input admittance varies rapidly with frequency, displaying a response similar to that of a conventional tuned *LCR* circuit. Mechanically analogous to a mass supported by a spring.
Scattering matrix		A succinct mathematical expression defining the electrical properties of an n-port structure. It is an $n \times n$ matrix in which the elements s_{lm} define the output from a given port for any combination of inputs to the remaining ports.
Single-mode cavity		A cavity in which the dimensions are chosen relative to the generator wavelength so that only a preferred mode is excited. Other modes have resonant wavelengths (frequencies) far enough removed to have insignificant excitation at the generator wavelength. The cavity is necessarily small to achieve this.

Item	Symbol Units	Description
Stepped twist		A twist section of rectangular waveguide in which the angle of twist is developed as the sum of a series of intermediate sections, each $\lambda_g/4$ long.
Susceptance	B Ω	The quotient of current/voltage of an inductor or capacitor.
T-junction		A waveguide junction in the form of a 'T'. The E-plane T has the stub arm in the broad face of the waveguide, the H-plane T in the narrow face. The T-junction is a three-port structure.
TE mode		Alias H-mode.
TM mode		Alias E-mode.
Travelling-wave tube		An electronic-valve amplifier for generating microwave power. It is capable of giving amplification over a wide frequency band, but is very expensive, and has little application in industry.
VSWR	S	Voltage standing-wave ratio. The quotient V_{min}/V_{max} ($S<1$) or the inverse V_{max}/V_{min} ($S>1$) where V_{max} and V_{min} are, respectively, the maximum and minimum voltages along the transmission line, separated in position by a quarter-wavelength. Dimensionless.
Waveguide		A hollow metal pipe for the transmission of microwave power, usually rectangular, of 2:1 aspect ratio. Features very high efficiency and zero leakage.
Waveguide wavelength	λ_g	In a waveguide, the wavelength of a propagating EM wave. It is longer than the free-space wavelength.

F_0 values, Z values, L_V values and D values

The pasteurisation and sterilisation of foods by heat processing is critically dependent on temperature and exposure time. The relationship between bacterial survival and temperature and time is very nonlinear. Likewise, the organoleptic quality of food is damaged by excess exposure to high temperatures and time, and so overexposure must be avoided. The 'window' for safe treatment and minimal degradation of quality of the food is small. The above parameter values are used by food technologists in quantifying the properties of heat treatment processes for foods (Jackson and Lamb, 1981).

F_0 values
The F_0 value of a heating process is a measure of its lethality taking into account both temperature and time. Because of its nonlinear characeristic, it is expressed logarithmically.

The F_0 value of a heat process is derived from the D, Z, and L_V values which are characteristics of a bacterium; *clostridium botulinum* is taken as a convenient reference standard.

D values
The D value is the exposure time required to reduce the bacterial count by a factor of 10 when the food is held at a constant temperature. A plot of log D against temperature is usually a straight line.

Z values
From the plot of log D against temperature, the change in temperature to give a change of ×10 in D can be estimated. This temperature change is the Z value, which for *c. botulinum* is 10 deg C.

The Z value is not the same for all bacteria, but for the purpose of food sterilisation *c. botulinum* is used as the reference standard by bacteriologists and food technologists.

L_V values

Using the Z value, it is possible to calculate the relative rate of destruction L_V of a micro-organism at different temperatures.

Customarily, an arbitrary rate of 1 is defined for a temperature of 121°C. At other temperatures θ the destruction rate is given by

$$L_V = 10^{-(121-\theta)/Z} \tag{A4.1}$$

Table A4.1 gives values of L_V for selected values of temperature and Z value.

Table A4.1   Values of L_V against temperature and Z value

Z	Temperature θ, deg C					
	110	115	120	125	130	135
6	0.01	0.10	0.68	4.6	32	215
8	0.04	0.18	0.75	3.2	13	56
10	0.08	0.25	0.79	2.5	7.9	25
12	0.12	0.32	0.83	2.1	5.6	15
14	0.16	0.37	0.85	1.9	4.4	10
16	0.21	0.42	0.87	1.8	3.6	7.5

The F_0 value is the area under the curve of L_V against time, and has the dimension minutes, i.e.

$$F_0 = \int\limits_{Process\ time} L_V\ dt$$

Q-BASIC program for mode counting in multimode ovens

The simple program below gives the total number of modes in an empty rectangular multimode cavity, and their eigenvalues (mode numbers), having their resonant frequencies within a bandwidth of $\pm\, r\,\%$. The cavity size is normalised to the free-space wavelength. Both TE and TM modes are included.

```
10 REM Normalised Mode Structure of Rectangular Multimode Oven
30 REM Oven dimensions( a,b,d) are normalised with respect to excitation wavelength
31 REM so that width a is the non-dimensional factor g0=a/wavelength (freespace)
32 REM p and q are proportion scaling factors b = pa and d = qa
33 REM The spectral density is given as the number of modes resonant within
34 REM a bandwidth +/- r%
35 REM l,m,n are the eigenvalues of the modes
50 DATA 24.2
55 READ g0
60 DATA 0.2, .1
65 READ p, q
70 DATA 0.5
75 READ r
76 COUNT = 0
80 FOR l = 0 TO 2 * g0 STEP 1
90 FOR m = 0 TO 2 * p * g0 STEP 1
100 FOR n = 0 TO 2 * q * g0 STEP 1
110 g = .5 * ((l ^ 2) + ((m / p) ^ 2) + ((n / q) ^ 2)) ^ .5
130 IF l = 0 AND m = 0 GOTO 170
140 IF m = 0 AND n = 0 GOTO 170
150 IF l = 0 AND n = 0 GOTO 170
151 IF g > g0 * (1 - r / 100) AND g < g0 * (1 + r / 100) THEN PRINT g, l, m, n
154 IF g > g0 * (1 - r / 100) AND g < g0 * (1 + r / 100) THEN COUNT = COUNT + 1
155 PRINT COUNT
170 NEXT n
180 NEXT m
190 NEXT l
```

Definitions of moisture content

Many definitions quantifying the moisture content of materials are in use, and it is very important to specify which is used in specifying it. Some definitions are peculiar to a particular industry or process, while others are more appropriate to the quantity of water present. For example 'percentage solids' is used for slurries and liquids because it shows directly the quantity most of interest, but moisture content 'dry-weight basis' is used for nearly dry materials. All the definitions are interrelated, but their numerical values for a specific case are very different.

It is therefore imperative to define moisture content precisely for performance specification.

Consider a wet material with the following properties:

Weight of dry matter = M_D
Weight of liquid = M_W

(*a*) Moisture content on percentage-dry-weight basis (MC_{DWB}):

$$MC_{DWB} = \frac{M_W}{M_D} \times 100\,\%$$ (A6.1)

(b) Moisture content on percentage-wet-weight basis (MC_{WWB}):

$$MC_{WWB} = \frac{M_W}{M_W + M_D} \times 100\%$$ (A6.2)

(c) Moisture content on percentage-solids basis (MC_{SB}):

$$MC_{SB} = \frac{M_D}{M_W + M_D} \times 100\%$$ (A6.3)

Figure A6.1 is a plot of the above moisture-content definitions against the ratio of water weight to dry matter. For moisture content less than 5% DWB, the dry-and wet-weight bases are nearly equal. As the moisture content rises, the wet-weight basis tends to 100%, while the dry-weight basis continues to rise linearly. When the weights of water and dry matter are equal, the percent-solids and the wet-weight-basis figures are equal at 50%, while the dry-weight basis is 100%.

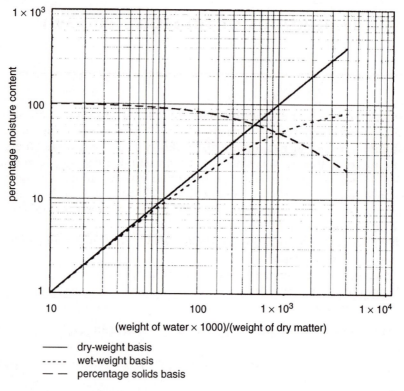

          —————   dry-weight basis
          - - - -   wet-weight basis
          — —   percentage solids basis

Figure A6.1   Moisture content on three definitions

Note that water may be present in a solid as free water, as chemically bound water (as water of crystallisation) and as adsorbed water where the water molecules have a chemical bond to the surface.

Most materials absorb water from the atmosphere. After completely drying, their moisture rises to an equilibrium level known as the regain, usually expressed as the percent regain, equivalent to the moisture content on the dry-weight basis. This is a common practice in the textile industry.

References

ALTMAN, J. L.: 'Microwave circuits' (Van Nostrand, New York, 1964)

ANSI: 'C95.1: Safety levels with respect to human exposure to radiofrequency electromagnetic fields 300 kHz to 199 GHz' (American National Standards Institute, New York, 1982)

BADEN-FULLER, A. J.: 'Microwaves' (Pergamon Press, 1979)

BARLOW, H. M., and CULLEN, A. L.: 'Technique of microwave measurements' (Methuen, 1948)

BENGTSSON, N. E., and RISMAN, P. O.: 'Dielectric properties of foods at 3 GHz by a cavity perturbation method. Parts 1 and 2', *J. Microw. Power*, 1971, **6** (2)

BOUCHEROT, P.: 'Concerning systems in which Vl/I2 is constant', *Rev. Gen. Electr.*, 1919, **5**, p. 203

BOWS, J.: 'Applications of microwave fringing field devices'. Proceedings of international conference on *Microwave and high frequency heating*, Cambridge, UK, 1995

BSI: 'BS 9030 1971: Specifications for magnetrons of assessed quality: generic data and methods of test' (British Standards Institution, London). Confirmed 1986. Including Amendments 1 and 2

BSI: 'BS 5175 1976: Specification for safety of commercial electrical appliances using microwave energy for heating foodstuffs' (British Standards Institution, London). Confirmed 1990. Includes Amendments 1–4

BSI: 'BS 7699: Specification for safety in electroheat installations' (British Standards Institute, London)

BSI: 'BS EN 60519: Safety in electroheat installations' (British Standards Institute, London)

BRYANT, G. H.: 'Principles of microwave measurements' (Peter Peregrinus Ltd. on behalf of the IEE, London, 1993)

BURFOOT, D., FOSTER, A. M., SELF, K. P., WILKINS, T. J., and PHILIPS, I.: 'Reheating in domestic microwave ovens: testing uniformity and reproducibility'. University of Bristol Food Refrigeration and Process Engineering Centre, Food Safety Dirctorate, Ministry of Agriculture, Fisheries and Food. Microwave Science Series Report 3, 1989/1991

CARSLAW, H. S., and JAEGER, J. C.: 'Conduction of heat in solids' (Oxford University Press, 1959)

CARTER, G. W.: 'The electromagnetic field in its engineering aspects' (Longmans, Green & Co, London, 1954)

CARTER, R., and WANG, P.: 'Design of helix applicators for microwave heating'. Proceedings of international conference on *Microwave and high frequency heating*, Cambridge, UK, 1995

CENELEC: 'ENV 50166-2: Human exposure to electromagnetic fields. High frequency

(10 kHz to 30 GHz)' (European Committee for Electrotechical Standardisation, Brussels, 1995)

CHAMBERS, D.: 'Designing high power SCR resonant convertors for very high frequency operation'. Proceedings of *PowerCon 9,* Washington DC, USA, 1982

CHAMBERS, D., and SCAPELLATI, C.: 'New high frequency high voltage power supplies for microwave heating applications'. Proceedings of *29th Microwave Power Symposium (IMPI),* 1994

CISPR: 'CISPR 11: Limits and methods of measurement of electromagnetic disturbance characteristics of industrial, scientific and medical (ISM) equipment, including amendments 1 and 2 and corrigendum 1' (International Special Committee on Radio Interference, 1990)

CISPR: 'CISPR 16: Radio disturbance and immunity measuring apparatus and methods. Part 1: Radio disturbance and immunity measuring apparatus. Part 2: Methods of interference and immunity measurements' (International Special Committee on Radio Interference)

COLLINS, G. B.: 'Microwave magnetrons' (McGraw-Hill, 1948)

COTTON, H.: 'Electrical technology' (Pitman, 1946)

CROMPTON, J. W.: 'A contribution to the design of multielement directional couplers', *IEE Proc. C,* 1957, Monograph 230R, **104**, (6), pp. 398-402

DAVIES, E. J.: 'Conduction and induction heating' (Peter Peregrinus Ltd. on behalf of the IEE, London, 1990)

DIBBEN, D. C., and METAXAS, A. C.: 'Time domain finite element analysis of multimode microwave applicators loaded with high and low loss materials'. Proceedings of international conference on *Microwave and high frequency heating,* Cambridge, UK, 1995

DISTEFANO, J. J., STUBBERUD, A. R., and WILLIAMS, I. J.: 'Feedback and control systems' (McGraw-Hill, New York, 1990)

ELECTRONIC TEMPERATURE INSTRUMENTS LTD.: '1997 catalogue' (Worthing, Sussex, UK, 1997)

ENGELDER, D. S., and BUFFLER, C. R.: 'Measuring the dielectric properties of food products at microwave frequencies', *Microw. World,* 1991, **12** (2)

ENTWISTLE, K. M.: 'The fracture of concentrically loaded square ceramic plates', *J. Mater. Science,* 1991, **26**, pp.1078-1086

FROHLICH, H.: 'Theory of dielectrics' (Oxford University Press, UK, 1958)

GABRIEL, C., and GRANT, G. H.: 'Dielectric sensors for industrial microwave measurement and control' . Proceedings of *High Frequency and Microwave Conference,* KEMA, Netherlands, 1989

GALLONE, G., LUCARDESI, P., MARTINELLI, M., and ROLLA, P. A.: 'A fast and precise method for the measurement of dielectric permittivity at microwave frequencies', *J. Microw. Power Electromagn. Energy,* 1996, **31** (3)

GURWITZ, D.: 'Resonant inverter'. US patent 5 371 668, 1993

GURWITZ, D., and MORRIS, P.: 'Resonant inverter'. International patent application WO 91/20172, 1991

HARVEY, A. F.: 'Microwave engineering' (Academic Press, New York, 1963)

HICKS, J.: 'Comprehensive chemistry' (Macmillan Press, 1971)

HORNER, F., TAYLOR, T.A., DUNSMUIR, R., LAMB, J., JACKSON, W.: 'Resonance methods of dielectric measurements at centimetre wavelengths', *J. IEE Part III,* 1946, **93**, (21), pp. 53-68

IEC: 'IEC 519 1982: Safety in electroheat installations. Part 6: specifications for safety in industrial microwave heating equipment' (International Electrotechnical Commission, Geneva) '

IEEE: 'C9.5.1 IEEE standard for safety levels with respect to human exposure to radiofrequency electromagnetic fields, 3 kHz to 300 GHz' (Institute of Electrical and Electronic Engineers, New York, 1992)

INTERNATIONAL MICROWAVE POWER INSTITUTE: 'Safety from microwave hazards in using microwave power equipment—an IMPI policy statement', *J. Microw. Power,* 1975, **10**, pp. 333-341

INTERNATIONAL NON-IONISING RADIATION COMMITTEE OF THE INTERNATIONAL RADIATION PROTECTION ASSOCIATION: 'Guidelines on limits of exposure to radiofrequency electromagnetic fields in the frequency range from 100 kHz to 300 GHz', *Health Phys.*, 1988, **54**, pp. 115-123

ISHITOBI, Y., and TOGAWA, M.: 'A practical diagram for computing long-line effects for CW magnetron', *J. Microw. Power*, 1978, **13** (2), pp. 131-137

JACKSON, A. T., and LAMB, J.: 'Calculations in food and chemical engineering' (Macmillan Press, 1981)

JAHNKE, E., and EMDE, F.: 'Tables of functions' (B. G. Teubner, Leipzig, 1933)

JAMES, H. M., NICHOLS, N. B., and PHILLIPS, R. S.: 'Theory of servomechanisms' (McGraw-Hill, New York, 1947)

JIA, X., and JOLLY, P.: 'Simulation of microwave field and power distribution in a cavity by a three dimensional finite element method', *J. Microw. Power Electromagn. Energy*, 1992, **27** (1), pp. 11-22

JOOS, G.: 'Theoretical physics' (Blackie, 1958)

KAYE, G. W. C., and LABY, T. H. (Eds.): 'Tables of physical and chemical constants, 15th edn.' (Longmans, 1986)

KEMP, P.: 'Alternating current electrical engineering' (Macmillan, 1949)

KENNINGTON, P. B., and BENNETT, D. W.: 'Living in harmony—industrial microwave heating and cellular radio', *IEE Eng. Sci. & Educ. J.*, 1993, **2** (5), pp.233-239

LAMONT, H. R. L.: 'Waveguides' (Methuen's Monographs on Physical Subjects, Methuen, London, 1942)

LAST, B. A.: 'Survey to determine the operational characteristics for industrial microwave heating apparatus for use in the 896/915 MHz band'. Radiocommunication Division, UK Radio Investigation Service, 1987

LATHAM, C., KING, and RUSHFORTH, L.: 'The magnetron' (Chapman & Hall, 1952)

LAU, R.: 'Modelling and microwave heating characteristics of foods and food packaging systems using computer simulation'. Proceedings of the 8th Microwave Association Conference, Leatherhead Food Research Association, Surrey, UK, 1995

LAU, R. W. M., and SHEPPARD, R. J.: 'The modelling of biological systems in three dimensions using the finite difference time domain method', *Phys. Med. Biol.*, 1986, **31** (11), pp. 1247-1256

LINES, A. W., NICOLL, G. R., and WOODWARD, A. M.: 'Some properties of waveguides with periodic structures', *Proc. IEE*, Radio Section paper 941, 1949

LOMER, P. D., and CROMPTON, J. W.: 'A new form of hybrid junction for microwave frequencies', *IEE Proc. B*, 1957, **103** (15)

LYTHALL, B. W.: 'Frequency instability of pulsed magnetrons with long waveguides', *Proc. IEE*, Rad Section, July 1946

MALONEY, E. D., and FAILLON, G.: 'A high power klystron for industrial processing using microwaves', *J. Microw. Power*, 1974, **9** (3)

MARCUVITZ, N.: 'Waveguide handbook' (Peter Peregrinus Ltd. on behalf of the IEE, London, 1986) first published in 1951 by McGraw-Hill

MAXWELL, E.: 'Conductivity of metallic surfaces', *J. Appl. Phys.*, July 1947

MEDICAL RESEARCH COUNCIL: 'Exposure to microwave and radiofrequency radiations'. Press notice, January 1971

MEREDITH, R. J.: 'Serpentine applicator'. British Patent 1 277 247, 1972

MEREDITH, R. J.: 'A three axis model of the mode structure of multimode cavities', *J. Microw. Power Electromagn. Energy*, 1993, **28** (4)

MEREDITH, R. J., SHUTE, M. R., and DAINES, P.: 'Combined microwave biscuit oven'. British Patent Application 9408975.2SR776, 1994

METAXAS, A. C.: 'Design of a TM_{010} resonant cavity as a heating device at 2.45 GHz', *J. Microw. Power*, 1974, **9** (2)

METAXAS, A. C.: 'Foundations of electroheat' (Wiley, 1996)

METAXAS, A. C., and MEREDITH, R. J.: 'Industrial microwave heating' (Peter Peregrinus Ltd. on behalf of the IEE, London, 1993) first published 1983

MILLER, S. J.: 'The travelling wave resonator and high power testing', *Microw. J.*, September 1960

MORENO, T.: 'Microwave transmission design data' (McGraw-Hill, New York, 1948)

MUMFORD, W. W.: 'Some technical aspects of microwave radiation hazards',

Proc. Inst. Radio Eng., 1961, **49**

NRPB: 'Advice on the protection of workers and members of the public from the possible hazards of electric and magnetic fields below 300 GHz: a consultative document' (National radiological protection board, HMSO, London, 1986)

NRPB: 'Guidance as to restrictions on exposures to time varying electromagnetic fields 1988: recommendations of the International Non-Ionising Radiation Committee, document GS 11' (HMSO, London, 1989)

NRPB: 'Restrictions on human exposure to static and time-varying electromagnetic fields and radiation'. *Documents of the NRPB*, **4** (5), pp. 7–63 (HMSO, London, 1993)

OSBORNE, W. C., and TURNER, C. G.: 'Woods practical guide to fan engineering' (Woods of Colchester, Colchester, UK, 1967)

PAOLINI, F.: 'Calculation of power deposition in a highly overmoded rectangular cavity with dielectric loss', *J. Microw. Power Electromagn. Energy*, 1989, **24** (1)

PIERCE, J. R.: 'Travelling-wave tubes' (Van Nostrand, New Jersey, 1950)

PIPES, L. A.: 'Applied mathematics for engineers and physicists' (McGraw-Hill, 1946)

PURANEN, L., and JOKELA, K.: 'Radiation hazard assessment of pulsed microwave radars', *J. Microw. Power Electrom. Energy*, 1996, **31** (3)

RAGAN, G.L.: 'Microwave transmission circuits' (McGraw-Hill, New York, 1948)

RAMO, S., WHINNERY, J. R., and VAN DUZER, T.: 'Fields and waves in communication electronics' (Wiley, 1965)

ROBERTS, S., and VON HIPPEL, A.: *J. Appl. Phys.*, 1946, **17**, p. 610

ROLLASON, E. C.: 'Metallurgy for engineers, 4th edn.' (Edward Arnold, 1973)

ROLPH, P. M. (Ed.): 'Fifty years of the cavity magnetron'. Proceedings of symposium, School of Physics & Space Research, University of Birmingham, UK, 1990

RYDER, J. D.: 'Electronic fundamentals and applications' (Sir Isaac Pitman & Sons, London, 1960)

SAY, M. G.: 'Performance and design of alternating current machines' (Sir Isaac Pitman & Sons, London, 1948)

SCHELKUNOFF, S. A.: 'Electromagnetic waves' (Van Nostrand, 1943)

SILVER, S.: 'Microwave antenna theory and design' (Peter Peregrinus Ltd. on behalf of the IEE, 1984) first published in 1949 by McGraw-Hill

SLATER, J. C.: 'Microwave transmission' (McGraw-Hill, 1942)

SLATER, J. C.: 'Microwave electronics' (Van Nostrand, 1950)

STRATTON, J. A.: 'Electromagnetic theory' (McGraw-Hill, 1941)

STUCHLY, M. A.: 'Potentially hazardous microwave radiation sources—a review', *J. Microw. Power*, 1977, **12** (4)

STUCHLY, M. A., and STUCHLY, S. S.: 'Coaxial line reflection methods for measuring the dielectric properties of biological substances at radio and microwave frequencies', *IEEE Trans.*, 1980, **IM-29** (3)

SUZUKI, T., OSHIMA, K., NAGAO, T., OKAKURA, T., and HOSEGAWA, K.: 'A travelling wave resonator for industrial microwave heating', *J. Microw. Power*, 1975, **10** (2)

SWIFT, E., and JONES, P. L.: 'The radio frequency assisted air float dryer', *Paper Technol. Indust.*, October 1983

THOMSON, J., and CALLICK, E. B.: 'Electron physics and technology' (English Universities Press, 1959)

TINGA, W., and XI, W.: 'Design of a new high-temperature dielectrometer system', *J. Microw. Power Electromagn. Energy*, 1993, **28** (2), pp.93–103

TURNER, R. F. B., VOSS, W. A. G., TINGA, W. R., and BALTES, H. P.: 'On the counting of modes in rectangular waveguides', *J. Microw. Power*, 1984, **19** (3)

TWISTLETON, J. R. G.: 'The measurement of magnetron frequency pulling', *Proc. IEE*, 1957, **104**

TWISTLETON, J. R. G.: 'Twenty kilowatt 890 MHz continuous-wave magnetron', *Proc. IEE*, Jan. 1964, **3** (1)

TWISTLETON, J. R. G.: 'Some magnetron and microwave developments carried out in the BTH/AEI Research Laboratory from 1949 to 1963'. Lecture to the IEE Rugby Centre, March 1997

VAN KOUGHNETT, A. L., and DUNN, J. G.: 'Doubly corrugated chokes for microwave heating systems', *J. Microw. Power,* 1973, **8** (1), p. 101

VAN KOUGHNETT, A. L., DUNN, J. G., and WOODS, L. W.: 'A microwave applicator for heating filamentary materials', *J. Microw. Power,* 1974, **9** (3)

VON HIPPEL, A. R.: 'Dielectrics and waves' (MIT Press, 1954)

VON HIPPEL, A. R. (Ed.): 'Dielectric materials and applications' (MIT Press, 1966)

WELLER, B. F.: 'Radio technology' (Chapman & Hall, 1946)

WILLIAMS, G. W.: 'Power electronics' (Macmillan Education Ltd., 1987)

YOUNG, J. F.: 'The Boucherot effect', *Wireless World,* August 1952

Index